Detection of Chemical, Biological, Radiological and Nuclear Agents for the Prevention of Terrorism

NATO Science for Peace and Security Series

This Series presents the results of scientific meetings supported under the NATO Programme: Science for Peace and Security (SPS).

The NATO SPS Programme supports meetings in the following Key Priority areas: (1) Defence Against Terrorism; (2) Countering other Threats to Security and (3) NATO, Partner and Mediterranean Dialogue Country Priorities. The types of meeting supported are generally "Advanced Study Institutes" and "Advanced Research Workshops". The NATO SPS Series collects together the results of these meetings. The meetings are co-organized by scientists from NATO countries and scientists from NATO's "Partner" or "Mediterranean Dialogue" countries. The observations and recommendations made at the meetings, as well as the contents of the volumes in the Series, reflect those of participants and contributors only; they should not necessarily be regarded as reflecting NATO views or policy.

Advanced Study Institutes (ASI) are high-level tutorial courses intended to convey the latest developments in a subject to an advanced-level audience

Advanced Research Workshops (ARW) are expert meetings where an intense but informal exchange of views at the frontiers of a subject aims at identifying directions for future action

Following a transformation of the programme in 2006 the Series has been re-named and re-organised. Recent volumes on topics not related to security, which result from meetings supported under the programme earlier, may be found in the NATO Science Series.

The Series is published by IOS Press, Amsterdam, and Springer, Dordrecht, in conjunction with the NATO Emerging Security Challenges Division.

Sub-Series

A.	Chemistry and Biology	Springer
B.	Physics and Biophysics	Springer
C.	Environmental Security	Springer
D.	Information and Communication Security	IOS Press
E.	Human and Societal Dynamics	IOS Press

http://www.nato.int/science
http://www.springer.com
http://www.iospress.nl

Series A: Chemistry and Biology

Detection of Chemical, Biological, Radiological and Nuclear Agents for the Prevention of Terrorism

Mass Spectrometry and Allied Topics

edited by

Joseph Banoub

Fisheries and Oceans Canada, Science Branch, Special Projects and
Memorial University of New Foundland, St. John's, Canada

Springer

Published in Cooperation with NATO Emerging Security Challenges Division

Proceedings of the NATO Advanced Study Institute on
Detection of Chemical, Biological, Radiological and Nuclear Agents
for the Prevention of Terrorism
Castelnuova Berardenga, Italy
25 May–2 June 2013

Library of Congress Control Number: 2014952891

ISBN 978-94-017-9247-9 (PB)
ISBN 978-94-017-9237-0 (HB)
ISBN 978-94-017-9238-7 (e-Book)
DOI 10.1007/978-94-017-9238-7

Published by Springer,
P.O. Box 17, 3300 AA Dordrecht, The Netherlands.

www.springer.com

Printed on acid-free paper

Preface

This NATO-ASI (Advanced Study Institute) entitled Detection of Chemical, Biological, Radiological and Nuclear Agents for the Prevention of Terrorism was held successfully at the Certosa di Pontignano – Università degli Studi di Siena, Castelnuovo Berardenga (SI), Italy, from May 25 to June 2, 2013. It was a great success from both scientific and organizational points of view.

We were extremely lucky to have an excellent organizing committee which facilitated all the tasks associated with this NATO-ASI. The excellent efforts of Professors Richard Caprioli (Vanderbilt University), Gianluca Giorgi (University of Siena) and Mokhtar El Essassi (Mohammed V University, Agdal) are sincerely acknowledged.

We are indebted to Professor Caprioli who was instrumental in inviting the best CBRN lecturers and industrial partners and, in addition, let us avail of Ms. Maureen Casey, Managing Editor of the *Journal of Mass Spectrometry*, for all the organizational work.

We were privileged and honoured to have as lecturers for this NATO-ASI a succession of world-renowned scientists, namely Professor Richard Caprioli; Stanley Cohen, Professor of Biochemistry, Director of the Mass Spectrometry Research Center, VICC Member and Chief Editor of the *Journal of Mass Spectrometry*, from Vanderbilt University; Professor Alvin Fox, Department of Pathology, Microbiology and Immunology, USC School of Medicine, Columbia, USA, who is also Editor in Chief of the *Journal of Microbiological Methods* and joint Editor in Chief of *Molecular and Cellular Probes*; Professor Guenter Allmaier, Institute of Chemical Technologies and Analysis, Vienna University of Technology; Professor Mark Duncan of the School of Medicine, University of Colorado, USA; Professor Gianluca Giorgi, University of Sienna; Professor Giovanni Sindona, University of Calabria; Dr. Steve Lammert, Torion Technologies Development, Utah, USA; and Dr. Oliver Terzic from the Organization for the Prohibition of Chemical Weapons, to present a series of lectures on CBRN agents detection.

The main objective of this NATO-ASI was to provide the attendees with the latest developments necessary to successfully understand the CBRN agents and their associated biotechnologies. Furthermore, this NATO Advanced Study Institute was designed to provide advanced training for doctoral and postdoctoral candidates in state-of-the-art technologies for bio-detection, and the meeting's primary focus was on mass spectrometry (including chromatographic and electrophoretic separation) and comparisons of spectroscopic, immunological and molecular analyses of chemical, biological, radiological and nuclear agents.

The NATO-ASI participants were taught how CBRN agents are easy to manufacture, conceal and release. This lack of watchfulness makes prediction of biological and chemical bioterrorism threats very difficult. Notwithstanding that the ability to rapidly detect, identify and monitor CBRN agents is imperative for the efficient use of both military and civilian defence resources. This detection knowledge allows the severity and extent of a hazard to be assessed so that areas which are clean and/or contaminated.

Additionally, the NATO-ASI participants were also introduced to the rapid developments in biotechnology, genomics and xenobiotics which could be used as CBRN agents and hence have severe implications for international peace and security.

Topics on the fields of microbiology, immunosensors and immunology of lethal bacteria, viruses, fungi, prions and spores were discussed. Special emphasis was placed on Gram-positive *Bacillus anthracis*, smallpox and polio viruses, SARS, Ebola and Marburg viruses, novel flu viruses, etc.

The relaxed atmosphere of this venue stimulated discussions between faculty and trainees concerning additional details of the lecture subjects and advanced protocols and technologies.

The main scientific consensus proposed by our NATO-ASI participants was as follows: While there is no single detection method that provides rapid and accurate detection of CBRN agents, the onus is on NATO countries to keep abreast of the powerful, new state-of-the-art detection technologies that can help protect the public from emerging food and environmental chemical/biological threats. There is a need to make sure that all NATO countries use the same protocols and identical instrumentation for the detection of CBRN agents.

Finally, this NATO-ASI has contributed to the critical assessment of existing knowledge on new and important detection technologies. It helped to identify directions for future research and to promote closer working relationships between scientists from different professional fields. In addition, it facilitated employment for postgraduate participants from different countries.

Agdal, Rabat, Morocco — Mokhtar El Essassi
St. John's, Canada — Joseph Banoub

Contents

Chapter 1
Biological Agents and Bioterrorism

Mauro Bologna

Abstract For this very stimulating course, I want to share with you some of my studies and even some of my scientific and phylosophical considerations on biological agents living in the environment and their relations with humans, in the very wide concepts of ecological relationships, parasitism, immunolgical defenses and infectious disease mechanisms. All these concepts must be studied and considered in the event of criminal use of biological agents (bioterrorism) aimed at harming human populations in time and in geographical space.

Keywords Biological weapons • Toxins • Poisons • Parasitism • Immunology • Micro-biome • Fear as weapon

1.1 Definitions and History of Bioterrorism

Practice of bioterrorism goes back to very remote times of human existence and indeed to pre-hystorical conflicts between humans, when groups of people devised deliberately to use biological agents for conflict sustainment (damaging weapons) and for propagation of fear in the enemy populations.

Since the early times of combat between humans, some biological means have been used: for instance through the contamination of water wells in conflict areas, tainted with rotting animal remains, or through hunting-fighting arrowheads dipped in toxic plant extracts or venomous substances, before throwing them at the living target.

Usable biological agents for harm are in fact all the pathogens of biological nature, like microrganisms (bacteria, viruses, fungi, prions), toxins, animal and plant venoms, together with the related carriers (fomites, instruments) or vectors (insects, etc.).

NATO Science for Peace and Security Program
Pontignano (Siena) May 2013

M. Bologna, M.D. (✉)
Professor of General Pathology, Department of Medicine, Health, Life and Environmental Sciences, Medical School and Biological Sciences School, University of L'Aquila, L'Aquila, Italy
e-mail: mauro.bologna@univaq.it

J. Banoub (ed.), *Detection of Chemical, Biological, Radiological and Nuclear Agents for the Prevention of Terrorism*, NATO Science for Peace and Security Series A: Chemistry and Biology, DOI 10.1007/978-94-017-9238-7_1

Terrorism is the use of violence to condition societies or governments in their political choices.

Bioterrorism is the use (or menace of use) of biological agents to enact terrorism events and induce generalized *fear* concerning negative consequences in target populations.

Here follow some examples of bioterrorism enacted in various ages:

(a) use of poison darts/arrows (primitive populations): mostly for hunting, but also for battles against enemies.

From here derived many of the advances of toxicology, the science of toxic substances (the word "toxicology" derives from the greek words "toxon" = arch and "farmakon" = poison; toxicon farmakon = poison for arch hunting):

(b) from such practice derives the knowledge we have of stricnin, curare, ouabain, aconite, other plant/animal poisons primarily devised for hunting and fighting.

(c) impingement of darts in decomposing cadavers or putrefaction soil (or manure >> tetanus) before throwing at enemies (New Guinea, tribal combats; Sciites 400 b.C.)

(d) last but not least, the use of fear that humans have of beasts. Here comes the example of Hannibal (from Carthago), leading the ships of Prusia I, king of Bithynia (West Turkey) in a battle of year 184 b.C. against Eumene II (Attalides, Pergamon); he won that naval battle because he managed to throw canisters full of reptiles at the enemy ships, causing terror and uncoordinated reactions leading to his victory. He therefore used fear as a weapon: snakes were not even harmful (not poisonous), but big was the surprise and reactions were out of control!

(e) In recent history, we can see that First World War (also named the war of Chemistry) contributed to the development and use of Nervine gases, chemical weapons banned everywhere but still existing in some countries. Second World War (also named the war of Physics) led to the development and use of the atomic bomb.

ThePeace Treaties: *Geneva Protocol* ruled against chemical weapons (1925) and was followed later by additions concerning bacteriologic war. Not all the states however subscribed it. Most states anyway have banned, in time, chemical and bacteriological weapons by 1975.

Today, how scared should we be of biological and chemical terrorism? Well, since these are lethal and cheap weapons, they are of considerable concern, because they may be seen as the atomic bomb of poors and represent remarkable threats to peace in local conflicts and in terroristic attacks worldwide. We should all know more on the subject and do extensive prevention.

Albert Einstein once said "I do not know by what weapons the Third World War will be fought, but for sure the Fourth will be fought with stones". Well, we do not want any more World Wars, for sure, period.

1.2 Microbes and Humans Interacting

Life began on Earth with unicellular beings: primitive bacteria and algae, ~3.5 billion years ago (Archean Age). Biological evolution of species produced today's forms of life, as we observe them. Millions of species co-exist and share the stage (biosphere).

Humans (we) pretend to have absolute priority, but … share the stage in such a crowded environment on earth means to learn, respect, understand and prevent.

On this planet we have millions of different species of live beings, with variable proportions in the biosphere and in different ecosystems: they are all co-existing, interacting and competing for food and survival. This encompasses the very universal phenomena of competition and of parasitism.

One of the most recent and precise evaluations of the number of species existing on planet earth (2012), but still very approximate and provisional, (well illustrated and summarized in National Geographic, 2013) finds evidence of more than 5,000 species of mammals, 10,000 species of birds, 12,000 species of reptiles, 15,000 species of amphibians, 45,000 species of fish, 150,000 species of crustaceans, 200,000 species of mollusks, 600,000 species of aracnids, five million species of insects and many, many millions (unestimable indeed) species of bacteria, viruses and other microrganisms. Are we many on Earth? Is there enough work for the immune system of each living multicellular organism to distinguish "self" from possibly harmful "not self"?

Species of microrganisms ascertained as pathogenic for humans are indeed a very small fraction of the existing species: we come to know them better because we study the diseases connected with them, but we ignore a lot about the great number of other, presumably innocuous species.

In general, different species interacting may set a parasitic relationship, in which the larger animal (*the host*) may receive harm (food loss or disease) and the smaller one (*the parasite*) may get advantages (more food, protection). Both must preserve their identity and prevent contamination by foreign genetic material (immunologic surveillance, bilaterally). Some parasites can even live within the hosts (endoparasitism), like some bacteria and all viruses.

Three types of interactions may occur between a microrganism and a human host: (a) symbiotic relationship, in which the microrganism and the host both benefit; (b) commensal relationship, in which the microrganism gains but the host suffers no harm; and (c) a true parasytic relationship, in which the microrganism gains and the host is harmed.

Symbiosis offers frequently mutual advantages and remains very stable in time. Pathogens are a minimal part of existing microrganisms. We humans host some advantageous bacterial populations (intestine, surface germs on the skin, commensal germs on the mucosae): we indeed are also *made of the germs living in/on our body. Indeed,* only one cell out of ten in our body is a human cell: the rest are bacteria (the so-called micro-biome). Among these, we count billions of bacteria in the intestines, useful for many functions (vitamin production, competition with pathogens, contribution to metabolism, etc.).

The balance between host and parasites depends on two basic forces, an aggressive force by the parasite depending on survival/proliferation/invasion capacity of the parasite itself and a defensive force by the host depending on the immune mechanisms (phagocytosis, cellular and humoral immune reactions). In this balancing of opposite forces the parasitic relationship is played by the contendents. If we have prevalence of parasite, we may have disease (and eventually death) of the host, but if we have prevalence of host defence we may have control (and eventually elimination) of parasites.

1.3 Main Diseases of Interest in the Field

In the light of recent concern and interest about the potential for biological terrorism (biofarware) there are several diseases and bacterial toxins that must be considered in particular, like anthrax [1, 2], smallpox [3, 4], plague [5], botulinum toxin [6], and tularemia [7]. A very detailed discussion of such diseases and other infectious diseases with similar risks in terms of bioterrorism goes beyond the scopes of this concise chapter, but some features of these and other infectious diseases representing important threats in the biofarware field will be mentioned.

In this respect, we may distinguish in time diseases which are:

1. old diseases which are disappearing and sometimes returning, like smallpox and polio virus infections (which are either extinct or close to be eradicated, thanks to planetary vaccination programs);
2. diseases still active at present times, like carbuncle (anthrax), plague, tularemia, tetanus, botulinum, TBC, etc.;
3. new diseases, which are appearing/spreading, like SARS (SevereAcute Respiratory Syndrome) and its more recent variety of MERS (Middle-East Respiratory Syndrome), infections by Ebola/Marburg viruses, hantavirus, filovirus, novel Flu virus strains, etc.

Now we will summarize the essential facts about some of these diseases. For a more complete medical reference to all of them, see for instance the Merck Manual of diagnosis and Therapy [8].

1.3.1 Smallpox (Variola)

Smallpox is a highly contagious disease (incubation 10–12 days) caused by the smallpox virus, an orthopoxvirus. It causes death in up to 30 % of infected subjects. Indigenous infection has been eradicated (last case, Ethyopia, 1990 – WHO). The main concern for outbreaks of smallpox is today from bioterrorism.

Smallpox is characterized by severe constitutional symptoms (fever, headache, extreme malaise) and a characteristic pustular rash. Treatment is supportive; prevention involves vaccination, which, because of its risks (eczema, encephalitis, etc.), is done selectively.

Pathogenesis of smallpox demonstrates that the virus is transmitted from person to person by direct contact or inhalation of droplet nuclei. Clothing and bed linens can also transmit infection. Most contagions are in the first 7–10 days after the skin rash appears. Once crusts form, infectivity declines. The virus invades the oropharyngeal and respiratory mucosa, multiplies in regional lymphnodes, causing viremia and localization in small blood vessels of the skin (rash) and rarely in CNS (encephalitis).

Officially, smallpox is dead on Earth. There are no longer cases detected in the world population since 1990, but can we destroy the samples of smallpox virus existing in some virology laboratories around the world? Certainly not [3], because we could no longer prepare vaccine doses without live virus samples to start from. And without vaccine, a small amount of wild virus could ignite a wide epidemic killing a large proportion of the human population, since the vaccination is no longer mandatory in any country and a large percentage of young populations have no longer been vaccinated after the early 1990s.

1.3.2 Poliomyelitis (Infantile Paralysis)

Poliomyelitis is an acute infection caused by a poliovirus. Manifestations include a nonspecific minor illness (abortive poliomyelitis), sometimes aseptic meningitis without paralysis (nonparalytic poliomyelitis) and, less often, flaccid weakness of various muscle groups (paralytic poliomyelitis). Diagnosis is clinical, although laboratory diagnosis is possible. Treatment is supportive. Vaccination is available, still mandatory in many countries, although soon legislations may change. Childhood vaccination produces immunity in 95 % of recipients. Declared cases worldwide have diminished remarkably, but some areas with particularly poor sanitary services or with conflicts preventing health services to operate are recording increased numbers of cases recently (Syria, 2013; China 2013).

Polioviruses have three serotypes. The virus enters the mouth via the fecal-oral route, then enters the lymphoid tissues of the GI tract. If not contained, infection may enter the CNS with significant damage in spinal cord and brain, specifically to nerves controlling motor and autonomic function (breathing). Spreading is through the enteric route. Vaccine is live, attenuated virus, able to immunize many contacts respect to the vaccinated subjects (community vaccination strategies; problems in nomad populations).

1.3.3 Anthrax (Carbuncle)

Anthrax is caused by *Bacillus anthracis*, toxin producing, encapsulated, aerobic or facultative anaerobic organisms. Anthrax, an often fatal disease of animals, is transmitted to humans by contact with infected animals or their products (woolsorter's disease).

In humans, infection tipically occurs through the skin. Inhalation infection is less common; oropharingeal, meningeal and GI infections are rare. For inhalation and GI infections, nonspecific local symptoms are typically followed in several days by severe systemic illness, shock and often death. Empyric treatment is with cyprofloxacin or doxycycline. A vaccine is available (antitoxin).

Pathogenesis of anthrax takes place since *Bacillus anthracis* readily forms spores when germs encounter dry environment -a condition unfavorable for growth. Spores resist destruction and can remain viable in soil, wool, and animal hair for decades. Spores germinate and multiply in favourable conditions (wet skin, tissue, blood) and can give human disease by contact (papules, black eschars, contagious also via fomites) ingestion (raw meat > fever, nausea, vomiting, diarrhea), and inhalation (flu-like illness, respiratory distress, cyanosis, shock, coma).

Of note is the anthrax bioterrorist attack through mailings (using spores in powder form) that took place in the USA in 2001 (US Postal Service, Washington DC), event that highly sensitized the public to the global theme of bioterroristic attacks.

1.3.4 Plague (Pestis, Black Death)

Plague is caused by *Yersinia pestis* (formerly named *Pasteurella pestis*). Short bacillus with hairpin shape, infects wild rodents and can infect humans via tick bites. Symptoms are either severe pneumonia or massive lymphadenopathy with high fever, often progressing to septicemia. Diagnosis is epidemiologic and clinical, confirmed by culture and serologic testing. Treatment is with streptomycin or doxycycline. Unfortunately, a vaccine is not available for plague.

1.3.5 Tularemia

Tularemia is a febrile disease caused by *Francisella tularensis*; it may resemble typhoid fever: symptoms are a primary local ulcerative lesion, regional lymphadenopathy, profound systemic symptoms, and, occasionally, atypical pneumonia. Diagnosis is primarily epidemiologic and clinical and supported by serologic tests. Treatment is with streptomycin, gentamycin and other antibiotics.

1.3.6 Tetanus

Tetanus is an acute poisoning from a neurotoxin produced by *Clostridium tetani*. Symptoms are intermittent tonic spasms of voluntary muscles. Spasm of the masseters accounts for the name “lockjaw” (trismus). Incubation requires 2–10 days. Diagnosis is clinical. Treatment with immune globulin and intensive support. Only unbound toxin can be neutralized. A vaccine is available, with a good extent of preventive protection.

1.3.7 Botulism

Botulism is a neuromuscular poisoning due to *Clostridium botulinum* toxin. Botulism may occur without infection if toxin is ingested. Symptoms are symmetric cranial nerve palsies accompanied by a symmetric descending weakness and flaccid paralysis without sensory deficits. Diagnosis is clinical and by laboratory idenification of toxin. Treatment is with antitoxin and support therapies.

1.3.8 Tuberculosis (TBC)

TBC is a chronic, progressive infection by *Mycobacterium tuberculosis*, often with a long period of latency following initial infection. It occurs most commonly in the lungs, with productive cough, chest pain and dyspnea. Diagnosis is most often by sputum culture and smear. TBC can involve any tissue (organ disease). Treatment is with multiple antimicrobial drugs. Forms of multiresistant TB bacteria are becoming more and more frequent.

1.3.9 SARS

Coronavirus infections in humans most frequently cause common cold symptoms; however in 2002, a relatively new coronavirus caused an outbreak of Severe Acute Respiratory Syndrome (SARS), which was much more severe than other coronavirus infections. SARS is an influenza-like disease leading to progressive respiratory insufficiency with significant mortality rate. First detected in China (Guandong, 2002), the SARS epidemic spread to more than 30 countries. In mid-July 2003, there were >8,000 cases with >800 deaths (10 % mortality).

Then the outbreak subsided and no new cases have been identified from 2004 to 2012. In 2012 a new similar epidemic (sustained by the virus nCoV, novel coronavirus) started in Middle East (Arabia), with an estimated mortality above 40 %. Later the nCoV epidemic has been named MERS (Middle East Respiratory Syndrome) and is being studied as a new zoonosis trasmitted to humans from Dromedary camels. Studies are currently in progress, with great attention by the international sanitary authorities [9].

WHO in 2013 indeed alarmed many countries against the new SARS-like coronavirus responsible of MERS, that infected at the moment of this writing (December 2013; www.who.int/en) more than 160 persons (Arabia, Great Britain, France,

Germany, Tunisia, Italy, Abu Dhabi, United Arab Emirates, etc.) with reduced infective capacity as compared to SARS, but still highly lethal and communicable via close contacts (family members). The latest available numbers call for 163 ascertained diagnoses in humans, with 71 deaths (mortality, 43.5 %).

Updates can be found at the following web sites: www.who.int/en; www.cdc.gov and (recommendations for clinicians) emergency.cdc.gov, emphasizing the need to consider the novel (nCoV) coronavirus when treating patients with a severe respiratory illness who have recently traveled to the Arabian Peninsula (or close contacts of the travelers).

1.3.10 Ebola/Marburg Diseases

Marburg and Ebola are filoviruses that cause hemorrhage, multiple organ failure and high mortality rates. Diagnosis is with enzyme-linked immunosorbent assay, PCR or electron microscopy. Treatment is supportive. Strict isolation and quarantine measures are necessary to contain outbreaks. Incubation 5–10 days. Marburg virus has been identified in bats and in primates. Human to human transmission occurs via skin and mucous membranes contact (humans/primates).

Filoviruses can affect intestines (nausea, vomiting, diarrhea), respiratory tract (cough, pharingitis), liver (jaudice), CNS (delirium, stupor, coma), and cause hemorrhagic phenomena (petechiae, frank bleeding) with high mortality rates (up to 90 % with Ebola virus). Survivors recover very slowly and may develop long lasting complications (hepatitis, uveitis, orchitis) with only supportive care available: no specific antivirals nor vaccines are available fo filovirus infections.

1.3.11 Hantavirus, Lassa Fever, etc.

Bunyaviridae contain the genus Hantavirus (four serogroups, nine viruses) causing hemorrhagic fevers with renal and pulmonary consequences, starting with flu-like symptoms and evolving with severe renal and pulmonary consequences. Lethal in 10–15 % of cases.

Lassa fever is an often fatal arenavirus infection occurring mostly in Africa. It may involve multiple organs, except CNS. Treated with ribavirin. No vaccinations are available so far for hantavirus infections.

Outbreaks of such infections have been recorded in Nigeria, Liberia, central Africa, with some rare imported cases in the USA and in the United Kingdom.

The animal reservoir of such viruses is in wild african rats (*Mastomys natalensis*), frequently found in african houses. Direct human to human transmission is documented via urine, feces, saliva or blood. Mortality (up to 45 %) can be reduced by prompt ribavirin treatment. Universal hygiene precautions, airborne isolation and surveillance of contacts are essential.

1.3.12 Influenza Virus, with New Strains Continuously Appearing

Last but not least … we must mention now influenza! Flu viruses are in nature among the most rapidly changing (mutating) organisms through their ability to infect a variety of hosts: birds (migrating waterfowl -ducks-, stantial poultry -chickens-), mammals (pigs, felines) and humans. In South East Asia (mostly in China, but also in Viet-Nam, Laos, Thailand, etc.) it is very common to have mixed farms of pigs, poultry and ducks, attended by humans.

Every year, new strains appear in SE-Asia, favoured by the recyprocal passage between migrating birds (mostly fowl), pigs and chickens, with exposure of many humans in farms, markets, rooster fighting sports, and food preparation places.

A common say in China tells that "Anything with four legs (except chairs) and anything that flies (except airplanes), can be eaten". With this phylosophy, there is generally a lot to be desired in food safety and in general hygienic prevention in such geographical areas.

After the avian flu H5N1 of 2005–2006, highly lethal but unable to give human to human contagion, new combinations of flu strains are expected and feared, with high lethality and high human to human transmissibility.

On this widely interesting theme for the world diffusion of new virus strains with pandemic potential, I wrote in 2010 together with the colleague virologist Aldo Lepidi a book entitled "Pandemics – virology, pathology and prevention of influenza" (Bollati Boringhieri publisher, Turin, Italy, 2010) [10].

In summary, we can see that a continuous surveillance is being devoted worldwide to the appearance of new strains of influenza viruses, in order to isolate as soon as possible potentially pandemic new strains and to prepare biological stocks suitable for massive vaccine preparations in due time to prevent the global spreading of potentially lethal new variants of the influenza viruses. Examples in time recall the cases of the highly lethal pandemics known as "Spanish flu" in 1917–1918 (in excess of 40 million deaths worldwide), "Asian flu" in 1956 (in excess of 100,000 deaths worldwide) and "Hong Kong flu" in 1978 (in excess of 700,000 deaths worldwide). The basic question is: when the new pandemic will strike ? Sometimes soon, as international experts say. The so called "Avian flu" came close to that, but sometimes in the future new mutations may emerge with the potential of being much worse.

In conclusion of this wide although rapid overview of the most frequent or alarming causes of microrganism-related human diseses with potential interest for bioterrorism, I hope to have provided sufficient matter for discussion and for further future diffusion of medical and microbiological culture that may be useful for prevention and the betterment of human social relationships and for peace promotion.

References

1. Inglesby TV, Henderson DA, Bartlett JG, Ascher MS, Eitzen E, Friedlander AM, Hauer J, McDade J, Osterholm MT, O'Toole T, Parker G, Perl TM, Russell PK, Tonat K (1999) Anthrax as a biological weapon: medical and public health management. Working Group on Civilian Biodefense. JAMA 281(18):1735–1745
2. Inglesby TV, O'Toole T, Henderson DA, Bartlett JG, Ascher MS, Eitzen E, Friedlander AM, Gerberding J, Hauer J, Hughes J, McDade J, Osterholm MT, Parker G, Perl TM, Russell PK, Tonat K (2002) Anthrax as a biological weapon, 2002: updated recommendations for management. JAMA 287(17):2236–2252
3. Henderson DA, Inglesby TV, Bartlett JG, Ascher MS, Eitzen E, Jahrling PB, Hauer J, Layton M, McDade J, Osterholm MT, O'Toole T, Parker G, Perl T, Russell PK, Tonat K (1999) Smallpox as a biological weapon: medical and public health management. Working Group on Civilian Biodefense. JAMA 281(22):2127–2137
4. Lovinger S (2002) Addressing the unthinkable: preparing to face smallpox. JAMA 288(20):2530
5. Inglesby TV, Dennis DT, Henderson DA, Bartlett JG, Ascher MS, Eitzen E, Fine AD, Friedlander AM, Hauer J, Koerner JF, Layton M, McDade J, Osterholm MT, O'Toole T, Parker G, Perl TM, Russell PK, Schoch-Spana M, Tonat K (2000) Plague as a biological weapon: medical and public health management. Working Group on Civilian Biodefense. JAMA 283(17):2281–2290
6. Arnon SS, Schechter R, Inglesby TV, Henderson DA, Bartlett JG, Ascher MS, Eitzen E, Fine AD, Hauer J, Layton M, Lillibridge S, Osterholm MT, O'Toole T, Parker G, Perl TM, Russell PK, Swerdlow DL, Tonat K (2001) Botulinum toxin as a biological weapon: medical and public health management. JAMA 285(8):1059–1070
7. Dennis DT, Inglesby TV, Henderson DA, Bartlett JG, Ascher MS, Eitzen E, Fine AD, Friedlander AM, Hauer J, Layton M, Lillibridge SR, McDade JE, Osterholm MT, O'Toole T, Parker G, Perl TM, Russell PK, Tonat K (2001) Tularemia as a biological weapon: medical and public health management. JAMA 285(21):2763–2773
8. Porter RK (ed) (2011) The Merck manual of diagnosis and therapy, 19th edn. Merck Sharp & Dohme Corp., Whitehouse Station
9. Mitka M (2013) Deadly Mers coronavirus not yet a global concern. JAMA 310(6):569
10. Bologna M, Lepidi A (2010) Pandemie¸ Virologia, patologia e prevenzione dell'influenza (Pandemics – virology, pathology and prevention of influenza). Bollati Boringhieri Publisher, Turin, Italy. ISBN 978-88-339-2037-5

Chapter 2
Immunological Defence Mechanisms Against Biological Agents

Mauro Bologna

Abstract A short discussion about attack strategies by microorganisms and immunological defence mechanisms by human body: facts and concepts for preventing disease and biological warfare.

Keywords Immunology • Microbes • Prevention • Vaccines

2.1 Immunology, How It Works

Immunology is a rather young science studying biological mechanisms existing in biological evolution since the earliest multicellular forms of life (sponges) and becoming progressively more complex in mollusks, insects, fish, birds up to mammals and humans. Immunology is a little more than 50 years old, with this specific name, although the first informations on the existence and validity of immune defence go back indeed to Edward Jenner with his "vaccination" practices in the 1790s.

That first "immunology" experiment is something done a little more than two centuries ago that is not ethically feasible today [1] …

Jenner, noticing that english milkmaids who caught cowpox did not get human smallpox, voluntarily and deliberately exposed his gardener's son (James Phipps) to cowpox (causing a fever illness); after recovery he challenged him with smallpox verifying his attained "immunity".

Jenner is therefore considered the father of "immunology". No mechanistic explanations of that experiment were possible at that time: only about 100–150 years later we started discovering antibodies and immune system functions allowing us to

NATO Science for Peace and Security Program
Pontignano (Siena) May 2013

M. Bologna, M.D. (✉)
Professor of General Pathology, Department of Medicine, Health,
Life and Environmental Sciences, Medical School and Biological Sciences School,
University of L'Aquila, L'Aquila, Italy
e-mail: mauro.bologna@univaq.it

J. Banoub (ed.), *Detection of Chemical, Biological, Radiological and Nuclear Agents for the Prevention of Terrorism*, NATO Science for Peace and Security Series A: Chemistry and Biology, DOI 10.1007/978-94-017-9238-7_2

understand what was biologically happening in the milkmaids and in the "vaccinated" individuals like Jenner's gardener's son.

The immune system is in fact like an "eye within" the body, vigilating that nothing extraneous is biologically active in each individual organism, recognizing effectively "self" from "not self", ensuring that replication of self cells is not contaminated by "foreign" biological agents and killing any extraneous biological entity through the action of soluble "light" weapons (antibodies) and of "heavy" killers (immune cells and macrophages, interacting) [1].

The key steps of an immune reaction are substantially three: (a) foreign particles inglobated by macrophages are degraded and "presented" to T-Helper lymphocytes which in turn start an elaborated attack, mostly through either (b) soluble weapons (antibodies, humoral response) or (c) cellular weapons (killer T-lymphocytes, cellular or cytotoxic response) or both (b+c), depending on the case.

Recently, it has been shown that activated T-lymphocytes tend to aggregate, like a swarm of bees, exchanging informations, useful to coordinate the immune response (i.e. to elaborate coordinated defence plans) [1].

When invaders are present in body fluids (extracellular, like most bacteria) they can be attacked with antibodies (specific surface recognition).

When invaders are instead intracellular (viruses, some bacteria like TBC) they are attacked with special killer cells (cytotoxic T-Ly) destroying self cells harboring the intruders, with their content [1].

Viruses have also a transit in body fluids, so antibodies are also produced against them.

In most responses, both humoral (antibodies) and cellular attacks are actuated. The immune response, moreover, is specific and potentiated by memory: therefore a second or further encounter with the same foreign agent (antigen) produces a stronger and quicker defense reaction [1].

2.2 Microbes, How They Prevail (Attack Strategies)

If germs make it to penetrate in the body, they enter an hostile territory, where they try to survive through strategic actions: rapid proliferation, identity disguise, toxic products (toxins), resistance to attacks, niche hiding, etc.

If microbes prevail, they can multiply and pass to other individuals.

The capacity to hide from immune attacks is exemplified by *Listeria monocytogenes*, which is able to escape vacuoles when internalized. The bacterium so hides and passes from cell to cell by endocytosis going undetected by antibodies.

Also tuberculosis (TBC) represents a great challenge, because the organisms (*Mycobacterium tuberculosis*) can survive even after they have been internalized (phagocytosed) by macrophages. The mycobacteria can indeed hide inside the bone marrow stem cells and evade the immune reactions for many years [1].

But fortunately, highly pathogenic germs are very few.

Millions of species of microorganisms exist. Most of them are not pathogenic for humans, because the immune system easily destroys them. The very limited number of effective pathogens is characterized by special properties making them resistant to the immune attacks.

Very virulent germs tend to disappear in nature, because they kill their sensitive hosts (natural selection): the killer disappears together with the killed. However, many virulent germs tend in time to attenuate their aggressiveness (which might make them extinguish), allowing milder forms of infection to emerge, that do not kill the hosts and guarantee survival of both contenders (parasite and host).

The attack strategies by germs are mostly implemented by rapid proliferation, which is a key factor for microorganisms: immune response indeed requires some time (7–10 days) to be strong. Another feature of attack by germs is their capacity to elaborate toxins (blocking key functions of the host): movement (tetanus), water balance (cholera), etc.

Further strategies are immune evasion (antigenic change – flu viruses), complement inhibition, resistance to phagocytosis, or even T-Ly elimination (HIV) [1].

2.3 Prevention to Favour Defence and Immunity Success

To control infections and favour immune responses, we may try:

1. to reduce microorganism proliferation (through the use of bacteriostatic drugs) or kill the germs (disinfection, use of bactericidal drugs);
2. to control the vectors diffusing the infection (insects, arthropods, birds, rats, etc.);
3. to immunize preventively the potential hosts (vaccinations): this requires public health planning, technology, costs and time;
4. to administer preformed antibodies (serotherapy): this also requires technology, costs and time.

Actions against germs (point 1, above) implies the use of chemotherapeutic agents (antibiotics) having a multitude of action mechanisms and being suitable under precise conditions; germs tend to develop resistance to them; therefore, fewer and fewer effective antibiotics are available today, also because the discovery of new molecules in this category has been scarce, lately. Hygienic measures are very effective, well known and can be enacted at low cost, whenever possible.

Campaigns against vectors (insecticides, biological competitors – fungi, bats, genetics -) can also be very effective (if they are done with a good biological knowledge of the ecological system) [1, 2].

But the best possible actions are the use of vaccines (if available and if time allows) and of immune sera (if available).

A vaccine is always preferable, because induces an active response and an advantageous state of immune memory in the individual, with minimal side effects (active immunization). A serum specific for a given antigen can be life-saving (serotherapy, immediately active), but has some side effects (since it introduces heterologous proteins -for instance, horse immunoglobulins, which in turn will be immunologically eliminated-) and does not have a lasting protection (passive immunization).

Modern life has created some extra occasions for germs and some "new" diseases [2]: air conditioning apparatuses for instance (if not properly cared and cleaned) are a new ideal environment for bacteria; because of dirt and humidity they can foster growth of airborne bacteria never seen before as human pathogens: *Legionella pneumophyla*. The story of Legionaries disease (a fatal lung infection by Legionella) is very instructive for microbiologists and epidemiologists. Human behaviours (homosexual intercourses, exchange of syringes among intravenous drug users and frequency of air travel) have extended the contagion of formerly rare infections like HIV: also in this case the facts are dramatically instructive for medicine and epidemiology. Transfusions of unscreened blood have also diffused hepatitis and HIV viruses. Centralized processing of foods has sometimes diffused a contagion of food-borne infections (*E. coli*) [1].

Airplanes have replaced ships for human travel and are at the center of attention for human communicable diseases spreading (Influenza, SARS, etc.). Many more people are traveling today to remote and tropical areas than before (forests, wilderness): this can expose more populations (even at home, on the return) to rare insects and microbes (see cases of malaria, Ebola virus, Marburg virus, etc.). In addition, economic development expands contacts: mining, forestry, agriculture in new tropical areas with recent deforestation.

And more, the increasing number of subjects with immunodeficiency diseases or immunosuppressant therapies (for transplants) increases the probability of new communicable agents to infect people, survive, and propagate in modern societies (opportunistic infections, with possible mutations in progress).

This entire panorama increases the variety of new emerging infections: here is a variety of bacterial and viral pathogens of recent discovery: Rotavirus, *Criptosporidium parvum*, *Legionella pneumophyla*, Ebola virus, Hantaan virus, *Campylobacter jejuni*, HTLV virus, HIV virus, *Helicobacter pylori*, Herpesvirus-6 and -8, virus Guanarito, virus Sabia, nCoV-MERS virus, to name just some of the most relevant and recent [2].

So, we have a lot to learn about new infections and a lot to do to control communicable diseases, also because of the continuously changing human ecosystem, in consequence of the ever changing human activities and the rapidity of genetic changes taking place in many microorganisms.

2.4 Vaccines: How to Prepare Them

What is a vaccine? In a dangerous disease caused by microorganisms and "foreign" not-self substances (an "antigen"), it may be possible to raise a vaccine, that is an innocuous preparation of that "antigen" which can induce the production of an immune response and create immune memory in the subjects we want to protect.

A vaccine can be developed in many ways: killed (denatured) or non replicating pathogens (viruses), recombinant protein antigens, live, attenuated (less harmful) strains of pathogens (for instance cowpox virus to protect against smallpox ...)

Vaccines can be very effective: cases of smallpox (extinction of the disease: non more cases of smallpox exist on earth for the success of global vaccine campaigns) and of polyo (near extinction of disease, except in some nomad populations and in critical areas of today's world, like Syria and also, surprisingly, of remote China provinces) are very instructive on the subject [2].

Requirements for vaccine preparation are the following. (a) An innocuous preparation of the antigen (+/– adjuvants); (b) a suitable biologic model in which to test the effectiveness of the immunization procedure; (c) a reduced series of administrations able to induce a long lasting response (number of boosters required): polio (very long lasting), as opposed to tetanus (lasts only some years, many boosters required); (d) an easy and affordable procedure for preparing quickly a large amount of product to use in case of necessity [2].

With such common and general requirements, the most used vaccinations include

1. killed whole germ preparations (influenza; polio type Salk; Hepatitis A; rabies; pertussis; cholera);
2. live attenuated germ preparations (tuberculosis; mumps-parotitis-rubella; polio type Sabin; yellow fever; variola; typhus and lately also against Ebola and Marburg viruses);
3. purified component vaccines (meningitis – subunits of Haemophylus influenzae b; acellular pertussis vaccine; tetanus-diphtheria anatoxins);
4. recombinant DNA vaccines (surface antigen of hepatitis B);
5. naked DNA vaccines (against many different germs, not yet widely used);
6. vaccines in genetically modified plants (against many different germs, not yet widely used) [2].

Technology of vaccine preparations is continuously evolving, in order to improve our capacity to prepare quickly enormous amounts of vaccinating doses from microorganisms soon after their isolation in newly appearing forms or strains. For viruses, for instance, an important step forward has been the technology of reproducing the organisms on cell cultures rather than on fertilized chicken eggs.

2.5 Vaccination Campaigns: How Best Use Vaccines

(supplies/demands)

Mass vaccination campaigns have taken place several times in history, like those of 1947 against smallpox, in which the citizens of New York City stood in long lines to be vaccinated at a rate of eight persons per minute, and like those of several countries against polio (after the Sabin vaccine introduction, in 1960), with periodic vaccination days organized to reach many secluded groups of citizens living in remote areas [2].

The realities however are that the existence of a vaccine is not enough to prevent a worldwide pandemic; more testing is needed before a new vaccine may be offered to the public; for vaccines prepared on chicken eggs, the successful mass production depends on the availability of the eggs, indeed; obstacles include organizational system for distribution and timely supply of doses, where and when they are needed.

Good news for scientists and for biologists is that since the September 2001 terroristic attacks in the USA (anthrax spores in the mailings), the employment needs of biologists have increased remarkably (production of vaccines, enactment of protection plans, treatment schemes for infections, etc.) [1, 2].

The field of immunology, in particular for vaccine research and development, continues to be of high impact in modern medicine and of high relevance in the contrast of bioterrorism.

To conclude, science is here to help; to be of use; to spread not infections but knowledge, across different cultures and nationalities.

Our role of scientists is mostly that of being "pontifices" (latin "pontes facientes", that means "bridge makers" across cultures and nationalities). For this reason it has been a real pleasure to exchange here our knowledge and to discuss it together, among so many different people and so wide scientific expertises.

References

1. Murphy K (2011) Janeway's immunobiology, 8th edition. Garland Science. ISBN-10: 0815342438
2. Abbas AK, Lichtman AH (2010) Basic immunology, 3rd edn. Saunders. Kindle Edition. ISBN-10: 141605569X

Chapter 3
Mass Spectrometry and Tandem Mass Spectrometry: An Overview

Gianluca Giorgi

Abstract In this chapter, a brief overview of fundamentals in mass spectrometry and of its applications in homeland security, forensic sciences and explosive detection, are presented.

This chapter presents the general features of mass spectrometry and the main concepts in ionization techniques and analyzers, with particular attention to resolution, one of most important and "hot" topic in modern mass spectrometry.

In addition, ambient mass spectrometry, a mass spectrometric analysis is discussed. This ambient ionization requires either no or minimal effort for sample preparation, using direct sampling. An overview of the wide range of ionization methods described in the literature, such as DART, DESI, PS is also presented.

Tandem mass spectrometry (MS/MS), a methodology used both for identifying and characterizing analytes is discussed, as well as it uses for quantitative purposes. Different MS/MS experiments, collision-induced dissociations (CID), interactions of ions with electrons (ECD, ETD), photons (IRMPD) and with surfaces (SID) are also briefly described.

3.1 Mass Spectrometry: General Features

Mass spectrometry (MS) constitutes an important and very powerful methodology for obtaining information about identity, structure and quantity of analytes, widely applied in different fields, ranging from food, environment, chemistry, clinics, biosciences, proteomics, homeland security, and many others [1–5].

Mass spectrometry is characterized by some peculiar features, and in particular: (i) high specificity and selectivity: a mass spectrum is related to the chemical structure of the molecule; (ii) high sensitivity: the attomol (10^{-11}) range and lower can be reached; (iii) speed: a mass spectrum can be acquired in a time frame of msec.

G. Giorgi (✉)
Department of Biotechnology, Chemistry & Pharmacy, University of Siena,
Via Aldo Moro, I-53100 Siena, Italy
e-mail: gianluca.giorgi@unisi.it

J. Banoub (ed.), *Detection of Chemical, Biological, Radiological and Nuclear Agents for the Prevention of Terrorism*, NATO Science for Peace and Security Series A: Chemistry and Biology, DOI 10.1007/978-94-017-9238-7_3

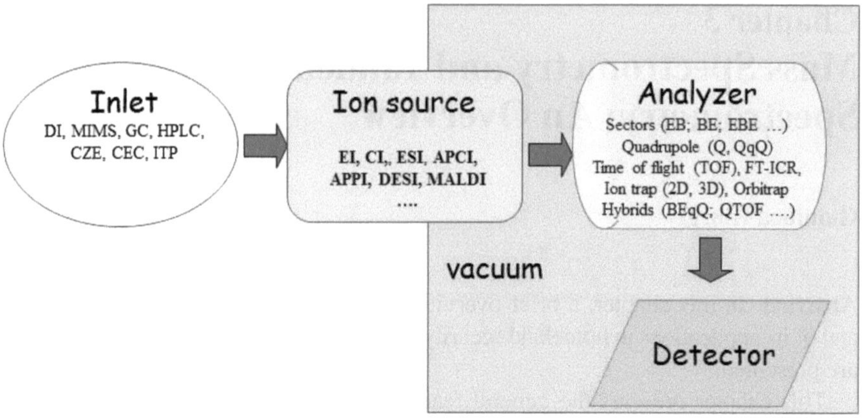

Fig. 3.1 The diagram of a mass spectrometer

Mass spectrometry is based on the production of ions in the gas phase, occurring in the ion source, on their separation, in the analyzer, according to their mass/charge (m/z) ratios, and finally detected by a detector (Fig. 3.1).

3.2 Ionization Techniques

Mass spectrometry can study wide classes of molecules, ranging from small volatile up to polar and very large molecules. All cannot be ionized by just one ionization technique, and a wide range of ionization techniques are nowadays available (Table 3.1). The choice of the most suitable will be based on the physico-chemical properties of the analyte under investigation, its molecular weight, volatility, polarity, etc.

Ionization techniques can be grouped into *hard* and *soft*. In the case of hard ionizations, mostly represented by electron ionization, a large amount of energy is deposited onto the molecule. It produces a molecular ion with a high energy excess that causes its abundant fragmentation and, as a consequence, lot of information on structural characterization of the analyte. If the energy excess is too large, the molecular ion might be undetectable thus not allowing molecular weight determination of the analyte.

Most of the ionization techniques used in modern mass spectrometry, such as electrospray (ESI), matrix assisted laser desorption ionization (MALDI), and all ionization techniques used in ambient mass spectrometry, can be categorized as *soft* ionizations.

In all these cases, the mass spectra will be characterized by the presence of protonated/deprotonated molecules and their adducts, while the fragmentation, is any, is limited to very few ions. This allows to determine the molecular weight of the analyte but no any information on its structure can be obtained.

Table 3.1 Main ionization reactions occurring in mass spectrometry[a]

Ionization reaction	Ionization technique
$M_g + e^- \rightarrow M^{+\bullet} + 2e^-$	EI
$M + H^+ \rightarrow [M + H]^+$	CI, ESI, APCI, APPI, MALDI, ……
$M + nH^+ \rightarrow [M + nH]^{n+}$	ESI
$M - H^+ \rightarrow [M - H]^-$	CI, ESI, APCI, APPI, MALDI, ……
$M - nH^+ \rightarrow [M - nH]^{n-}$	ESI

[a]Abbreviations: M_g molecule in the gas phase, *EI* electron ionization, *CI* chemical ionization, *ESI* electrospray, *APCI* atmospheric pressure chemical ionization, *ESI* electrospray, *APPI* atmospheric pressure photoionization, *MALDI* matrix assisted laser desorption ionization

3.3 Analyzers and Resolution

Once the ions are formed in the ion source, they are accelerated towards the mass analyzer where their separation according to their *m/z* ratios occurs.

Analyzers can be divided into two main groups: those based on separation *in space* (sectors, quadrupole, time of flight) and those separating ions *in time* (ion traps, Orbitrap, FT-ICR) (Table 3.2).

One of the most important performances of analyzers is their resolving power, *i.e.* the ability to distinguish between ions differing in the quotient mass/charge by a small increment. Resolving power and resolution are the most debated terms in mass spectrometry.

Two main definitions can be used for resolution (Fig. 3.2): (i) valley definition: it considers two peaks of equal height in a mass spectrum at masses *m* and ($m + \Delta m$) be separated by a valley (generally defined at 5–10 %) which, at its lowest point, is at x % of the height of either peaks; (ii) Peak width definition: for a single peak made up of singly charged ions at mass *m* in a mass spectrum, the resolution may be expressed as $m/\Delta m$, where Δm is the width of the peak at a height which is a specified fraction of the maximum peak height. Generally the 50 % peak height is considered, so a resolution at full width at half maximum (FWHM) is defined.

Depending on the analyzer, resolving power and resolution can be constant or they vary in the scan. In particular, in quadrupole and ion trap analyzers, the peak width remains constant over all the scan. It follows that if the FWHM is 0.5 u, at *m/z* 40 the resolution will be 80 (i.e. 40/0.5), but at *m/z* 1,000 it will be 2,000.

In the case of TOF and sector analyzers, resolution remains constant overall the scan, so if $R_{FWHM} = 20{,}000$, at *m/z* 40 the peak with will be 0.002 u, but at *m/z* 4,000 it will be 0.2 u. For FTICR, at constant detection time, the resolving power is inversely proportional to *m/z*. Thus if at *m/z* 100 the resolving power is 1×10^6, it will be 1×10^5 at *m/z* 1,000. In the Orbitrap analyzer, at constant detection time, the resolving power is inversely proportional to the square root of *m/z*. Thus if the resolving power is 100,000 at *m/z* 100, at *m/z* 1,000 it will be 31,622 (i.e. $100{,}000(100/1{,}000)^{1/2}$).

Table 3.2 Analyzers and their main features

Analyzer	Force	Separation based on	*m/z* range	Resolving power	Mass accuracy
Ion separation *in space*					
Sectors (EB, BE)	Magnetic and electrostatic fields	Ion momentum and kinetic energy	10,000	10,000	<1 ppm
Quadrupole	Electric field and radiofrequency	Stability/instability	2,000–4,000	Unit (0.2 u FWHM)	No (>20 ppm)
Time of flight		Speed	>100,000	>10,000	2–5 ppm
Ion separation *in time*					
3D and 2D ion trap	Electric field and radiofrequency	Frequency of the orbits	<6,000	<500	No
Fourier Transform Ion Cyclotron Resonance (FT-ICR)	Electric field, radiofrequency, magnetic field	Frequency of the orbits	>10,000	>100,000	<1 ppm
Orbitrap	Electric field	Frequency of the harmonic oscillations	<6,000	100,000	2–5 ppm

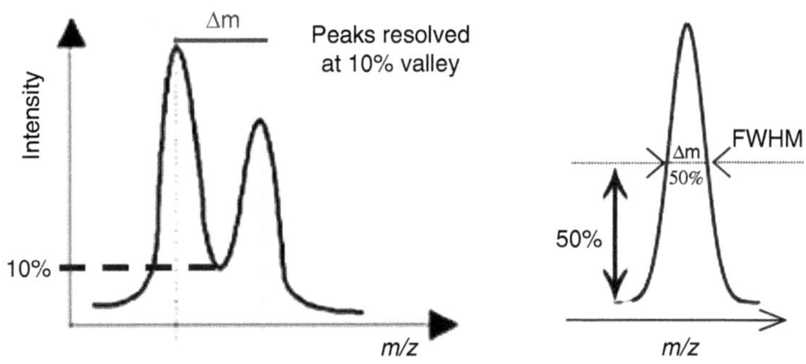

Fig. 3.2 Definitions of resolution: valley definition (*left*) and full width half maximum (FWHM, *right*)

3.4 Ambient Mass Spectrometry

Ambient mass spectrometry is defined as a mass spectrometric analysis with no or minimal effort for sample preparation, using direct sampling and ionization at ambient conditions [6]. It follows that the samples are analyzed in their native state and ionization occurs at atmospheric pressure, externally to the mass spectrometer. The formed ions are then introduced into the mass spectrometer [7].

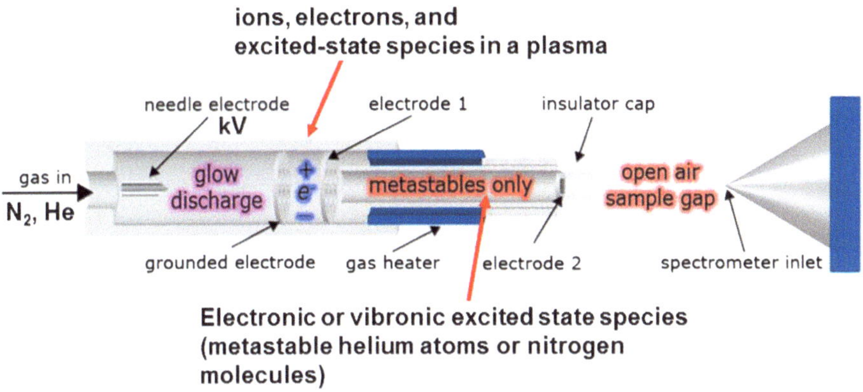

Fig. 3.3 The Direct Analysis in Real Time (DART) ionization (Adapted from Jeol, USA)

Decades of ambient ionization methods have been developed in the last years [6–15], some of them are based on direct ionization of the sample, such as paper spray [16], low temperature plasma (LTP) [17], others on desorption processes (desorption electrospray ionization (DESI) [18], femtosecond laser desorption ionization (fs-LDI) [19], desorption atmospheric pressure photoionization (DAPPI) [20], nanospray desorption electrospray ionization (nano-DESI) [21]); some others are based on two ionization steps, such as direct analysis in real time (DART) [22], laser ablation electrospray ionization (LAESI) [23, 24], also known as laser electrospray ionization mass spectrometry (LEMS) when fs-LDI is combined with ESI [25], and many others [12].

Ambient mass spectrometry ionization techniques are generally soft ionizations and they yield minimal or none fragmentation of the analytes. For example, ionization by the spray-based ambient ionization methods such as DESI produces an energy deposition of 2 eV in typical cases, which is similar to the internal energy of electrospray ionization [26].

The DART ionization [22] has been introduced in 2005 and it has been defined as the first open-air, ambient ion source for mass spectrometry. A schematic drawing of the ion source is reported in Fig. 3.3.

The key processes taking place in DART ionization are (i) a thermo-desorption of condensed-phase analytes by a stream of a hot gas which carries active species (metastable helium atoms or nitrogen molecules) derived from a plasma discharge, and (ii) an APCI-like ionization, enabling acquisition of respective mass spectra of the analytes.

DART is a fast and easy way to analyze samples with a minimal sample preparation for most samples. It can tolerate “dirty” or high-concentration samples and without contamination and it is suitable for obtaining fast fingerprinting of materials, but it cannot ionize large biomolecules, such as DNA and proteins. An application of DART in the study of explosives is reported in Fig. 3.4.

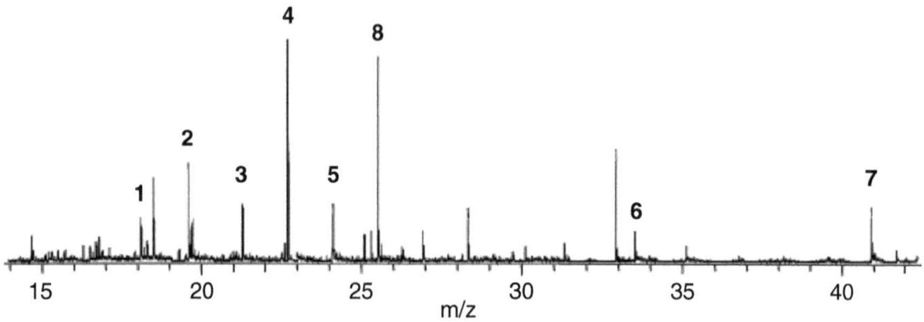

Fig. 3.4 Detection of explosives in muddy water by DART ionization: *1* dinitrotoluene (DNT); *2* amino-DNT; *3* trinitrobenzene; *4* trinitrotoluene (TNT); *5* tetryl; *6* RDX (TFA adduct); *7* HMX (TFA adduct); *8* palmitate (in pond water) (Reprinted from Ref. [22] with permission. Copyright 2005, American Chemical Society)

DESI ionization has represented the first application of ambient mass spectrometry to explosives detection done by Cooks and coworkers in early 2005 [27]. DESI uses charged solvent droplets, generated by a pneumatically assisted electrospray emitter posed at a given incident angle, for sampling and ionizing of analytes on sample surfaces [28]. The charged solvent droplets usually have diameters of less than 10 μm and velocities of about 100 m s^{-1}. DESI has excellent performances in surface analysis and it can be also used for imaging mass spectrometry and for performing chemical reactions.

The use of DESI and a miniaturized mass spectrometer in the analysis of explosives is reported in Fig. 3.5.

Spectra obtained by different ambient mass spectrometry techniques in the characterization of the radical anion and deprotonated anion of TNT are shown in Fig. 3.6.

Different ionization and ambient mass spectrometry techniques find several applications in homeland security, in forensic applications [37], in the direct analysis of microorganisms [38, 39].

3.5 Tandem Mass Spectrometry

As reported above, most of the ionization techniques used in modern mass spectrometry are *soft* ionizations. They allow to determine the molecular weight of an analyte but, without fragmentation in the ion source, it is not possible to have information about its structure. If fragments are not produced in the source, it is possible to form them in the analyzer region. This is one of the main goals of tandem mass spectrometry.

The concept of tandem generally recalls in our mind tandem bikes in which at least two cyclists ride the bike. Similarly, in tandem mass spectrometry experi-

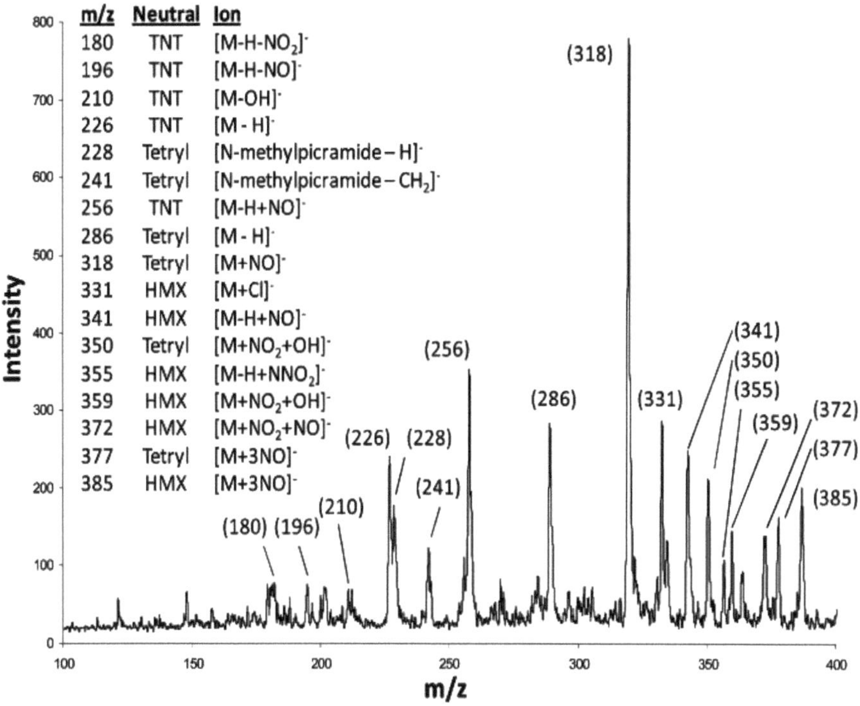

Fig. 3.5 Negative ion mass spectra of 1 µg of the mixture of the explosives Tetryl, TNT, and HMX deposited on frosted glass in a total area of 1 cm^2 obtained using DESI on the Mini 10 mass spectrometer. N-Methylpicramide is a decomposition product from Tetryl (Reprinted from Ref. [29] with permission. Copyright 2010, American Chemical Society)

ments, two (MS/MS or MS^2) or more (MS^n) sequential stages of mass analysis (which can be spatially or temporally separated) are used in order to examine selectively the decomposition of given ions in a mixture of ions.

Different kinds of experiments can be done by tandem mass spectrometry (Fig. 3.7): the most common is the *product ion scan*. This is carried out by selecting ions with a given *m/z* value detecting selectively their *product ions* formed owing to induced dissociations. A second kind of experiments is the *precursor ion scan*. In this case, all precursors of a given fragment ion are detected. This approach is particularly useful when a common fragment ion, for example *m/z* 97 ($[HSO_4^-]$) in the case of sulphated compounds, is produced from different compounds of a mixture.

The third kind of experiment, i.e. *neutral loss scan*, is the detection of fragmentation pathways involving elimination of neutral species, such as H_2O, CO, CO_2, HCN.

In tandem mass spectrometry, ion decomposition can be obtained by four main mechanisms: (a) collisions of ions with a gas, generally nitrogen, helium, or argon: collision induced dissociation (CID), collision activated dissociation (CAD) experiments; (b) interactions of ions with electrons: electron capture dissociation (ECD)

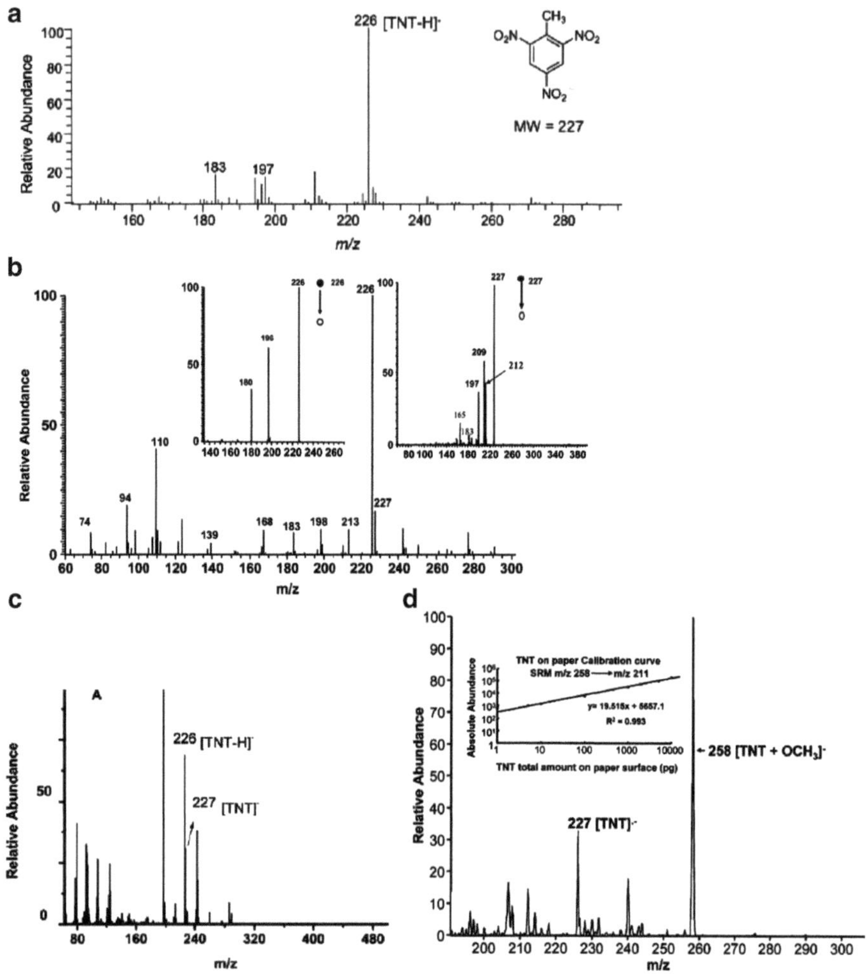

Fig. 3.6 Mass spectra recorded using different ionization techniques, showing the varied ratio of the radical anion of TNT (*m/z* 227) to the deprotonated TNT anion (*m/z* 226). (**a**) Deprotonated molecules of TNT detected using Low Temperature Plasma (*LTP*) Desorption (Reprinted from Ref. [30] with permission. Copyright 2009, Royal Society of Chemistry). (**b**) A typical mass spectrum recorded using Surface Desorption Atmospheric Pressure Chemical Ionization (*DAPCI*) with $[M]^{-\bullet}$ and $[M\text{-}H]^-$ MS/MS spectra in the insets (Reprinted from Ref. [31] with permission. Copyright 2007, Wiley). (**c**) A typical mass spectrum recorded using Dielectric Barrier Discharge Ionization (*DBDI*), showing predominated fragments (*m/z* 197, 226) of TNT (Reprinted from Ref. [32] with permission. Copyright 2007, Wiley). (**d**) A typical mass spectrum recorded using Desorption Electrospray Ionization (*DESI*) and a calibration curve in the inset (Reprinted from Ref. [33] with permission. Copyright 2005, American Chemical Society). (**e**) A typical mass spectrum recorded using Extractive Electrospray Ionization (*EESI*). No fragment of TNT was observed using EESI. MS/MS spectra of the radical anion of TNT and of the adduct $[RDX + CH_3CO_2]^-$ in the insets (Reprinted from Ref. [34] with permission. Copyright 2009, Elsevier BV). (**f**) A typical mass spectrum recorded using Secondary Electrospray Ionization (*SESI*) and a calibration curve in the inset. No radical anion of TNT (*m/z* 227) was observed using SESI (Reprinted from Ref. [35] with permission Copyright 2009, Elsevier BV). (**g**) *Top panel* (*A*) TNT single photon laser ionization mass spectrum. *Lower panel* (*B*) shows an expanded view of the molecular anion region (Reprinted from Ref. [36] with permission. Copyright 2006, American Chemical Society)

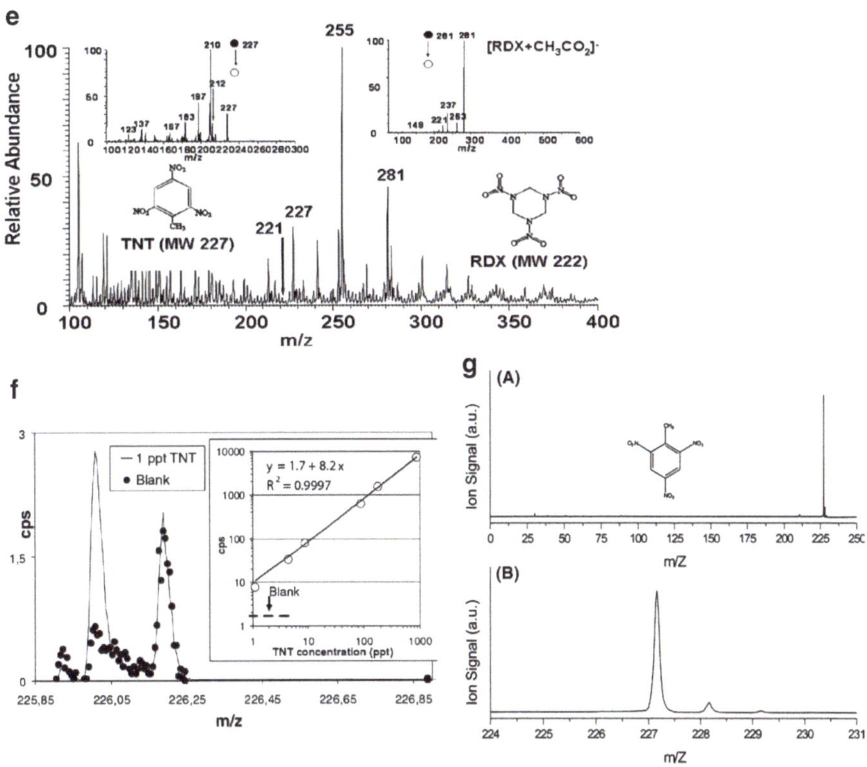

Fig. 3.6 (continued)

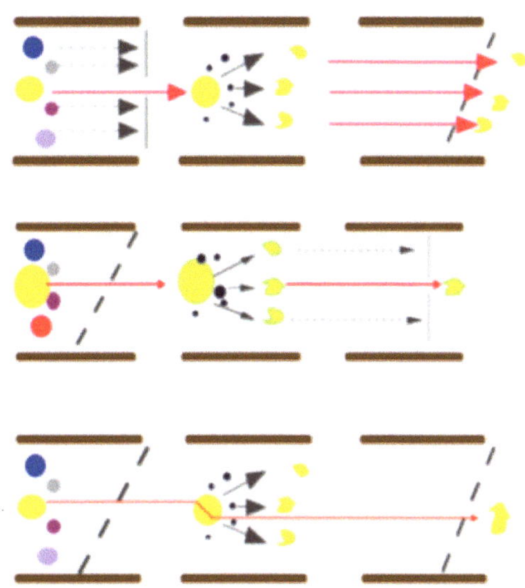

Fig. 3.7 Scan modes used in tandem mass spectrometry: *top*: product ion scan; *middle*: precursor ion scan; *bottom*: neutral loss scan

and electron transfer dissociation (ETD); (c) interactions of ions with photons by using infrared multiphoton dissociation (IRMPD); (d) interactions of ions with surfaces: surface-induced dissociation (SID).

(a) *Collision-induced dissociations* (CIDs): in CID experiments, a selected ion collides with a neutral target gas, generally helium, nitrogen or argon. This is accompanied by a conversion of part of the translational energy of the ion to internal energy with a consequent bond cleavage and dissociation.

The effects of the collisions, yielding to ion decomposition, depend on different parameters, such as collision energy (high (keV) collision energies, using tandem sector and time-of-flight instruments, or low (eV range) energies, in tandem quadrupole and ion trapping instruments), path length (ranging from 1 to several cm) and number of collisions (single or multiple collisions). A comparison between features of collision induced dissociations carried out in different analyzers, is reported in Table 3.3.

Table 3.3 Comparison between features of collision induced dissociations carried out in different analyzers: sectors (*top*), quadrupole (*middle*), ion trap (*bottom*)

Analyzer	Energy (eV)	Path length (cm)	Time scale (μs)	Number of collisions
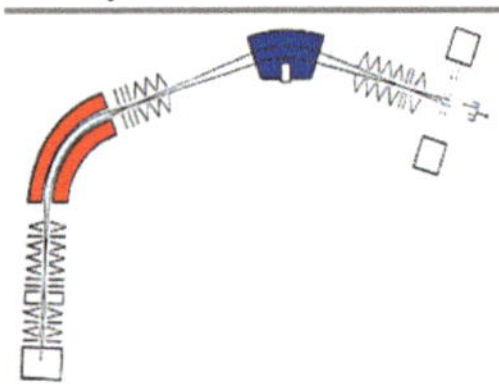	10^4	1	0.25	1–5
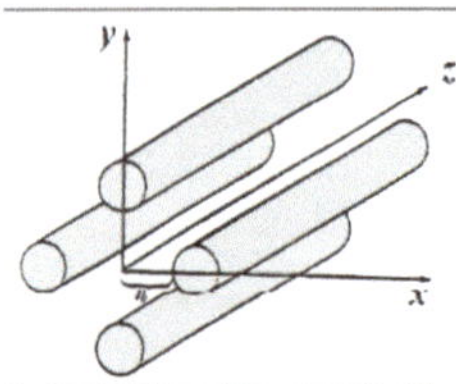	10–100	15	100	1–20
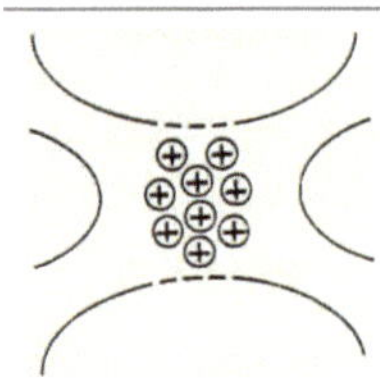	1–10	300–3,000	10^5–10^6	1–5

Full scan negative-ion mode and low energy CID spectra of the molecular anions produced by LTP ionization of different explosives are reported in Fig. 3.8.

(b) *interactions of ions with electrons*: electron capture dissociation (ECD) and electron transfer dissociation (ETD). The first method on interactions of ions with electrons has been electron capture dissociation [41]. ECD is generally used in Fourier transform ion cyclotron resonance (FT-ICR) mass spectrometers, due to the need of several milliseconds interaction between ions and electron, and because ECD efficiency is highest for low-energy electrons (<1 eV) which are difficult to provide in quadrupole ion traps where strong r.f. potentials affect the electrons' movement. Stabilization of the captured electron is faster than electron emission, which is usually on the time-scale of 10^{-14} s, so bond dissociation occurs faster than a typical bond vibration.

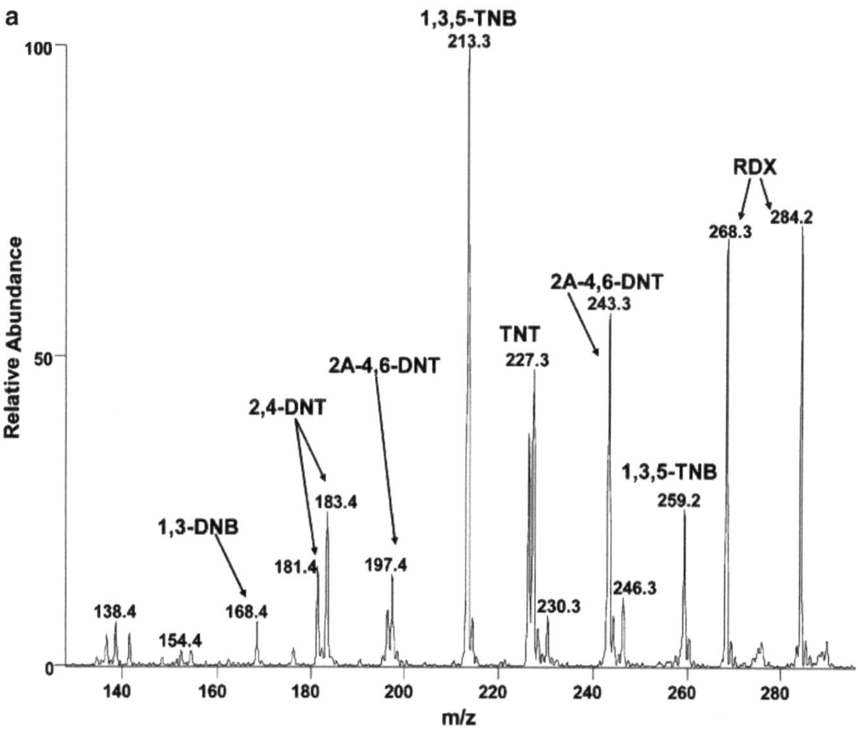

Fig. 3.8 (**a**) Mixture of explosives (30 ng each) examined from a heated (120 °C) glass slide. Full-scan negative-ion mode LTP-MS spectrum of 2,4-dinitrotoluene (2,4-DNT) (*m/z* 181 ($[M-H]^-$) and 183 ($[M+H]^-$)); 1,3-dinitrobenzene (1,3-DNB) (*m/z* 168 ($M^{-\bullet}$)); 1,3,5-trinitrobenzene (1,3,5-TNB) (*m/z* 213 ($M^{-\bullet}$) and 259 ($[M+NO_2]^-$)); 2-amino-4,6-dinitrotoluene (2A-4,6-DNT) (*m/z* 197 ($M^{-\bullet}$) and 243 ($[M+NO_2]^-$)); RDX (*m/z* 268 ($[M+NO_2]^-$) and 284 ($[M+NO_3]^-$)); and TNT (*m/z* 226 ($[M-H]^-$) and 227 ($M^{-\bullet}$)). (**b**–**e**) LTP-MS/MS product ion negative-ion mode spectra of the $M^{-\bullet}$ radical anions of selected explosives: (**b**) *m/z* 227 (TNT); (**c**) *m/z* 168 (1,3-DNB); (**d**) *m/z* 213 (1,3,5-TNB); and (**e**) *m/z* 197 (2A-4,6-DNT) (Reprinted from Ref. [40] with permission.. Copyright 2011, American Chemical Society)

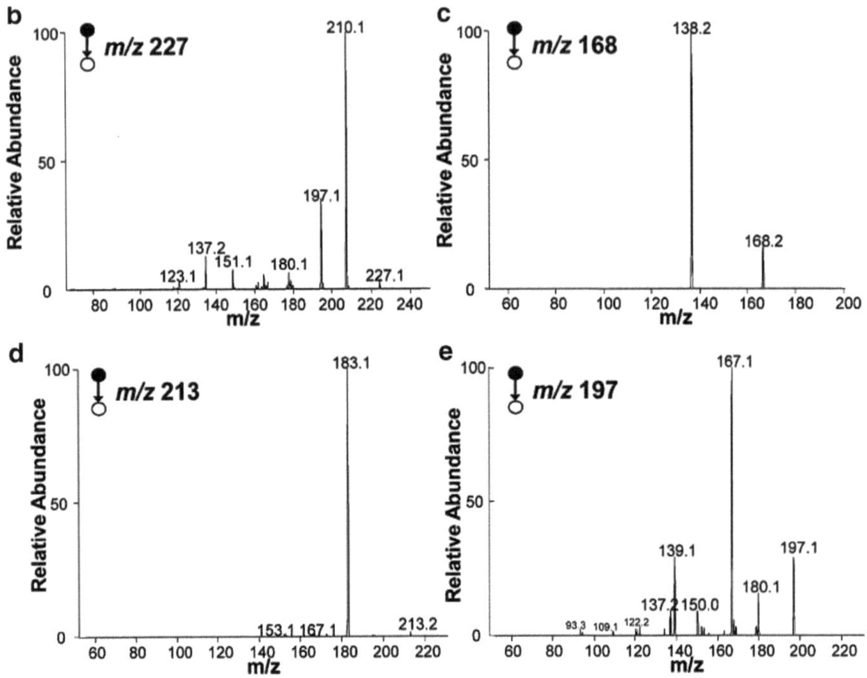

Fig. 3.8 (continued)

Electron transfer dissociation (ETD) [42] is based on ion-ion reactions, occurring in a quadrupole ion trap, between an electron-reach species, i.e. a radical anion of a polyciclic aromatic compound produced by chemical ionization, and a multiply charged ion.

In both ECD and ETD, the final result is the acquisition of an electron with a charge-state reduction of the ion and subsequent fragmentation. That is the reason why ECD and ETD can be applied to positive ions with at least a double charge and electrospray is the most common ionization techniques.

In both methods the occurrence of direct bond cleavages only is observed with no or minimal energy redistribution and randomization away from the reaction site. As a result, strong backbone N-Cα bonds of peptides are cleaved, forming *c* and *z*-type ions, complementary to *b* and *y* ions normally produced by low-energy CID.

(c) *interactions of ions with photons* by using infrared multiphoton dissociation (IRMPD) or ion spectroscopy.

Ions can also be excited and subsequently fragmented by the absorption of one or more photons generally produced by using lasers of different wavelengths. In infrared wavelengths, multiphoton processes are consequently needed to excite ions sufficiently for efficient fragmentation. IRMPD and ion spectroscopy are applied to trapped ions in ion trap or FT-ICR analyzers.

In IRMPD, precursor ions, that must be able to absorb energy in the form of photons, producing excited states above the threshold of its dissociation, are activated by a low-power (<100 W) continuous-wave CO_2 (10.6 μm) laser for a selected irradiance time (usually on the order of tens to hundreds of milliseconds), followed by the rapid redistribution of energy over all the vibrational degrees of freedom and consequent fragmentation.

IRMPD is a non-selective activation method and, as CID, results in cleavage sites at the most vulnerable bonds, such as peptide bonds, through rearrangement reactions usually yields rich fragmentation spectra. The activation is stepwise, by subsequent absorption of photons, and dissociation occurs by low-energy pathways, often the lowest that is available.

High power (up to 10 MW) free electron lasers with wide tunability (40–2,500 cm^{-1}), available only in specialized installations, allow to obtain a full infrared spectrum of the ions of interest.

(d) *interactions of ions with surfaces:* surface induced dissociation (SID) [43, 44]. SID involves a single step activation (time frame ~10^{-12} s) occurring when an ion collides on a target surface with or without self-assembled monolayer, causing dissociation of the impacting ions as well as the formation of other species, such as ion-surface reaction products, due to other side reactions.

The products of SID depend on the energy of the ion and on the nature of both the ion and the surface.

References

1. De Hoffmann E, Stroobant V (2007) Mass spectrometry: principles and applications, 3rd edn. Wiley, Chichester
2. Gross JH (2011) Mass spectrometry. A textbook, 2nd edn. Springer, Berlin
3. Dass C (2007) Fundamentals of contemporary mass spectrometry. Wiley-Interscience, New York
4. Desiderio DM, Nibbering NM (2009) In: Ekman R, Silberring J, Westman-Brinkmalm AM (eds) Mass spectrometry: instrumentation, interpretation, and applications. Wiley, Hoboken
5. Gross ML, Caprioli RM (eds) (2003) The encyclopedia of mass spectrometry, vol 1–10. Elsevier Science, Oxford
6. Venter A, Nefliu M, Cooks RG (2008) Ambient desorption ionization mass spectrometry. Trac 27:284–290
7. Cooks RG, Ouyang Z, Takáts Z, Wiseman JM (2006) Ambient mass spectrometry. Science 311:1566–1570
8. Van Berkel GJ, Pasilis SP, Ovchinnikova O (2008) Established, emerging atmospheric pressure surface sampling/ionization techniques, for mass spectrometry. J Mass Spectrom 43:1161–1180
9. Chen H, Gerardo G, Zenobi R (2009) What can we learn from ambient ionization techniques? J Am Soc Mass Spectrom 20:1947–1963
10. Alberici RM, Simas RC, Sanvido GB, Romao W, Lalli P, Benassi M, Cunha IBS, Eberlin MN (2010) Ambient mass spectrometry: bringing MS into the "real world". Anal Bioanal Chem 398:265–294

11. Huang MZ, Yuan CH, Cheng SC, Cho YT, Shiea J (2010) Ambient ionization mass spectrometry. Annu Rev Anal Chem 3:43–65
12. Huang MZ, Cheng SC, Cho YT, Shiea J (2011) Ambient ionization mass spectrometry: a tutorial. Anal Chim Acta 702:1–15
13. Ifa DR, Wu C, Ouyang Z, Cooks RG (2010) Desorption electrospray ionization and other ambient ionization methods: current progress and preview. Analyst 135:669–681
14. Weston DJ (2010) Ambient ionization mass spectrometry: current understanding of mechanistic theory; analytical performance and application areas. Analyst 135:661–668
15. Harris GA, Galhena AS, Fernández FM (2011) Ambient sampling/ionization mass spectrometry: applications and current trends. Anal Chem 83:4508–4538
16. Liu J, Wang H, Manicke NE, Lin JM, Cooks RG, Ouyang Z (2010) Direct analysis of biological tissue by paper spray mass spectrometry. Anal Chem 82:2463–2471
17. Liu YY, Ma XX, Lin ZQ, He MJ, Han GJ, Yang CD, Xing Z, Zhang SC, Zhang XR (2010) Imaging mass spectrometry with a low-temperature plasma probe for the analysis of works of art. Angew Chem Int Ed 49:4435–4437
18. Ifa DR, Wiseman JM, Song QY, Cooks RG (2007) Development of capabilities for imaging mass spectrometry under ambient conditions with desorption electrospray ionization (DESI). Int J Mass Spectrom 259:8–15
19. Coello Y, Jones AD, Gunaratne TC, Dantus M (2010) Atmospheric pressure femtosecond laser imaging mass spectrometry. Anal Chem 82:2753–2758
20. Pol J, Vidova V, Kruppa G, Kobliha V, Novak P, Lemr K, Kotiaho T, Kostiainen R, Havlicek V, Volny M (2009) Automated ambient desorption-ionization platform for surface imaging integrated with a commercial fourier transform ion cyclotron resonance mass spectrometer. Anal Chem 81:8479–8487
21. Laskin J, Heath BS, Roach PJ, Cazares L, Semmes OJ (2012) Tissue imaging using nanospray desorption electrospray ionization mass spectrometry. Anal Chem 84:141–148
22. Cody RB, Laramée JA, Durst HD (2005) Versatile new ion source for the analysis of materials in open air under ambient conditions. Anal Chem 77:2297–2302
23. Nemes P, Vertes A (2007) Laser ablation electrospray ionization for atmospheric pressure, in vivo, and imaging mass spectrometry. Anal Chem 79:8098–8106
24. Nemes P, Barton AA, Li Y, Vertes A (2008) Ambient molecular imaging and depth profiling of live tissue by infrared laser ablation electrospray ionization mass spectrometry. Anal Chem 80:4575–4582
25. Judge EJ, Brady JJ, Dalton D, Levis RJ (2010) Analysis of pharmaceutical compounds from glass, fabric, steel, and wood surfaces at atmospheric pressure using spatially resolved, nonresonant femtosecond laser vaporization electrospray mass spectrometry. Anal Chem 82:3231–3238
26. Nefliu M, Smith JN, Venter A, Cooks RG (2008) Internal energy distributions in desorption electrospray ionization (DESI). J Am Soc Mass Spectrom 19:420–427
27. Takáts Z, Cotte-Rodriguez I, Talaty N, Chen H, Cooks RG (2005) Direct, trace level detection of explosives on ambient surfaces by desorption electrospray ionization mass spectrometry. Chem Commun 1950–1952
28. Takats Z, Wiseman JM, Gologan B, Cooks RG (2004) Mass spectrometry sampling under ambient conditions with desorption electrospray ionization. Science 306:471–473
29. Sanders NL, Kothari S, Huang G, Salazar G, Cooks RG (2010) Detection of explosives as negative ions directly from surfaces using a miniature mass spectrometer. Anal Chem 82:5313–5316
30. Zhang Y, Ma XX, Zhang SC, Yang CD, Ouyang Z, Zhang XR (2009) Direct detection of explosives on solid surfaces by low temperature plasma desorption mass spectrometry. Analyst 134:176–181
31. Chen HW, Zheng J, Zhang X, Luo MB, Wang ZC, Qiao XL (2007) Surface desorption atmospheric pressure chemical ionization mass spectrometry for direct ambient sample analysis without toxic chemical contamination. J Mass Spectrom 42:1045–1056

32. Na N, Zhang C, Zhao MX, Zhang SC, Yang CD, Fang X, Zhang XR (2007) Direct detection of explosives on solid surfaces by mass spectrometry with an ambient ion source based on dielectric barrier discharge. J Mass Spectrom 42:1079–1085
33. Cotte-Rodriguez I, Takats Z, Talaty N, Chen HW, Cooks RG (2005) Desorption electrospray ionization of explosives on surfaces: sensitivity and selectivity enhancement by reactive desorption electrospray ionization. Anal Chem 77:6755–6764
34. Chen HW, Hu B, Hu Y, Huan YF, Zhou ZQ, Qiao XF (2009) Neutral desorption using a sealed enclosure to sample explosives on human skin for rapid detection by EESI-MS. J Am Soc Mass Spectrom 20:719–722
35. Martinez-Lozano P, Rus J, De La Mora GF, Hernandez M, De La Mora JF (2009) Secondary electrospray ionization (SESI) of ambient vapors for explosive detection at concentrations below parts per trillion. J Am Soc Mass Spectrom 20:287–294
36. Mullen C, Irwin A, Pond BV, Huestis DL, Coggiola MJ, Oser H (2006) Detection of explosives and explosives-related compounds by single photon laser ionization time-of-flight mass spectrometry. Anal Chem 78:3807–3814
37. Ifa DR, Jackson AU, Paglia G, Cooks RG (2009) Forensic applications of ambient ionization mass spectrometry. Anal Bioanal Chem 394:1995–2008
38. Fenselau C, Demirev P (eds) (2011) Rapid characterization of microorganisms by mass spectrometry, ACS symposium, vol 1065. American Chemical Society, Washington, DC
39. Chingin K, Liang J, Chen H (2014) Direct analysis of in vitro grown microorganisms and mammalian cells by ambient mass spectrometry. RSC Adv 4:5768–5781
40. Garcia-Reyes JF, Harper JD, Salazar GA, Charipar NA, Ouyang Z, Cooks RG (2011) Detection of explosives and related compounds by low-temperature plasma ambient ionization mass spectrometry. Anal Chem 83:1084–1092
41. Zubarev RA, Kelleher NL, Mc Lafferty FW (1998) Electron capture dissociation of multiply charged protein cations. A nonergodic process. J Am Chem Soc 120:3265–3266
42. Syka JE, Coon JJ, Schroeder MJ, Shabanowitz J, Hunt DF (2004) Peptide and protein sequence analysis by electron transfer dissociation mass spectrometry. Proc Natl Acad Sci U S A 101:9528–9533
43. Mabud MA, Dekrey MJ, Cooks RG (1985) Surface-induced dissociation of molecular ions. Int J Mass Spectrom Ion Proc 67:285–294
44. Wysocki VH, Ding JM, Jones JL (1992) Surface-induced dissociation in tandem quadrupole mass spectrometers: a comparison of three designs. J Am Soc Mass Spectrom 3:27–32

Chapter 4
Modern Sample Preparation Techniques for Gas Chromatography-Mass Spectrometry Analysis of Environmental Markers of Chemical Warfare Agents Use

Oliver Terzic and Pim de Voogt

Abstract The chapter introduces problematics of on-site chemical analysis in the investigations of past chemical warfare agents (CWA) events. An overview of primary environmental degradation pathways of CWA leading to formation of chemical markers of their use is given. Conventional and modern sample preparation approaches for on-site gas chromatography (GC) – mass spectrometry (MS) analysis of CWA and their degradation products in environmental sample matrices are presented. The advantages, disadvantages, and relative performance of the sample preparation techniques are discussed.

Keywords Chemical warfare agents • On-site analysis • Gas chromatography – mass spectrometry • Liquid phase microextraction • Solid phase microextraction • In-sorbent tube sample preparation • Thermal desorption

---

The views and recommendations in this article are those of authors and do not represent official OPCW or UvA/KWR Watercycle Research Institute policy.

O. Terzic (✉)
Inspectorate Division, Organisation for the Prohibition of Chemical Weapons (OPCW), Johan de Wittlaan 32, 2517 JR The Hague, The Netherlands
e-mail: Oliver.Terzic@opcw.org

P. de Voogt
Institute for Biodiversity and Ecosystem Dynamics, University of Amsterdam (UvA), POBox 94240, 1090 GE Amsterdam, The Netherlands

KWR Watercycle Research Institute, Nieuwegein, The Netherlands

J. Banoub (ed.), *Detection of Chemical, Biological, Radiological and Nuclear Agents for the Prevention of Terrorism*, NATO Science for Peace and Security Series A: Chemistry and Biology, DOI 10.1007/978-94-017-9238-7_4

4.1 Introduction

Recent developments in the Syrian civil war, where the government and antigovernment forces accused each other of using chemical weapons (CW), have revived interest of security communities in technicalities of investigation of chemical warfare agents (CWA) events.

A CWA can be defined as a substance intended for use in military operations or terrorist activity to kill, seriously injure, or incapacitate through its physiological effects. Traditionally, riot control agents, defoliants, smoke and incendiary weapons are excluded from this definition.

Almost any type of conventional military weapon system like artillery shells, aerial bombs, grenades, mines, rockets, and missiles can be used to deliver CWA to target. These systems, however, have to be specially designed to carry and disseminate chemical payload as a vapor or/and aerosol [1]. Additionally, CWA can be sprayed from air, land, and water vehicles or covertly used to contaminate food and water supplies. Non-traditional or improvised CWA dissemination means could be expected in terrorist attacks such as the Tokyo subway incident [2].

When a CWA is released in the environment, it becomes distributed among four major compartments: water, air (atmosphere), soil, and living organisms (Fig. 4.1).

Immediately following the release, distribution of CWA is largely determined by characteristics of delivery system. Agent vapors and small liquid/solid particles are carried by air currents, while any large particles and liquid drops fall out in a ballistic-like trajectory and are quickly deposited on the ground. A portion of the agent payload can be thermally degraded upon detonation of CW, if a high explosive charge ("burster") is used for the dispersal. Soon after the dispersion, the environmental fate of the CWA becomes principally governed by its physical and chemical properties, weather conditions and terrain, similar to other environmental contaminants [3–5]. The agent may be removed from the air by depositing and

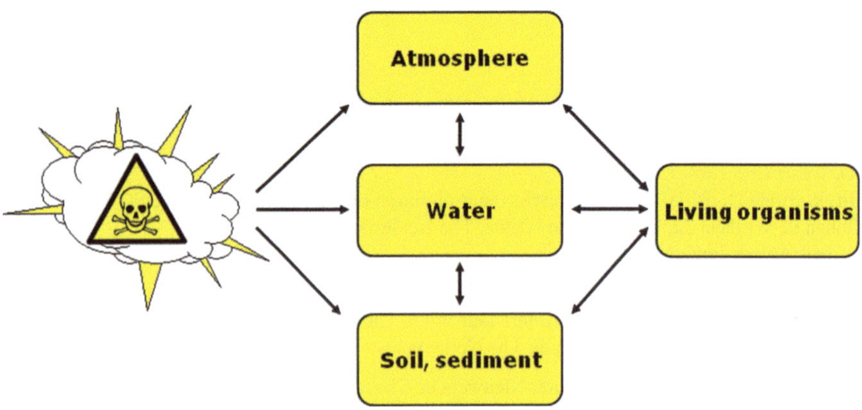

Fig. 4.1 Release of CWA in the environment

adsorbing to the ground or vegetation, or by wet precipitation. Rain is known to be efficient in cleansing the atmosphere. Sufficiently volatile agents may be re-released from surfaces or water bodies to the atmosphere for further cycles of travel and present a hazard until sufficiently diluted or degraded. High winds and strong turbulence reduce the concentration and increase the area coverage by more quickly carrying away and diffusing the agent. The agent adsorbed onto solid particles or dissolved in water can move vertically through the soil profile or across surface in run-off. A number of different types of chemical reactions may occur in the environmental compartments and at their borders transforming originally released CWA and producing new compounds. Depending on a time period elapsed from the release, the CWA degradation products may be the only chemical markers of the event remaining on the field.

A field investigation of a prior CWA event may provide very diverse sampling opportunities and sample types. The sampling priorities are always samples likely to contain intact "neat" agents, like munitions contents, residues or fragments from munitions and other delivery systems. Environmental samples, such as soil, concrete chips, other types of solid secondary fragmentation, liquids or water collected from the point of a munitions impact, contaminated paints, chips of plastics or rubber, dry or wet swabs from munitions and possibly contaminated surfaces, equipment or clothing from affected personnel, they all may contain only traces of "neat" agents and/or their degradation products. Additionally, such samples usually have a more complex matrix, producing interferences in the analytical procedure. Air sampling may have a limited, if any value when applied in investigation of past events. Body fluids such as human blood or urine, vegetation, dead animals or insects at the point of impact are more likely to contain metabolites, protein and DNA adducts identified as specific long term or short term CWA exposure markers [5]. Fur of exposed animals or human hair may retain traces of chemical warfare agents and their degradation products.

The principal analytical technique used for the identification of CWA and their degradation products is gas chromatography-mass spectrometry (GC-MS). Liquid chromatography-mass spectrometry (LC-MS) has increasing use, but it is still more limited to applications in analysis of some less reactive CWA and polar degradation products in aqueous samples/water extracts, as well as exposure markers in biomedical samples [6, 7]. Beside versatility, the major advantage of GC-MS over current LC-MS instrumentation is that GC-MS is fieldable. Sample transport to an off-site laboratory for analysis can be a complicated and expensive task due to possible toxicity of the sample. It is also likely to be a time-consuming process. In the majority of cases an on-site analysis capability is a clear advantage providing relatively fast information on presence/absence of chemical hazards and reducing the number of samples to be sent off-site through initial screening. A range of on-site GC-MS instrumentation is now commercially available (Fig. 4.2). Inficon has developed a man-portable, self contained Low Thermal Mass (LTM) GC-Quadrupole MS instrument capable of sampling and on-spot analysis [8]. Torion® has introduced an even lighter compact, field LTM GC-Toroidal Ion Trap MS [9]. The Griffin™ 460 mobile LTM GC-Cylindrical Ion Trap MS enables multi-modal

Fig. 4.2 Examples of on-site GC-MS instrumentation

sample introduction [10]. Agilent Technologies has invested into a rugged version of bench top instrumentation through its LTM GC -5975T Quadrupole MS [11]. These are only few examples of the on-site GC-MS equipment available on the market. Even standard laboratory bench top GC-MS instruments can be used in an on-site laboratory. The Organisation for the Prohibition of Chemical Weapons (www.OPCW.com) has been shipping worldwide the bench top GC-MS instruments, using them successfully in wide range of environments and conditions.

Once the representative sample is obtained, success of the analysis depends greatly on sample preparation. Neat agent samples typically require only dilution with an organic solvent. With the environmental samples, the analytes must be extracted from the sample matrix. The environmental degradation of CWA can yield polar products with insufficient volatility for GC analysis or products which elute poorly on GC. The analytical task is complicated and this is reflected in the sample preparation procedures, which may be time consuming.

This article gives an overview of environmental chemical markers of CWA use, presents conventional sample preparation approaches for GC-MS analysis of CWA and their degradation products in environmental sample matrices as well as some modern, alternative sample preparation techniques. The advantages, disadvantages, and relative performance of the sample preparation techniques are discussed. Special attention is given to their potential for application in on-site analysis.

Compared to the off-site analysis, sample preparation procedures used in an on-site laboratory aim at reduction of the amount and weight of sample, solvents, reagents, waste generated and equipment used. Another desired feature of candidate on-site sample preparation technique is reduction in overall analysis time, providing fast information and allowing for higher sample throughput.

4.2 An Overview of Environmental Chemical Markers of CWA Use

Chemical warfare agents are commonly classified into blister, nerve, choking, vomiting, and blood agent categories based on their effect on humans. Major CWA of concern in terms of military and terrorist capacity and past use are the blister and nerve agents. For these reasons, in the present review the emphasis will be on the latter groups of CWA. The choking, blood, and vomiting agents are generally considered obsolete CWA.

4.2.1 Blister Agents

Blister agents, or vesicants, are probably the best known CWA. As the name suggests, blister agents can cause large and often life-threatening skin blisters which resemble severe burns [12]. Three families of blistering compounds that dominated military application are: sulfur mustards, nitrogen mustards and Lewisites.

4.2.1.1 Sulfur Mustards

Term "sulfur mustards" is commonly applied for a family of chlorinated thioethers with blistering capabilities. Depending on the manufacturing process, several family members can be present in weapon grade agent along with number of other impurities. The three best known compounds are shown in Fig. 4.3. The discussion in this article refers to bis(2-chloroethyl)sulfide, with NATO designation HD, which is usually the major component of all weapon grade sulfur mustard agents. The crude, undistilled form of bis(2-chloroethyl)sulfide has NATO designation H. When produced in a mixture with bis(2-chloroethylthioethyl)ether, or agent T, it has NATO designation HT. Sesquimustard, or agent Q, is normally present as an impurity of bis(2-chloroethyl)sulfide. It can be also prepared in a mixture with H to depress the freezing point of the former. The mixture receives then the NATO designation HQ. A more complete list of common bis(2-chloroethyl)sulfide impurities is given elsewhere [13].

The chemistry of HD in the environment is very complex. The most important environmental degradation pathways are shown in Fig. 4.4.

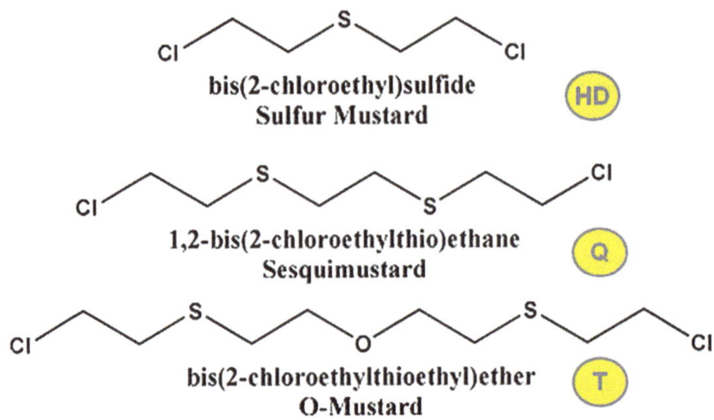

Fig. 4.3 The best known members of sulfur mustards family of CWA, their chemical names, synonyms and NATO codes

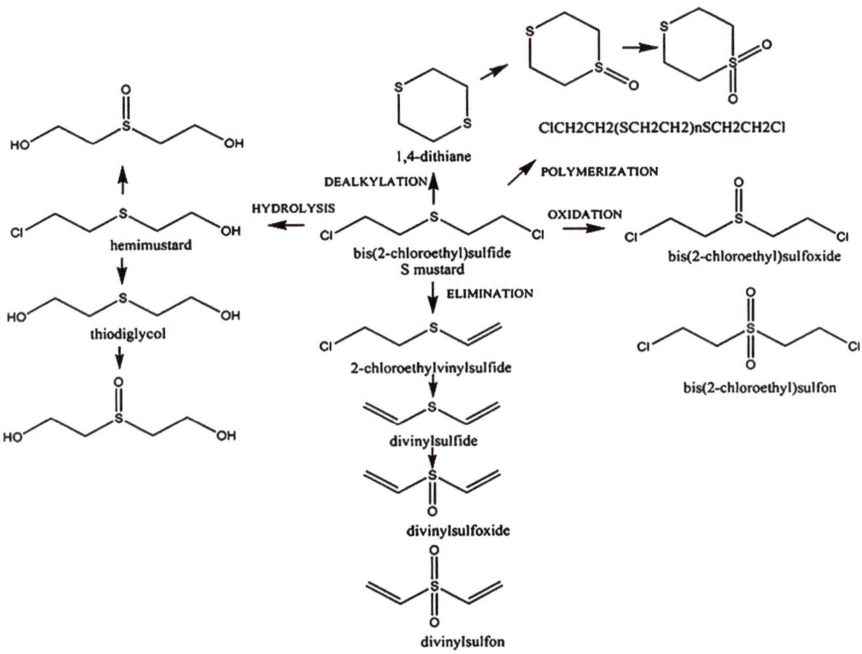

Fig. 4.4 Primary degradation pathways of HD in the environment

1,4-Dithiane is a common impurity of HD, but it is also created by thermal decomposition of the agent during detonation of the CW. Persistency of HD in the environment depends greatly on the environmental conditions and matrices. It is known that HD can persist for weeks in soil, if the ambient temperature and humidity are low [14]. In one occasion, soil samples were collected from a

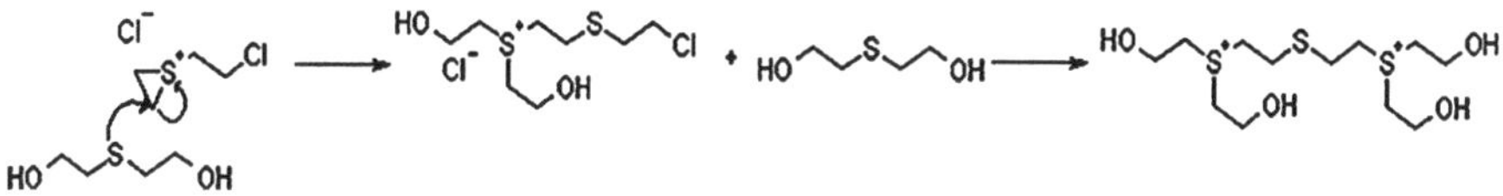

Fig. 4.5 Hydrolysis of HD with formation of sulfonium ion aggregates in water

bomb crater 10–12 weeks after an alleged CW event. Beside the intact HD, a number of degradation products from Fig. 4.4 were detected using GC-MS [15]. In water, HD can have a peculiar fate. It dissolves slowly, but once dissolved hydrolyses rather rapidly. However, sulfonium ion aggregates can form around the undissolved, bulk HD, shielding it from further dissolution (Fig. 4.5). In this way, HD can persist under water and retain blister properties for decades [16, 17]. Especially persistent is viscous mustard, containing polymer thickeners such as polystyrene or montan wax. Thiodiglycol is the main hydrolysis product of HD before mineralisation. It is miscible with water, resistant to hydrolysis and photolysis, but susceptible to microbial degradation. A caution is required when taking thiodiglycol as a sole marker for prior presence of HD. Thiodiglycol is a common commercial product, in use as a solvent in antifreeze solutions, in dyestuffs for printing, and as a costabilizer in the production of polyvinyl chloride.

All of the sulfur containing HD related chemicals can have the reaction to the sulfoxide or the sulfone. However, these are typically minor pathways. The other sulfur mustards family members follow similar degradation pathway as HD: the hydrolysis of the terminal chlorine produces the corresponding chloroalcohols, while the subsequent hydrolysis gives the diols. In all cases the sulfones and sulfoxides should be expected to form.

4.2.1.2 Nitrogen Mustards

A number of chlorinated ethyl amines was tested as potential CWA candidates in between two world wars [18]. The three tertiary alkyl amines that have found military application are shown in Fig. 4.6. Tris-(2-chloroethyl)amine or HN3 is relatively persistent in the environment, while 2,2′-dichlorotriethylamine (HN1) and bis-(2-chloroethyl)methylamine (HN2) are moderately persistent. All three compounds are unstable in sunlight undergoing photolytic degradation with hydroxyl radicals in the atmosphere.

Hydrolysis is the major degradation pathway of nitrogen mustards in soil and water, especially under weakly alkaline conditions. Environmental degradation pathways of nitrogen mustards are shown in Fig. 4.7.

The intermediates in nitrogen mustards hydrolysis are ionic species that can form byproducts. In the case of HN1, the dimeric salt is a relatively minor byproduct. In the case of HN2, the reaction leading to formation of the dimer can be very fast, even explosive under certain conditions. The ultimate end point of the hydrolysis for

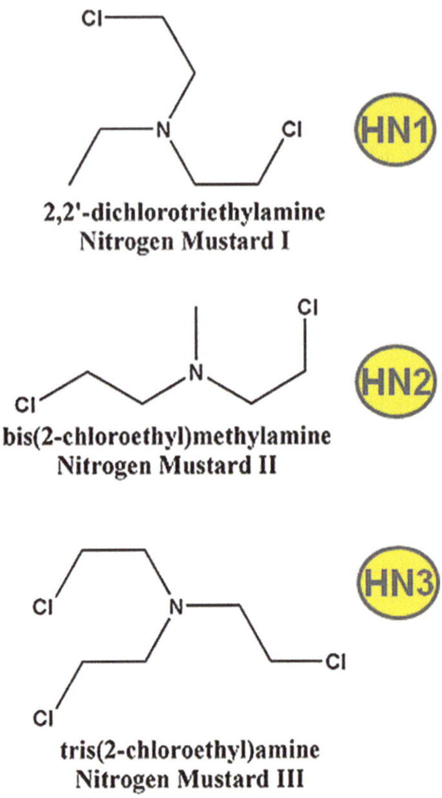

Fig. 4.6 Nitrogen mustards family of CWA, their chemical names, synonyms and NATO codes

all cases is corresponding ethanolamine. It is important to note that diethanolamine (hydrolysis product of HN1) and triethanolamine (hydrolysis product of HN3) have industrial uses, so that their presence in the environment is not unique to the CWA contamination.

HN3 reacts with hydrochloric acid giving odorless, rhombic plates of tris(2-chloroethyl) ammonium chloride. This is a water soluble compound, pretty stable and has the same toxic properties as a free amine. As such, it is a potential contaminant for water and food supplies.

4.2.1.3 Lewisites

Weapons-grade Lewisite presents a mixture of predominantly trans isomers of 2-chlorovinyldichloroarsine (L1), bis(2-chlorovinyl)chloroarsine (L2), and tris(2-chlorovinyl)arsine (L3). The chemical structures of Lewisites are shown in Fig. 4.8. A typical composition of the mixture is 90 % L1, 9 % L2 and 1 % L3. A catalytically mediated conversion of L1 stored in metal containers and/or its thermal dissociation after deployment of the weapon can result in relatively higher

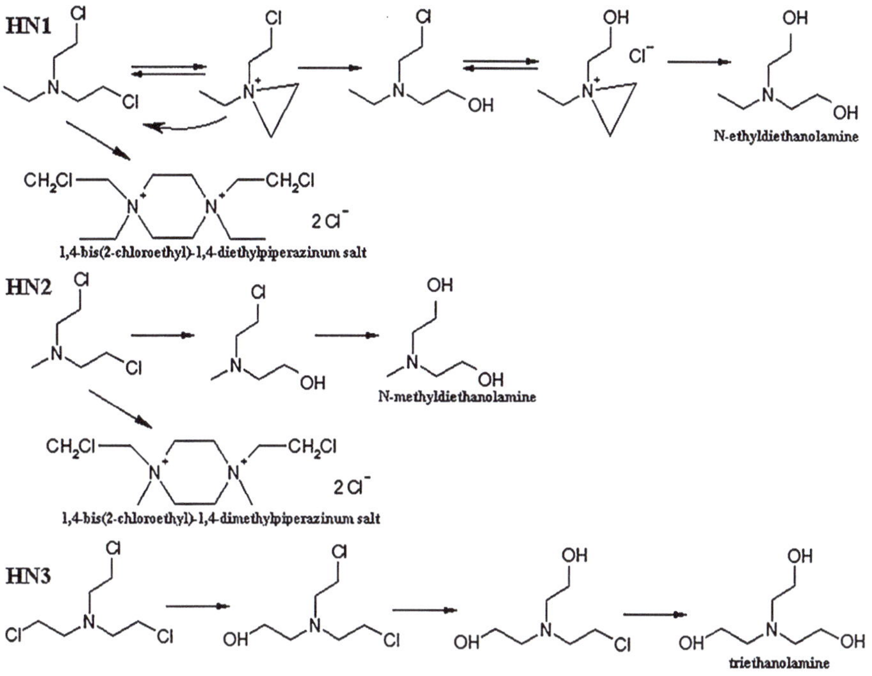

Fig. 4.7 Environmental degradation pathways of nitrogen mustards

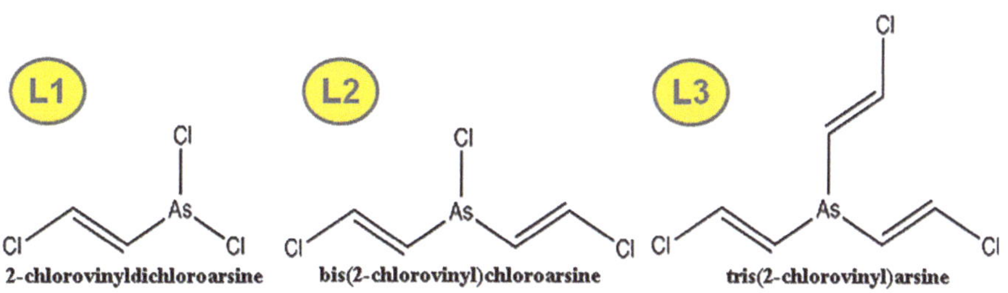

Fig. 4.8 Lewisites, their chemical names and NATO codes

percentages of L2 and L3 [18]. Although the mixture has only a limited solubility of 0.5 g/l in water, hydrolysis is the primary degradation pathway of L1 and L2 in the environment. The hydrolysis mechanisms are complex and in water, depending on the pH, may include several reversible reactions. These can be summarized as shown in the scheme in Fig. 4.9. Lewisite 3 is relatively inert, does not have blistering properties and does not react with nucleophiles such as water.

Both, the arsonous acid and the anhydride forms of L1 possess toxic/blistering properties. Since Lewisite and its immediate hydrolysis products are never found in the environment per se, they are good markers for prior CWA use. 2-Chlorovinylarsonous acid is considered the urinary metabolite/biomarker of L1 exposure [19].

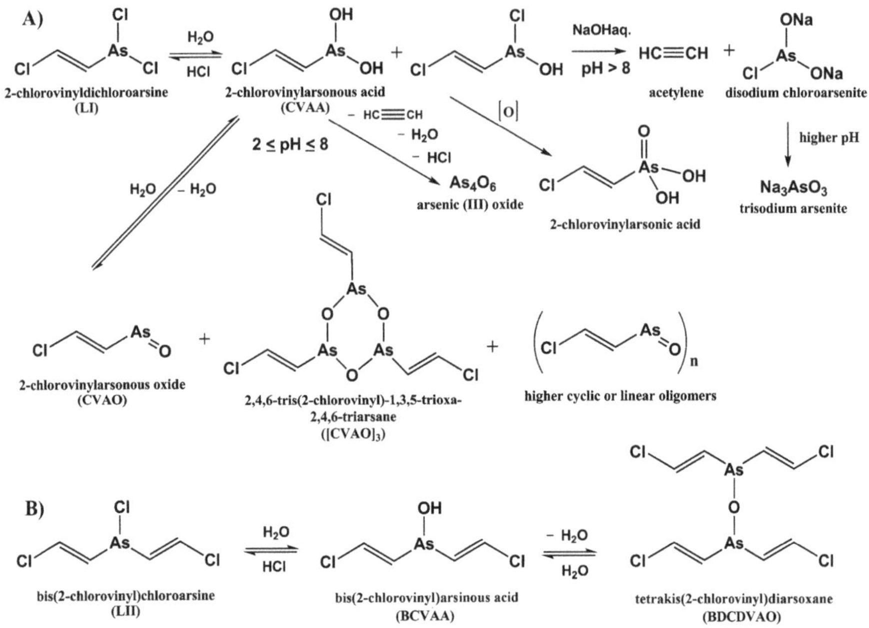

Fig. 4.9 Hydrolysis pathways of L1 (**a**) and L2 (**b**)

4.2.2 *Nerve Agents*

Nerve agents are primarily organophosphorus esters. They are generally divided into two groups, known by their code letters as G agents and V agents. The common characteristic of nerve agents from both groups is a C–P bond which is rarely found in the environment.

4.2.2.1 G Series of Nerve Agents

The principal G agents are GB (O-isopropyl methylphosphonofluoridate; sarin), GD (O-pinacolyl methylphosphonofluoridate; soman), GF (O-cyclohexyl methylphosphonofluoridate; cyclosarin), and GA (O-ethyl N,N-dimethylphosphoramidocyanidate; tabun). Their chemical structures are shown in Fig. 4.10.

The primary environmental degradation pathway of all G agents is hydrolysis. The principal hydrolysis product of GB, GD and GF is O-alkyl methylphosphonic acid (alternative name O-alkyl methylphosphonate), which slowly hydrolyses to methylphosphonic acid. The alkyl methylphosphonic acids are also the major urinary metabolites of nerve agents [20]. The hydrolysis reaction of GB is shown in Fig. 4.11 as an example. O-isopropyl methylphosphonic acid is chemically pretty stable and resistant to microbial degradation. Hydrolysis products (acids) and presence

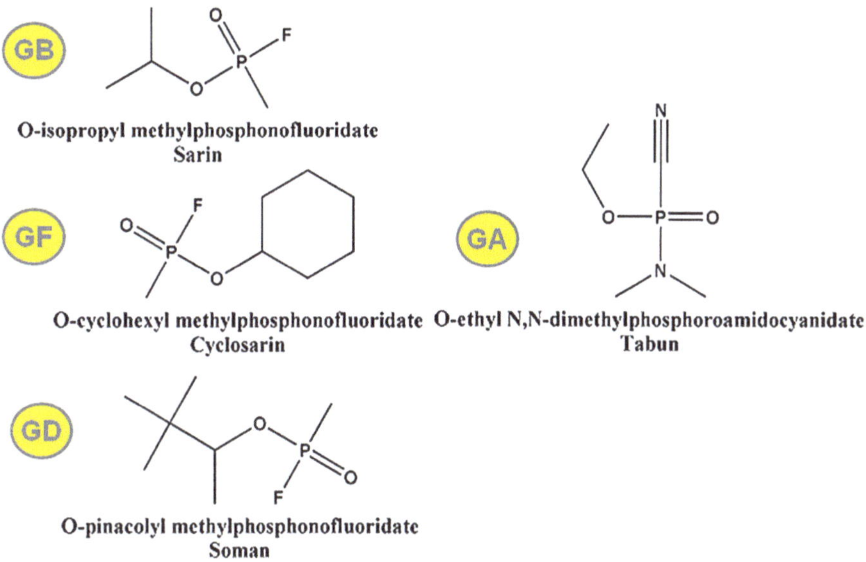

Fig. 4.10 Principal G series nerve agents, their chemical names and NATO codes

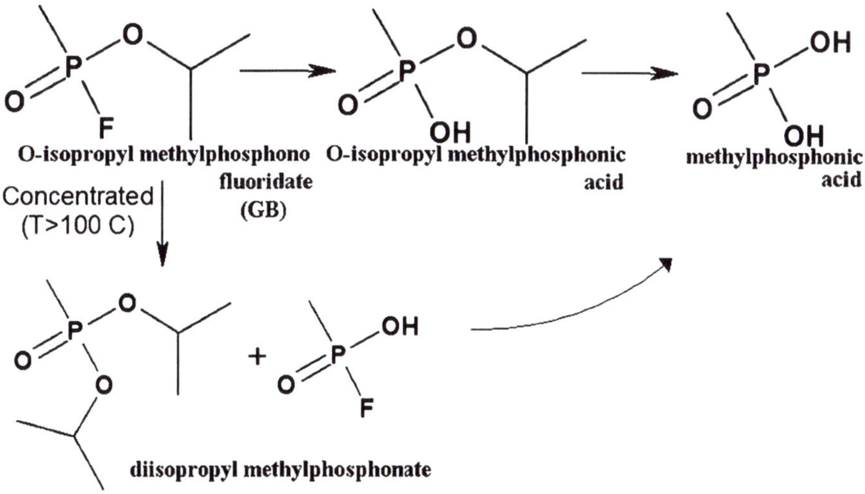

Fig. 4.11 Primary degradation pathway of GB

of cations, like Ca^{2+} and Mg^{2+} in seawater, increase the rate of hydrolysis. Diisopropyl methylphosphonate is miscible with water and potentially stable for months.

The reaction of GA with water is pH dependant [13] and as such it can proceed via different pathways (Fig. 4.12).

GA is toxic for aquatic organisms; however it is subject to microbial degradation in soil via O-dealkylation, C-dealkylation, nitrile hydrolysis and N-dealkylation.

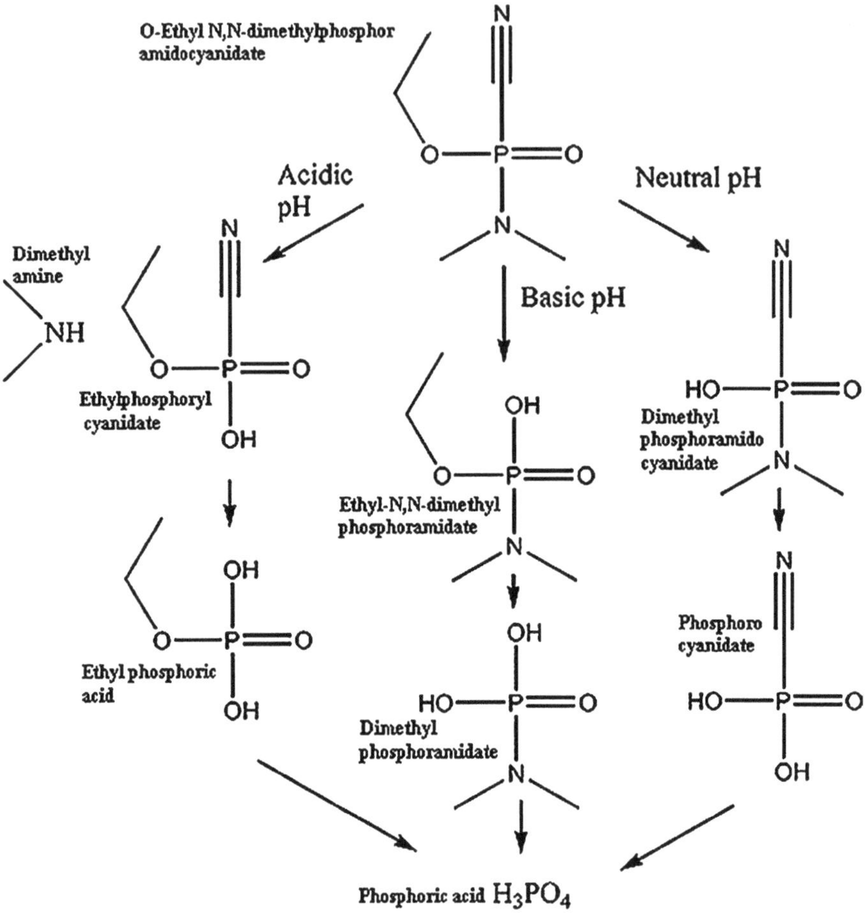

Fig. 4.12 Hydrolysis pathways of GA

4.2.2.2 V Series of Nerve Agents

The principal V agents are VX (O-ethyl S-2-diisopropylaminoethyl methylphosphonothiolate) and Russian VX or VR (O-isobutyl S-2-diethylaminoethyl methylphosphonothiolate). Their chemical structures are shown in Fig. 4.13.

V compounds are moderately soluble in water and relatively resistant to hydrolysis. Hydrolysis may occur by three pathways (P–S, P–O and S–C cleavage), depending on pH, temperature and concentration [21, 22]. These are shown in Fig. 4.14 with VX as an example. Hydrolysis chemistry of VR is similar to VX.

The intermediate known as the EA2192 (S-2(diisopropylaminoethyl)methylphosphonic acid) is a toxic compound, soluble and very stable in water. V compounds exposed to sunlight may undergo reversible photoisomerization to the corresponding phosphonothionates.

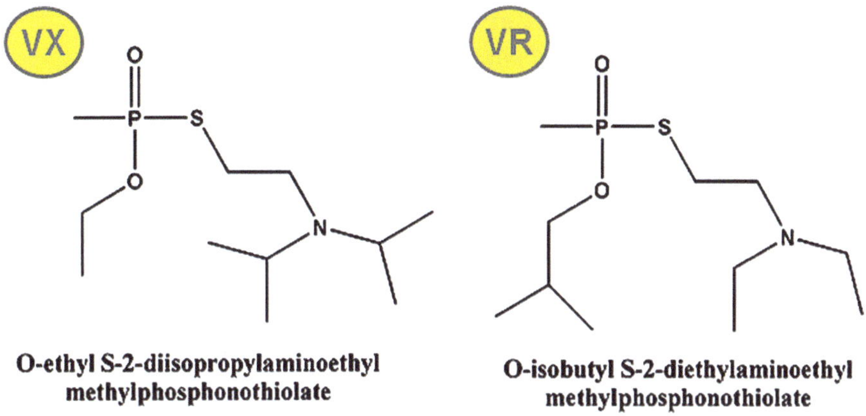

Fig. 4.13 Principal V series nerve agents, their chemical names and NATO codes

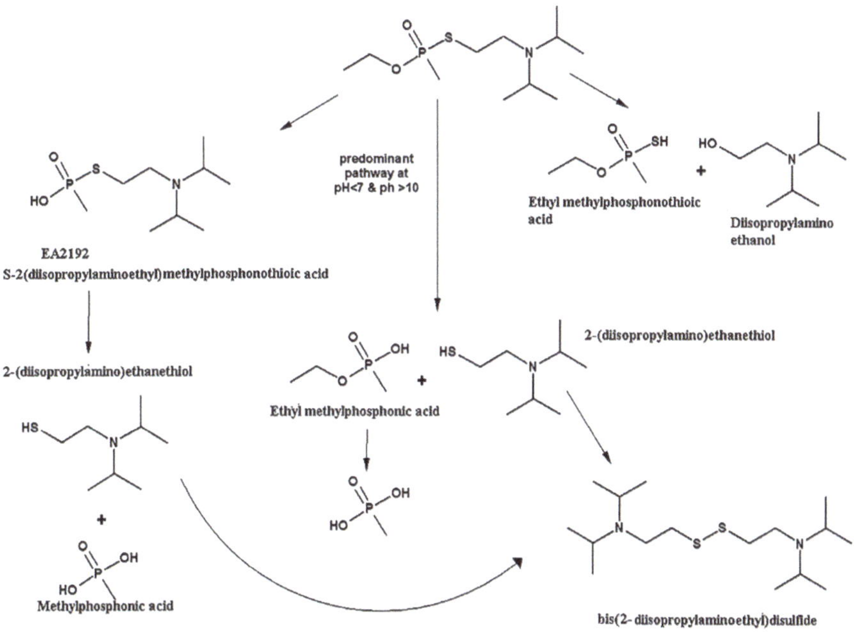

Fig. 4.14 Hydrolysis pathways of VX

Methylphosphonic acid is the final product of VX and VR degradation. It can persist for more than a decade in soil of testing sites. However, methylphosphonic acid should be taken cautiously as a sole chemical marker of CWA event. Its release to the environment is possible as degradation product of pesticides such as O,O-bis(2,4,5-trichlorophenyl) methylphosphonate, flame retardants such as dimethyl methylphosphonate, Fyrol 58 (Akzo Nobel), and Antiblaze 19 (Albright and Wilson).

4.3 On-site Sample Preparation for GC-MS Analysis in Investigation of Prior CWA Event

Chemical analysis in investigations of prior CWA events is focused on unambiguous identification of the CWA and their environmental chemical markers rather than on producing quantitative results. Since a wide range of compounds with widely differing polarity (Table 4.1), volatility and reactivity is targeted, sample preparation procedures are generic, aiming at isolation of groups of compounds with similar physical and chemical properties.

Conventionally, this is achieved utilizing solid–liquid and liquid–liquid extractions. The first step in the sample preparation is extraction of sample matrix with organic solvent to recover less polar analytes such as "intact" blister and nerve CWA. The extract is dried, concentrated if needed, and analyzed directly. High levels of hydrocarbons can be present in environmental matrices causing chromatographic interferences and thus, they have to be removed prior to the analysis. This can be achieved by adding an immiscible, relatively polar aprotic solvent such as acetonitrile, to the organic extract. The analytes, generally more polar than the hydrocarbons, are partitioning from nonpolar hydrocarbon solvent to the aprotic solvent [23]. The next step in the sample preparation is to extract the sample with D.I. water to recover more polar analytes, such are the majority of the CWA degradation products. A general flowchart of sample preparation for environmental samples is given in Fig. 4.15.

Less volatile and more polar compounds require derivatisation prior to GC-MS analysis. Trimethylsilyl (TMS) derivatisation with N,O-bis(trimethylsilyl)trifluoro-acetamide (BSTFA) or *tert*-butyldimethylsilyl (TBDMS) derivatisation with N-Methyl-N-*tert*-butyldimethylsilyl trifluoroacetamide (MTBSTFA) are commonly selected for on-site analysis (Fig. 4.16). This is due to their applicability to the broadest range of the target compounds [24]. The silylation reagents and the derivatives are sensitive to the presence of water; therefore water has to be removed prior to the derivatisation.

This can be achieved using laboratory equipment like a centrifuge vacuum evaporator or nitrogen evaporator (commercial examples are shown in Fig. 4.17). The same equipment can be used to support the derivatisation reaction that requires heating at about 60 °C for 30–45 min.

One portion of the aqueous sample/extract should be passed through the solid phase extraction cartridge containing a strong cation exchange resin prior to the evaporation and derivatisation. This is to remove cations such as Ca^{2+} and Mg^{2+} that form insoluble salts with alkylphosphonic or alkylthiophosphonic acids, reducing the silylation efficiency.

For analysis of Lewisite related compounds, an extractive derivatisation is performed using thiol solution in an organic solvent [24]. Two most commonly used thiols are n-butanethiol (BUSH) and 3,4-dimercaptotoluene (DMT). The latter reagent has a disadvantage in that it reacts with Lewisite 1 and Lewisite 2 species producing the same derivative (Fig. 4.18), making it impossible to differentiate between the parental agents.

Table 4.1 Names, CAS numbers and hydrophobicity parameters[a, b] of selected CWA and their environmentally most important transformation products

Name CWA transformation product	CAS	log P[a]	log K_{ow}[b]
bis(2-chloroethyl)sulfide	505-60-2	1.98	1.37
1,2-bis(2-chloroethylthio)ethane	3563-36-8	2.70	2.99
bis(2-chloroethylthioethyl)ether	63918-89-8	2.93	2.53
1,4-dithiane	505-29-3	0.112	0.77
thiodiglycol	111-48-8	–0.71	–0.77
divinyl sulfide	627-51-0	–	0.85
2,2′-dichlorotriethylamine	63978-54-1	1.42	2.02
bis(2-chloroethyl)methylamine	55-86-7	0.91	0.91
tris(2-chloroethyl)amine	555-77-1	1.31	2.27
N-methyldiethanolamine	105-59-9	–0.62	–1.5
triethanolamine	102-71-6	–1.11	–1.00
N-ethyldiethanolamine	139-87-7	–0.11	–1.01
2-chlorovinyldichloroarsine	541-25-3	–	2.56
bis(2-chlorovinyl)chloroarsine	40334-69-8	–	–
tris(2-chlorovinyl)arsine	40334-70-1	–	–
2-chlorovinylarsenous acid	159939-86-3	–	1.4 to 2.4
2-chlorovinylarsonous oxide	3088-37-7		–1.4 to –0.4
O-ethyl S-2-diisopropylaminoethyl methylphosphonothiolate (VX)	50782-69-9	2.05	2.09
O-isobutyl S-2-diethylminoethyl methylphosphonothiolate (VR)	159939-87-4	2.20	2.14
S-(2-Diisopropylaminoethyl) methylphosphonothioic acid	73207-98-4	–	0.96
ethyl methylphosphonothioic acid	18005-40-8	–	1.26
methylphosphonothioic acid	5994-73-0	–0.61	1.07
methylphosphonic acid	993-15-5	–1.25	–2.28
diisopropylaminoethanol	96-80-0	–	1.08
bis(2-diisopropylaminoethyl) disulfide	65332-44-7	–	3.48
O-isopropyl methylphosphonofluoridate (sarin)	107-44-8	0.51	0.30
O-cyclohexyl methylphosphonofluoridate (cyclosarin)	329-99-7	1.62	1.60
O-pinacolyl methylphosphonofluoridate (soman)	96-64-0	1.78	1.78
O-ethyl N,N-dimethylphosphoroamidocyanidate (tabun)	77-81-6	0.08	0.38
isopropylmethylphosphonic acid	1832-54-8	–0.49	0.27
diisopropylmethylphosphonate	1445-75-6	0.82	1.03
methylphosphonic acid	993-15-5	–1.25	–0.70

[a]log P = logarithm of the n-octanol/water partition coefficient calculated by freeware program ACDLabs

[b]log K_{ow} = logarithm of the n-octanol/water partition coefficient, either experimentally obtained [13] or calculated from the EPI Suite™ estimation program

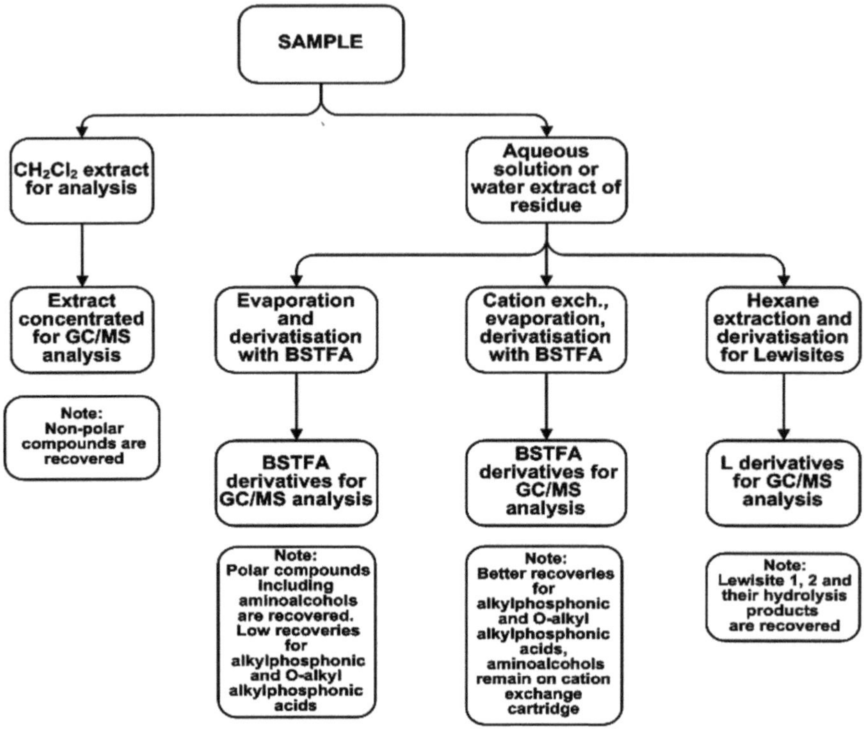

Fig. 4.15 General flowchart for conventional approach to environmental sample preparation for GC-MS determination of chemical warfare agents

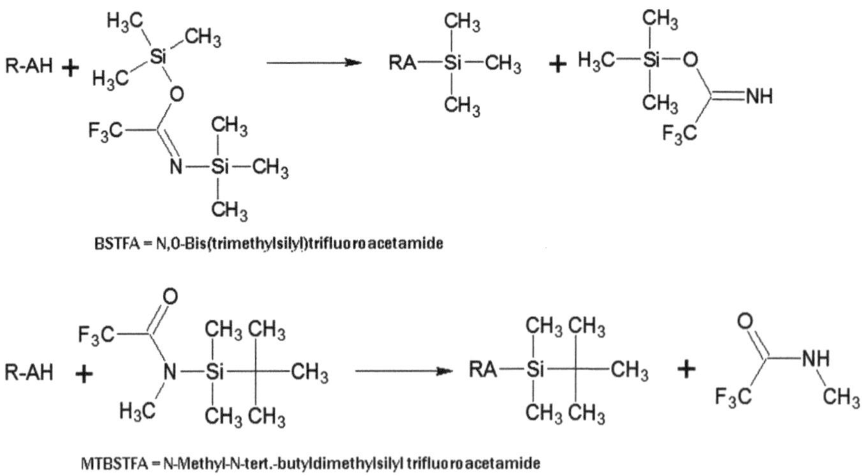

Fig. 4.16 Derivatisation with BSTFA and MTBSTFA, general reaction

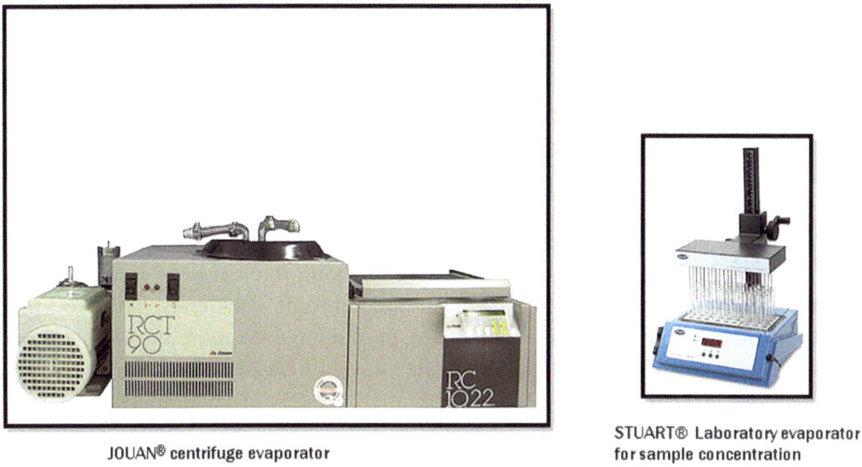

Fig. 4.17 Examples of commercially available equipment for evaporation of water and concentration

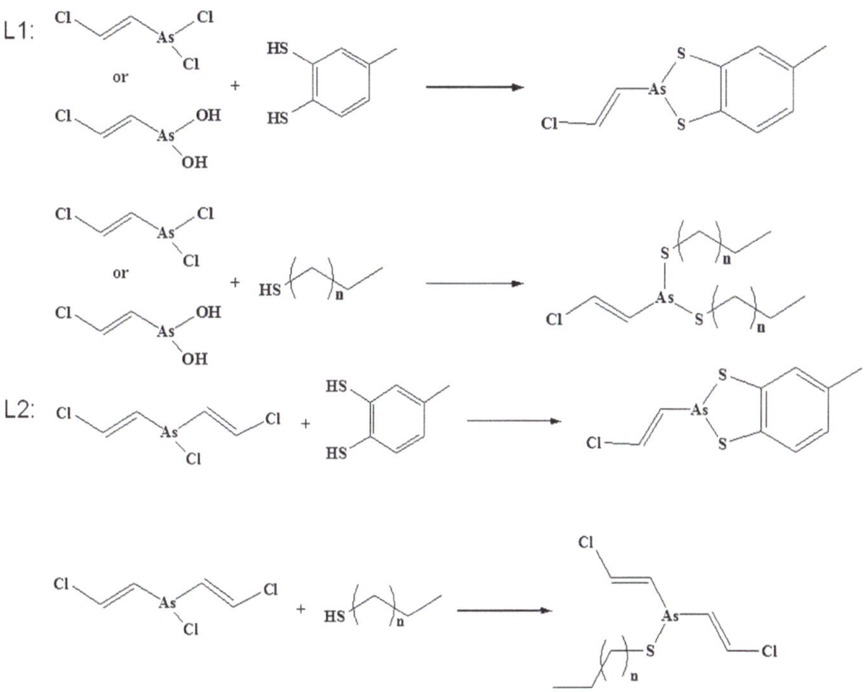

Fig. 4.18 Derivatisation of Lewisite related species with alkylthiols and 3,4-dimercaptotoluene

The presented scheme of liquid-liquid and solid-liquid extractions usually works fine in achieving the analytical aims when applied to samples containing μg/ml or μg/g (parts per million, ppm) analyte levels in moderately clean environmental matrices. However, there are many drawbacks in the approach, especially when utilized by an on-site, field laboratory. First of all, the procedure requires relatively large sample sizes/volumes of at least 5–10 ml or g that may not be available. The scheme relies on the use of generous amounts of organic solvents, like dichloromethane and hexane. The organic solvents have to be treated as dangerous goods, when transported to the field. The extractions are labor intensive and create multiple fractions that beside organic solvents can contain toxic chemicals. The procedure also consumes lots of laboratory glassware, producing even more hazardous waste. Dealing with the hazardous waste on-site may not be only expensive, but also problematic. The procedure cannot cope appropriately with samples containing soluble nonvolatiles (polymers), surfactants, finely suspended solids or emulsions. Such samples may be obtained from industrial discharges, polluted and natural water sources or CWA decontamination waste. If the emulsion is not already present in the samples, it may form during the extraction of a sample containing surfactants and fine solid particles. The boundary between the organic solvent and the sample will then have a layer with a cloudy or milky appearance which can be even greater than the solvent layer. If this emulsion cannot be broken, the analytical method may not be efficient. The aqueous samples or water extracts are especially problematic. Evaporation of water is a major time factor in analysis that reduces sample throughput, i.e. the number of samples that can be analysed. The time available for investigation of a CWA event may be restricted by numerous reasons, including security concerns. Evaporation requires heavy equipment (Fig. 4.17) that adds to the logistic burden of the on-site laboratory. With samples containing water soluble polymers like polyethylene glycols, it can be very difficult to completely remove water. Evaporation may also results in losses of volatile compounds like thiodiglycol.

In order to overcome all these drawbacks, a number of modern, solvent-minimized sample preparation techniques has been developed and tested.

4.4 Liquid Phase Microextraction Methods in Preparation of Aqueous Samples/Extracts for GC-MS Analysis of CWA and Related Compounds

Liquid phase microextraction (LPME) methods are a fairly recent development in sample preparation, where the traditional liquid-liquid extraction principle has been miniaturized by greatly reducing the acceptor-to-donor ratio [25]. The extraction normally takes place into a volume of several microliters of a water-immiscible solvent (acceptor phase) immersed in an aqueous sample containing analytes (donor phase). The LPME methods have a potential to integrate analyte extraction, concentration and sample introduction in a single step. They can be divided into three main categories: single-drop microextraction (SDME), hollow-fiber microextraction (HF-LPME), and dispersive liquid–liquid microextraction (DLLME).

4.4.1 Single-Drop Microextraction (SDME)

In SDME, a 1–3 μl volume of chloroform, trichloroethylene or toluene is drawn into a conventional, gas tight GC syringe. The syringe needle is pushed through a septum-closed sample vial until the tip of the needle is immersed in the sample solution in which the organic solvent used is immiscible. A drop of solvent is carefully suspended from the tip of the needle. Alternatively, the drop is suspended in the headspace above the sample allowing for use of more polar solvents, including D.I. water. Both techniques are illustrated in Fig. 4.19. The method requires optimization in respect to stirring rate, concentration of added salting out agent, and extraction time. After the optimized extraction time, usually no longer than 30 min, the microdrop is retracted into the syringe and transferred to the GC-MS for direct injection and analysis. Advantages of the approach are that it is simple, inexpensive, sensitive, relative fast and possible to automate. However, it can be applied only for relatively "clean" aqueous solutions not containing organic solvents, surface active substances and particulate material. This is due to possible drop instability, dissolution and dislodgment. Another major disadvantage is that SDME can be applied only for analysis of medium polarity to non-polar CWA and their degradation products [26, 27].

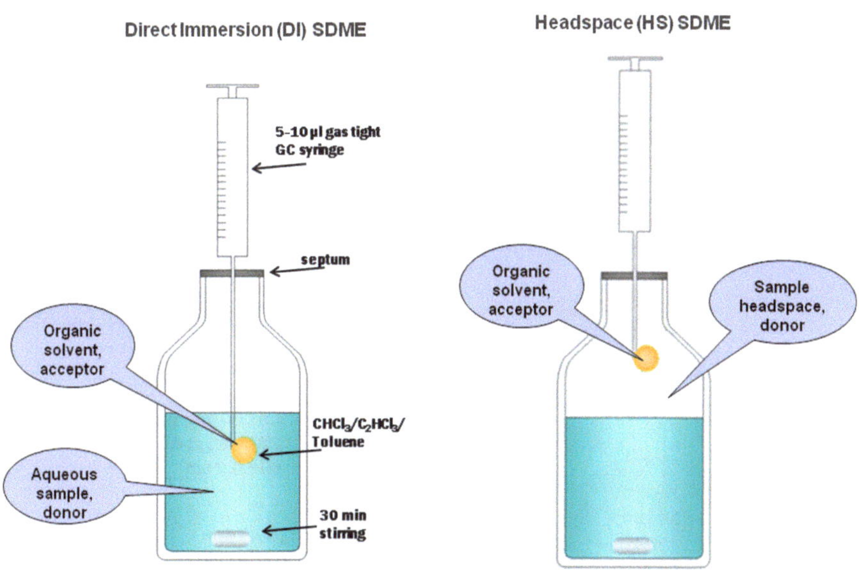

Fig. 4.19 Two modes of Single-Drop Microextraction (*SDME*)

4.4.2 Hollow-Fiber Microextraction (HF-LPME)

The drop instability from SDME has been circumvented in HF-LPME by containing the organic solvent within the lumen of a porous hydrophobic hollow fiber, typically made of polypropylene (Fig. 4.20). The fiber protects the drop against mechanical disturbance and also develops the surface area of the organic solvent in a contact with the sample for more efficient extraction. After conditioning, the fiber (1–3 cm in length) is briefly immersed in a solvent (chloroform or trichloroethylene) to immobilise the solvent in the pores prior to extraction. Several microliters of solvent are drawn into a GC syringe, before affixing the prepared fiber to the tip of the syringe needle. The fiber is completely immersed into the aqueous sample and the syringe plunger is carefully depressed to fill the hollow fiber with solvent. Extraction is carried out with stirring for 15–45 min. Upon completion of extraction, the organic solvent is withdrawn into the syringe and the hollow fiber is disposed. A volume of 1 μl of solvent is injected directly into a GC instrument for analysis.

As described, HF-LPME can be applied for the analysis of non-polar CWA [28, 29]. By co-injecting the extract with BSTFA derivatising reagent, it is possible to analyze

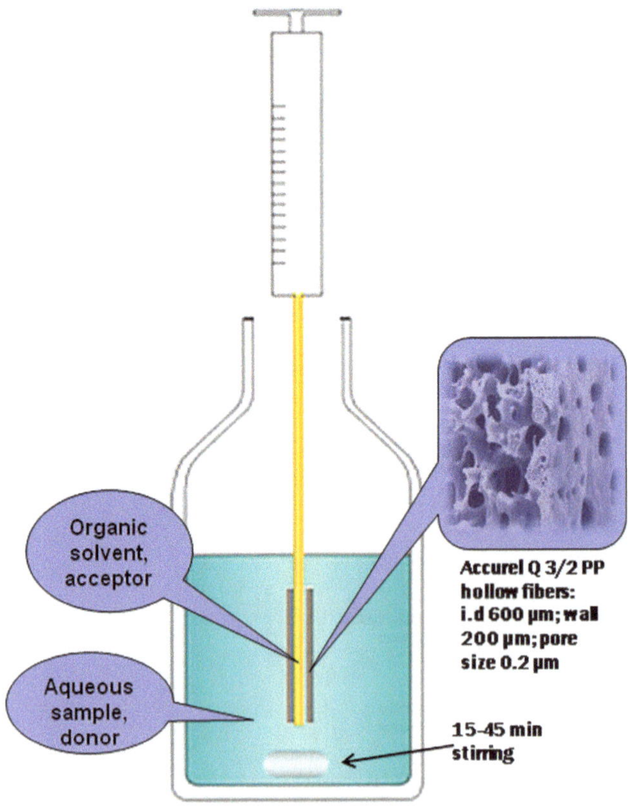

Fig. 4.20 The experimental set-up for hollow-fiber microextraction (*HF-LPME*)

the alkaline CWA related compounds, such as thiodiglycol and aminoalcohols [30]. Additionally, HF-LPME can be applied for analysis of acidic analytes, such as alkylphosphonic and alkylthiophosphonic acids, by using the mixture of chloroform and MSTBFA for simultaneous derivatization and extraction of the analytes [28]. Overall, HF-LPME has shown to be more rugged and having also a better extraction efficiency than SDME. It is an inexpensive, sensitive technique that uses minimum amounts of organic solvents and allows for in-situ derivatisation. However, concerns about robustness when applied to viscous and "dirty", especially organics rich environmental samples remain.

4.4.3 *Dispersive Liquid–Liquid Microextraction (DLLME)*

In this method, a mixture of extraction and dispersive solvents is rapidly injected into the aqueous sample creating a multiphase system of cloudy appearance. The extraction solvent is immiscible with water and has a higher specific gravity. The dispersive solvent is miscible with both, extracting solvent and water, playing a key role in formation of fine droplets of extraction solvent in aqueous samples. Due to the high surface area of the extracting solvent droplets, the extraction can be quite efficient and rapid. The sample is then centrifuged and the sedimented phase collected for direct analysis or further workup. As described (Fig. 4.21), the method can be applied for analysis of less-polar CWA and related compounds.

Palit and Mallard [31] explored a new approach, where extractive derivatization is accomplished with the dispersion of a mixture of 1-(heptafluorobutyryl)imidazole derivatizing reagent and dichloromethane as the organic phase in the aqueous sample, using acetonitrile as the dispersing solvent. The derivatization of the alcohols, degradation products of CWA, takes place at the interface between the analyte-containing aqueous phase and derivatization reagent-laden organic phase.

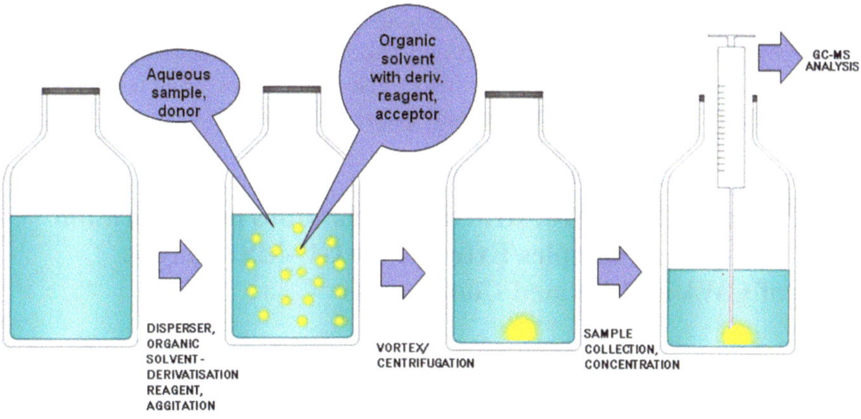

Fig. 4.21 Schematic representation of dispersive liquid–liquid microextraction (*DLLME*)

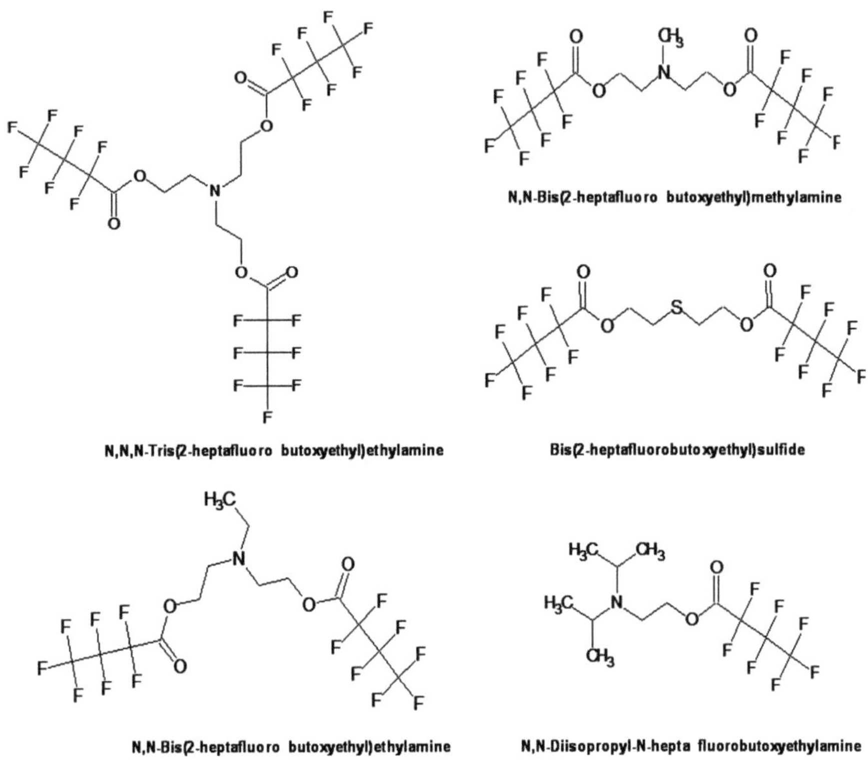

Fig. 4.22 Heptafluorobutyric derivatives of some CWA related alcohols

Amongst other, the approach enables analysis of ethanolamines (hydrolysis products of HN1, HN2 and HN3), thiodiglycol (hydrolysis product of HD), and dialkylaminoethanol (hydrolysis product of V-agents). Chemical structures of some heptafluorobutyryl derivatised analytes are shown in Fig. 4.22. The advantages of DLLME are simplicity of operation, high enrichment factor and very short extraction time. One limitation is that, for now, the method is restricted to less-polar CWA compounds and related alcohols only. Additionally, there are concerns about robustness of the method when applied to viscous samples, samples containing surfactants and organic solvents, and multiphase samples.

4.5 Solid Phase Microextraction Techniques in Preparation of Aqueous Samples/Extracts for GC-MS Analysis of CWA and Related Compounds

The first successful modern microextraction technique was solid-phase microextraction (SPME) developed by the group of Pawliszyn at University of Waterloo, Ontario, Canada [32]. Since then, this extraction technique has become the standard that other microextraction techniques have tried to reach and outperform.

4.5.1 *Solid-Phase Microextraction (SPME)*

The SPME technique uses a small, retractable, polymer-coated fiber installed in a holder with a stainless steel needle that serves for the fiber protection. A variety of SPME fibers are commercially available [33], with solutions for both manual and automated operation. For analysis of less-polar compounds and intact CWA, the 65 μm Polydimethylsiloxane/Divinylbenzene (PDMS/DVB) fiber [34] is placed in the sample solution or in the headspace of the sample solution for a period of time. The analytes diffuse by convection to the surface of the polymer coating (extracting phase) until equilibrium is achieved. Once withdrawn from the sample source, the fiber is inserted into the hot GC injection port and the sorbed analytes are thermally desorbed into the GC column.

For analysis of acidic CWA degradation products, the 75 μm Carboxen/Polydimethylsiloxane (CAR/PDMS) fiber is first exposed to the headspace of MTBSTFA [35]. The extraction is carried out from the acidified sample by stirring for 30 min. Sodium chloride salt is added to saturation for salting-out effect. The fiber is again exposed to the MTBSTFA headspace for 15 min prior to desorption in the GC injection port. The same procedure is followed for analysis of alkaline CWA degradation products, except that the sample pH is adjusted to 10, sodium sulfate is the preferred salting-out agent, and BSTFA is the preferred derivatisation reagent [36]. The SPME process is presented in Fig. 4.23.

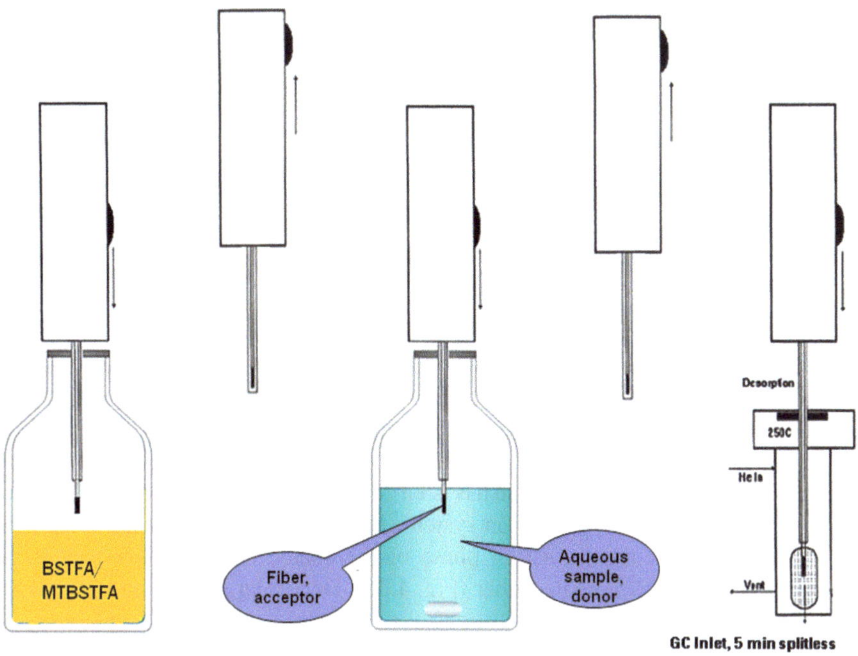

Fig. 4.23 Schematic representation of solid-phase microextraction (*SPME*)

The method has also been developed for the determination of 2-chlorovinylarsonous acid, the hydrolysis product of Lewisite 1, which combines dithiol derivatization with SPME [37].

As the sample-preparation technique, SPME has following attractive features:

- it is a rapid, simple, and sensitive method;
- it can be applied for neat CWA and degradation products;
- it has a small size, i.e. it is convenient for designing portable devices and for field sampling [38];
- it is versatile and it can be used for air/vapour sampling as well [39, 40].

However, there are some limitations of the method, as follows:

- conditioning of fibers requires 30 min–2 h time;
- fibers are relatively expensive and have short lifetimes;
- fibers swell in the presence of organic solvents;
- the needles can bend and the fibers are fragile;
- concerns about robustness when applied to 'dirty' and multiphase samples.

4.6 Fiber Assisted Microevaporation and Derivatization (FAMEAD)

In this technique, several drops of aqueous sample are carefully applied on an acryl SPME fiber. The water is removed by exposing the fiber to a gentle stream of nitrogen. The fiber is then immersed in a headspace of BSTFA for 15 min and analysed for TMS derivatives of CWA degradation products by thermal desorption in the GC inlet (Fig. 4.24). The method is simple and fast, but has similar limitations as the SPME technique.

4.6.1 Stir Bar Sorptive Extraction (SBSE)

The stir-bar sorptive extraction technique (SBSE) presents a variation of the SPME technique [41] that utilizes a glass stir bar (having a magnet within) coated with a polymeric sorbent. The stir bar is placed in the aqueous sample, stirred for an appropriate time (usually tens of min), removed, dried to remove any water droplets, and placed in the specially designed GC liner for thermal desorption of the sorbed analytes into the GC. The technique has been commercialized by Gerstel (Baltimore, Maryland) as the Twister extraction system (Fig. 4.25) [42]. The available surface area and volume of the extracting phase of the coated stir bar are substantially higher than of an SPME coated fiber (Fig. 4.26). Hence, larger amounts of analyte

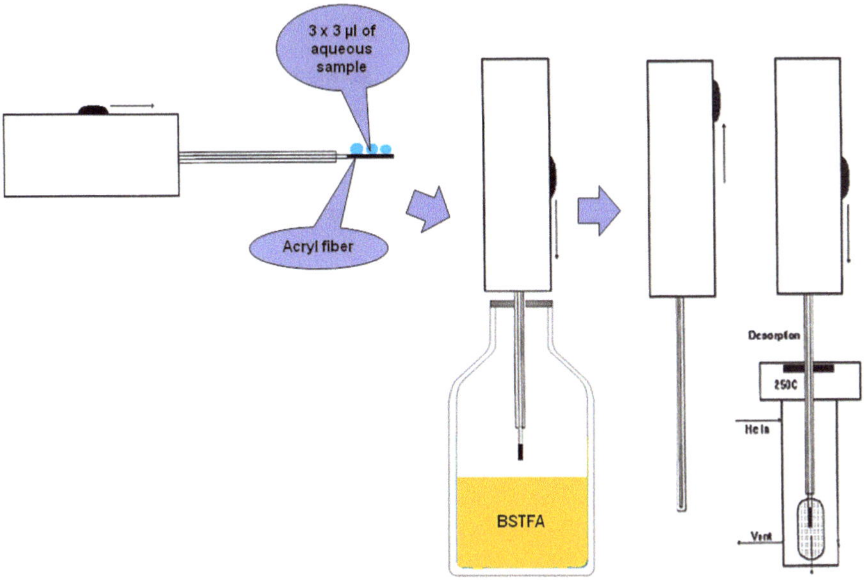

Fig. 4.24 Schematic representation of fiber assisted microevaporation and derivatization (*FAMEAD*)

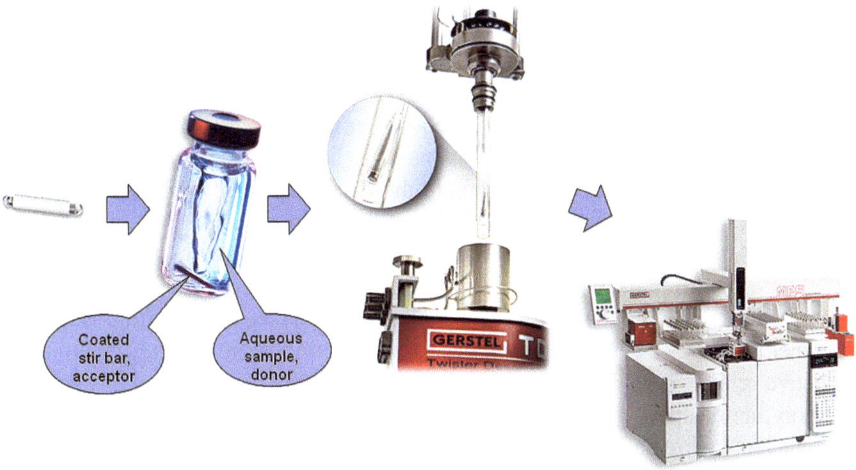

Fig. 4.25 Gerstel Twister SBSE extraction and automated analysis system (Adapted from http://www.gerstel.com/en/thermal-twister-desorption.htm)

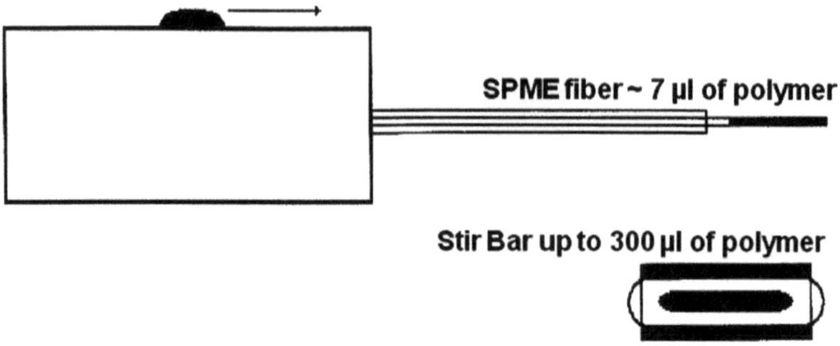

Fig. 4.26 SBSE vs. SPME

can be extracted and overall sensitivity is improved. The stir bars are more robust and resistant to braking compared to the SPME fibers, but they are commercially available with only few types of polymer coatings. This reduces the versatility of the method, especially in respect to more polar analytes [43].

The method was shown to be effective in analysis of intact GB, HD and VX in aqueous matrices [44]. In-situ derivatisation for some more polar compounds may also be possible [45].

4.7 Molecularly Imprinted Polymers (MIPs)

Molecularly imprinted polymers (MIPs) have synthetic recognition sites with a predetermined selectivity for the analyte(s) of interest [46]. They are prepared by mixing the template molecule with functional monomers, cross-linking monomers and a radical initiator in a proper solvent [47]. During polymerization, the complexes are formed between the template molecule and the functional monomers that are stabilized within the structure of forming rigid, highly cross-linked polymer. After polymerization, the template molecules are extracted out, leaving the three-dimensional cavities that are complementary in both shape and chemical functionality arrangement to those of the template. Thus, the resulting imprinted polymer possesses a permanent memory for the imprint species, ideally enabling the selective rebind of the imprint molecule from a mixture. The interactions between the polymer receptor site and the imprint molecule can either be non-covalent, such as ionic and hydrogen bonding, or reversible covalent [48]. The concept of molecular imprinting is illustrated in Fig. 4.27.

Although usually time consuming, synthesis of MIPs is a relatively straightforward and inexpensive procedure. The products are stable, reusable, and they can be used as an extraction phase in solid phase extraction (SPE), SPME, SBSE and HF-LPME. When applied in sample preparation for analysis of CWA degradation products in aqueous matrices, MIPs products bring selectivity only for limited number

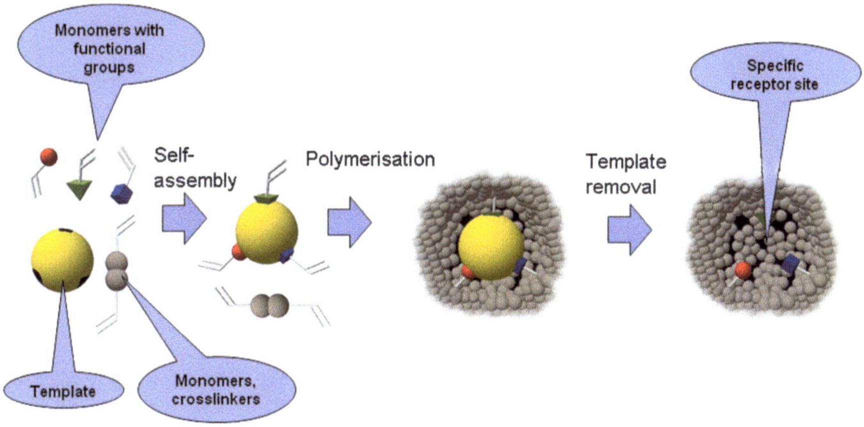

Fig. 4.27 Molecular imprinting

of analytes [49–52]. Other important downsides of MIPs use are bad reproducibility, loss of specificity and performance when directly applied at complex environmental samples [53].

4.8 In-Tenax Tube Sample Preparation for Thermal Desorption GC-MS

This novel method has been developed for on-site use by inspection teams of the OPCW. The method utilizes Tenax® TA packed tubes and in-tube silylation followed by thermal desorption GC–full scan MS for analysis of CWA and their degradation products in liquid samples/extracts [54]. The same Tenax tubes and equipment are used for sampling and analysis of air/vapour samples [55].

Tenax® TA comes commercially as the granular adsorbent consisting of macro-porous particles with specific surface areas between 20 and 35 m^2/g. The individual particles are constructed of a network of highly elongated and entangled 2,6-diphenyl-p-phenylene polymer material with open spacings that ranges from 100 to 300 nm, as shown in Fig. 4.28 [56, 57].

Granular Tenax® TA has an extremely low affinity for water, a high heat tolerance (>375 °C) and it is chemically inert. It has been shown an effective sorbent material for air/vapour sampling and thermal desorption GC–MS analysis of a wide spectrum of CWA [58–62].

The novel sample preparation method uses Tenax® TA as the depository for the more polar/less volatile CWA degradation products, a trap for more volatile/less polar analytes such as intact CWA, and a solid support for the derivatisation.

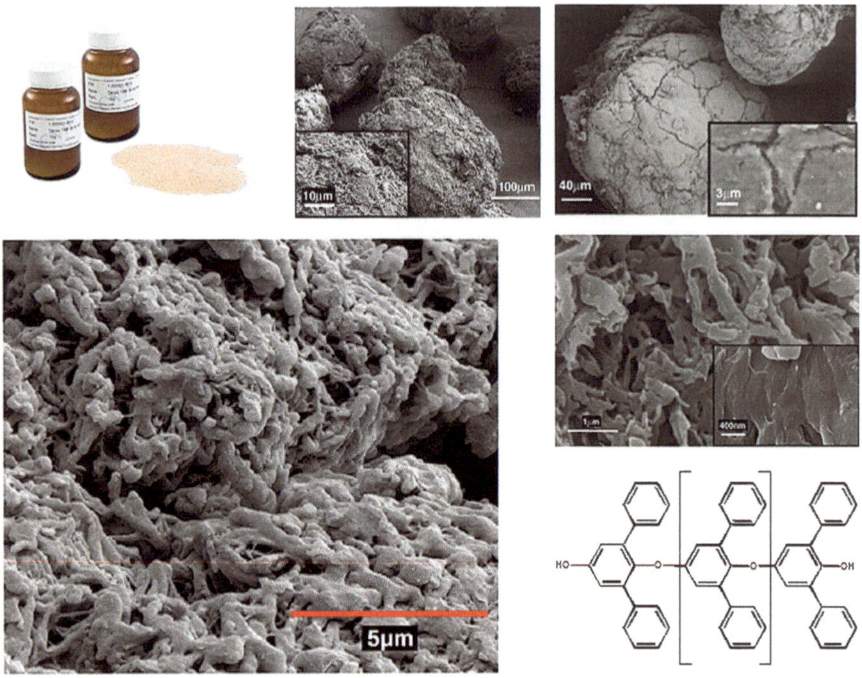

Fig. 4.28 Granular Tenax TA, SEM micrographs (Adapted from Alfeeli et al. [56, 57]) and its chemical structure

A small quantity of Tenax® TA 60/80 mesh is packed in a GC borosilicate glass liner as a body of a thermal desorption tube that fits in a Programmable Temperature-Vaporization (PTV) GC inlet (Fig. 4.29).

Several microliters (0.5–10 μl) of the liquid sample is injected into the tube using a conventional GC syringe and spiking attachment of the Six-Tube Conditioner Model 9600 (CDS Analytical Inc., Oxford, PA), as shown in Fig. 4.30a. Water (solvent) is removed using a programmable heater and a helium flow (Fig. 4.30b). Then, the TMS derivatisation is performed in-tube by injecting 3 μl of BSTFA. The less-/non-volatile analytes are converted into more volatile compounds which are retained by the Tenax bed (Fig. 4.30c). These are analyzed in the same GC run with the trapped less polar compounds after direct thermal desorption of the tube into a PTV GC inlet (Fig. 4.30d). The splitless operation of the inlet is used to transfer the maximum amount of analyte(s) into the GC column.

For analysis of the Lewisite related species, the pH of the aqueous sample is adjusted to ≤ pH 2 with HCl solution [63]. The sample injection is immediately followed by the BuSH derivatisation. The sample is analysed after water removal. The experimental set-up for Lewisites related species is shown in Fig. 4.31.

The in-Tenax tube sample preparation procedure is simple, easy to perform and very fast, with the whole operation completed within 9 min. The sample volume

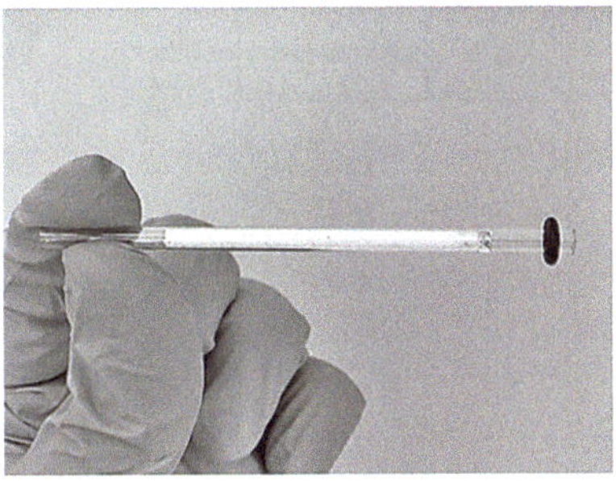

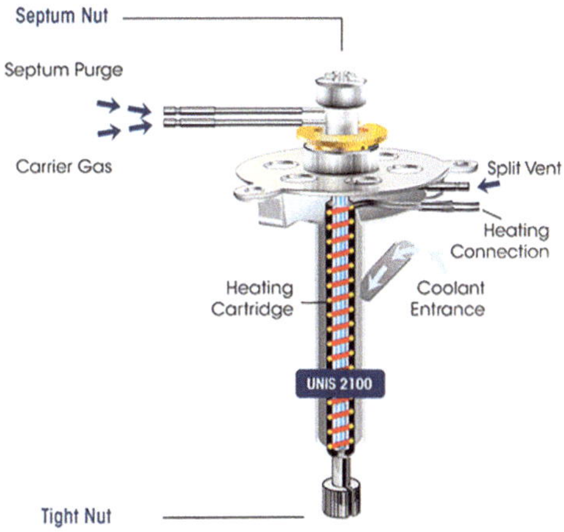

Fig. 4.29 Thermal desorption tube (*left*) for UNIS 2100 Programmed Temperature Vaporizing Inlet (*right*)

required is very small, ranging from 1 to 10 μl for aqueous solutions in low parts per million/high parts per billion concentration ranges, or 10 μl and more for lower parts per billion ranges using full scan MS. For the samples composed of two or more solvents of different polarity, or multiphase sample matrices, usually there is no need for a separation or an additional sample preparation step. The solvents are removed simultaneously leaving the previously distributed analytes concentrated in the tube. The procedure requires just a few μl of derivatising reagent in hexane. Consequently, the amounts of dangerous goods needed and hazardous waste generated by the laboratory are reduced to a bare minimum. The used TD tube is exchanged after the analysis, removing residual contamination deposited into the tube. This makes the system less susceptible for contamination coming from nonvolatiles. The equipment needs and related logistics are minimal as well. The same equipment set-up, reagent and TD tubes are used for the collection, preparation and analysis of air samples (Fig. 4.32). Moreover, a change from the PTV TD mode to a standard liquid injections mode is accomplished simply by loading of an adequate GC–MS method in the control software package. All these features make the method especially convenient for a forensic or environmental field laboratory use.

4.9 Conclusions

The majority of CWA is reactive and, hence, unstable in environmental matrices yielding products with widely differing polarity, volatility and reactivity from the parental agents. Therefore, chemical analysis in investigation of prior CWA use must target the CWA degradation products as well as the intact compounds.

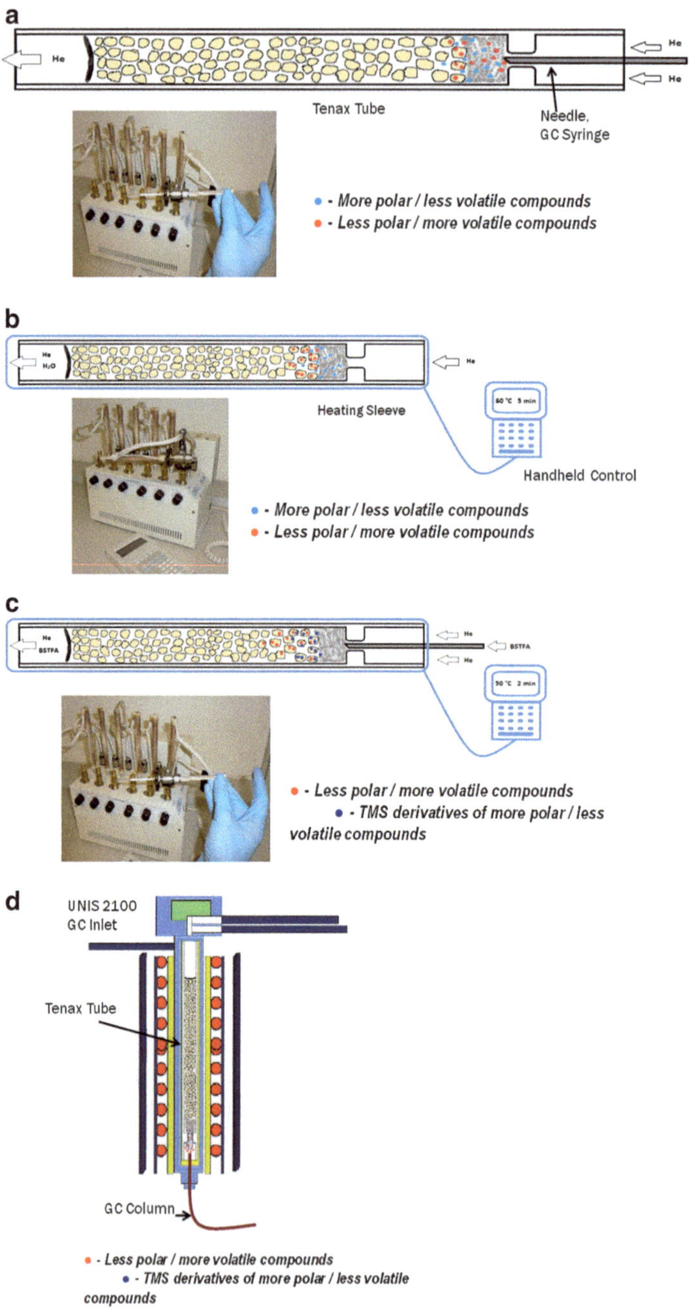

Fig. 4.30 In-Tenax tube sample preparation for thermal desorption GC-MS analysis

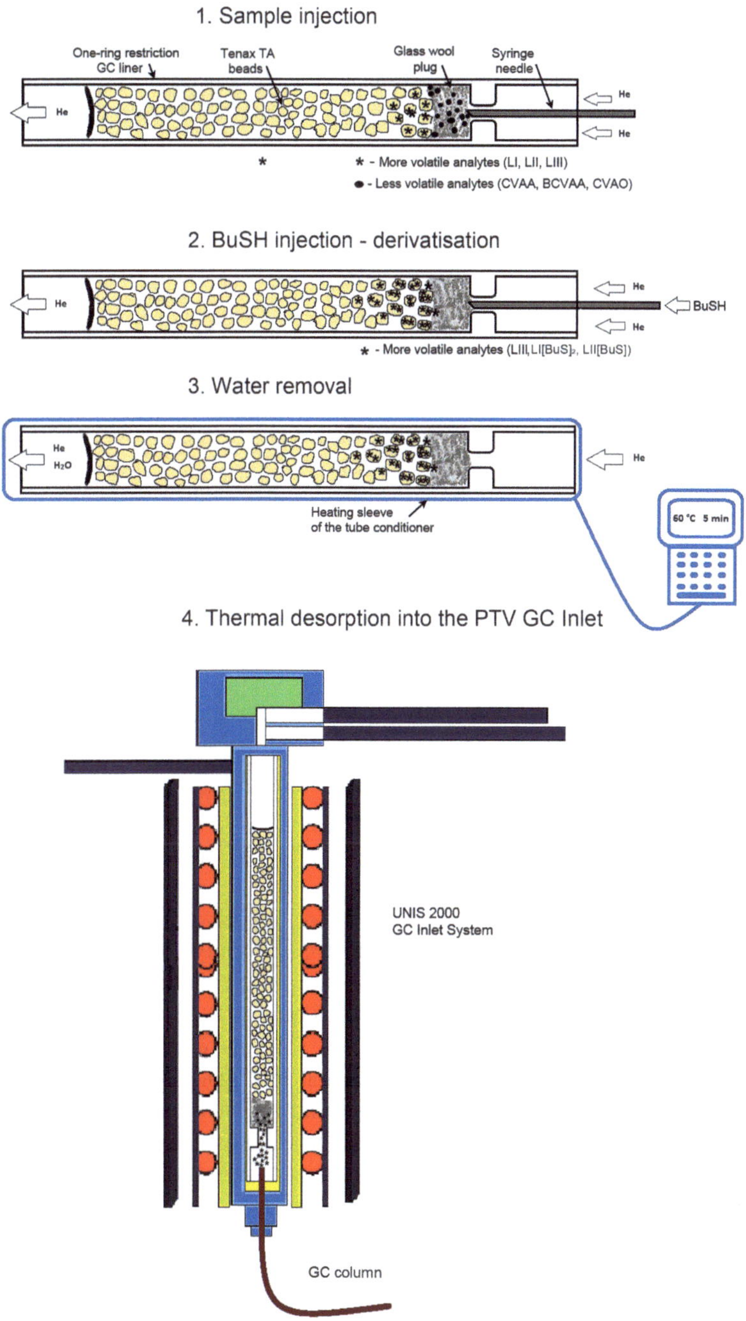

Fig. 4.31 In-Tenax tube sample preparation for analysis of Lewisite related species (Adapted from Terzic et al. [63])

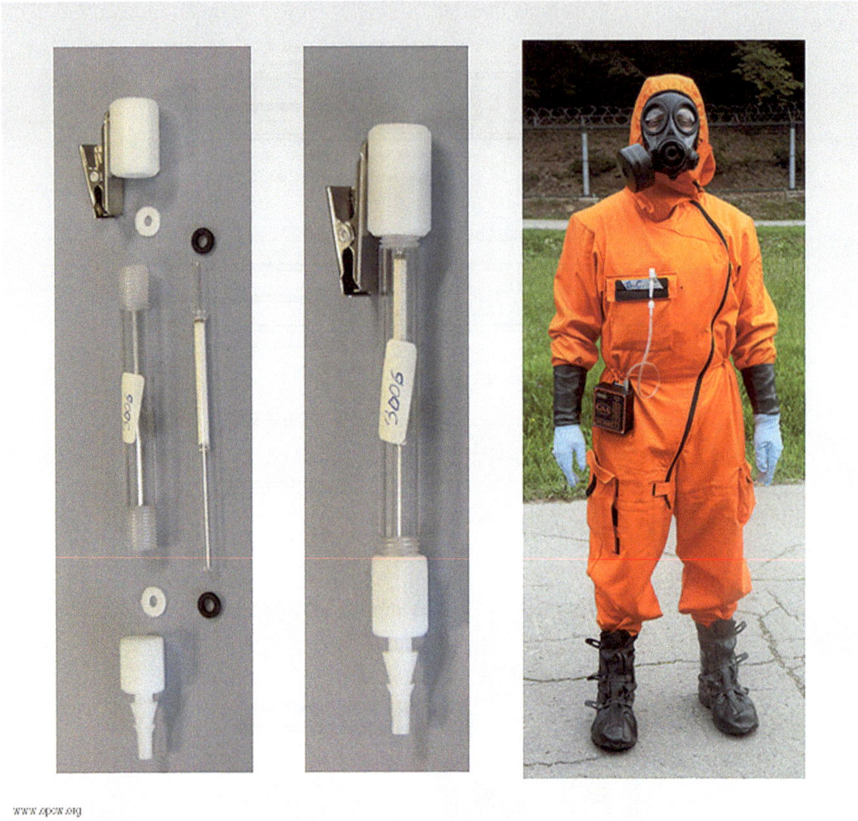

Fig. 4.32 The OPCW air/vapour sampling tube

Gas Chromatography – Mass Spectrometry is currently the primary analytical technique for this task. The conventional sample preparation approach for the on-site GC-MS analysis is based on liquid-liquid and solid-liquid extractions. This includes extensive use of harmful organic solvents, equipment, and leads to a time consuming, laborious sample preparation that generates lots of hazardous waste. Especially problematic are aqueous samples/water extracts and multiphase sample matrices. Modern sample preparation procedures look at simpler, faster and more robust approaches that require smaller sample size/volume, utilize less organic solvents, equipment and energy. Ideally, the modern methods should allow for automation. Methods based on liquid and solid microextractions have shown to be effective only when targeting a limited number of the analytes. Additionally, they have all shown to lack robustness when dealing with complex and multiphase sample matrices. Thus, they can be used only as complementary analysis to further screen. In-Tenax tube sample preparation for thermal desorption GC-MS analysis has been shown to be currently the most versatile and robust approach for on-site sample preparation and analysis of a whole range of the target analytes.

References

1. Sidell FR, Takafuji ET, Franz DR (eds) (1997) Medical aspects of chemical and biological warfare, 1st edn. United States Government Printing, Office of the Surgeon General, Borden Institute, Washington, DC
2. Okumura T, Takasu N, Lshimatsu S, Miyanoki S, Mitsuhashi A, Kumada K, Tanaka K, Hinohara S (1996) Report on 640 victims of the Tokyo subway sarin attack. Ann Emerg Med 28:129
3. De Voogt P, Jansson B (1993) Verticals and long range transport of persistent organics in the atmosphere. Rev Environ Contam Toxicol 132:1–27
4. Bartelt-Hunt SL, Knappe DRU, Barlaz MA (2008) A review of chemical warfare agent simulants for the study of environmental behavior. Environ Sci Technol 38:112–136
5. FM 3-6/FMFM 7-11-H/AFM 105-7, Field Behavior of NBC Agents (Including Smoke and Incendiaries), Department of the Army, Department of the Air Force, United States Marine Corps, Washington, DC, 3 Nov 1986
6. D'Agostino PA, Chenier CL (2006) Analysis of chemical warfare agents: general overview, LC-MS Review, InHouse LC-ESI-MS methods and open literature bibliography, Technical report DRDC Suffield TR 2006-022, Suffield, Canada
7. Bogusz MJ (ed) (2008) Handbook of analytical separations, vol 6, Forensic Science. Elsevier B.V, Amsterdam, pp 839–872
8. Inficon. http://www.inficonchemicalidentificationsystems.com/en/index.html. Cited 05 May 2013
9. Torion. http://www.torion.com/products. Cited 05 May 2013
10. FLIR. http://gs.flir.com/detection/chemical/mass-spec/griffin-460. Cited 05 May 2013
11. Agilent Technologies. http://www.chem.agilent.com/en-US/products-services/Instruments-LTMSystems/Mass-Spectrometry/5975T-LTM-GC-MSD-System. Cited 05 May 2013.
12. Gupta RC (ed) (2009) Handbook of toxicology of chemical warfare agents, 1st edn. Academic Press, San Diego, pp 98–99
13. Munro NB, Talmage SS, Griffin GD, Waters LC, Watson AP, King JF, Hauschild V (1999) The sources, fate, and toxicity of chemical warfare agent degradation products. Environ Health Perspect 107(12):933–974
14. Trapp R (1985) The detoxification and natural degradation of chemical weapon agents: SIPRI studies on chemical and biological weapons. Taylor & Francis Inc, Philadelphia, p 3
15. Hay A, Roberts G (1990) The use of poison gas against the Iraqi Kurds: analysis of bomb fragments, soil, and wool samples. JAMA 263(8):1065–1066
16. Theobald N, Ruhl NP (1994) Chemical warfare agent munitions in the Baltic Sea. Deut Hydrographische Z 46:121–131
17. Jorgensen BS, Olesen B, Berntsen O (1985) Mustard gas accidents on Bornholm, Ugeskr. Laeger 28(147):2251–2254
18. Franke S (1967) Lehrbuch der Militärchemie, Band I: Chemie der Kampfstoffe. Deutscher Militärverlag, Berlin
19. Black RM (2008) An overview of biological markers of exposure to chemical warfare agents. J Anal Toxicology 32:2–9
20. Reynolds ML, Little PJ, Thomas BFG, Bagley RB, Martin BR (1985) Relationship between the biodisposition of {3H} soman and its pharmacological effects in mice. Toxicol Appl Pharmacol 80
21. Epstein J, Callahan JJ, Bauer VE (1974) The kinetics and mechanisms of hydrolysis of phosphonothiolates in dilute aqueous solution. Phosphorus 4:157
22. Yang Y-C, Szafraniec LL, Beaudry WT, Rohrbaugh DK (1990) Oxidative detoxification of phosphonothiolates. J Am Chem Soc 1:6621
23. Vanninen P (ed) (2011) Recommended operating procedures for analysis in the verification of chemical disarmament. The Ministry for Foreign Affairs of Finland, University of Helsinki, Finland, pp 124–125

24. Black RM, Muir B (2003) Derivatisation reactions in the chromatographic analysis of chemical warfare agents and their degradation products. J Chromatogr A 1000:253–281
25. Sarafraz-Yazdi A, Amiri A (2010) Liquid phase microextraction. Trends Anal Chem 29:1
26. Palit M, Pardasani D, Gupta AK, Dubey DK (2005) Application of single drop microextraction for analysis of chemical warfare agents and related compounds in water by gas chromatography/mass spectrometry. Anal Chem 77:711–717
27. Park Y, Kim SK, Choi K, Son B, Park J, KangBull H (2009) Analysis of chemical warfare agents in water using single drop microextraction. Kor Chem Soc 30(1):49–52
28. Dubey DK, Pardasani D, Gupta AK, Palit M, Kanaujia PK, Tak V (2006) Hollow fiber-mediated liquid-phase microextraction of chemical warfare agents from water. J Chromatogr A 1107:29–35
29. Lee HSN, Basheer C, Lee HK (2006) Determination of trace level chemical warfare agents in water and slurry samples using hollow fibre-protected liquid-phase microextraction followed by gas chromatography-mass spectrometry. J Chromatogr A 1124:91
30. Lee HSN, Sng MT, Basheer C, Lee HK (2008) Determination of basic degradation products of chemical warfare agents in water using hollow fibre-protected liquid-phase microextraction with in-situ derivatisation followed by gas chromatography-mass spectrometry. J Chromatogr A 1196–1197:125–132
31. Palit M, Mallard G (2011) Dispersive derivatization liquid-liquid extraction of degradation products/precursors of mustards and V-agents from aqueous samples. J Chromatogr A 1218:5393–5400
32. Arthur CL, Pawliszyn J (1990) Solid phase microextraction with thermal desorption using fused silica optical fibers. Anal Chem 63:2145
33. Sigma-Aldrich. http://www.sigmaaldrich.com/analytical-chromatography/sample-preparation/spme/selecting-spme-fiber.html. Cited on 19 June 2013
34. Ng WF, Lakso H-Å (1997) Determination of chemical warfare agents in natural water samples by solid-phase microextraction. Anal Chem 69:1866–1872
35. Sng MT, Ng WF (1999) In situ derivatisation of degradation products of chemical warfare agents in water by solid-phase microextraction and gas chromatographic-mass spectrometric analysis. J Chromatogr A 832:173–182
36. Vanninen P (ed) Recommended operating procedures for analysis in the verification of chemical disarmament, The Ministry for Foreign Affairs of Finland, University of Helsinki, 2011, pp 163
37. Szostek B, Aldstadt JH (1998) J Chromatogr A 2(807):253–263
38. Torion. http://www.torion.com/products/22. Page visited on 19 June 2013
39. Popiel S, Sankowska M (2011) Determination of chemical warfare agents and related compounds in environmental samples by solid-phase microextraction with gas chromatography. J Chromatogr A 47(1218):8457–8479
40. Schneider JF, Boparai AS, Reed LL (2001) Screening for sarin in air and water by solid-phase microextraction-gas chromatography/mass spectrometry. J Chromatogr Sci 10(39):420–424
41. Baltussen HA, Sandra PJF, David F, Cramers C (1999) Stir bar sorptive extraction (SBSE), a novel extraction technique for aqueous samples: Theory and principles. J Microcolumn Sep 10(11):737–747
42. Gerstel Application Note 8/2001
43. Blasco C, Fernändez M, Picó Y, Font G (2004) Comparison of solid-phase microextraction and stir bar sorptive extraction for determining six organophosphorus insecticides in honey by liquid chromatography-mass spectrometry. J Chromatogr A 1030:77–85
44. Stuff JR, Dupont Durst H (2002) Stir bar microextraction for detection of CW agents in liquid matrices. In: Unclassified report ECBC-TR-233. Edgewood Chemical Biological Center, Aberdeen Proving Ground (APG), MD, USA, May 2002
45. Kawaguchi M, Sakui N, Okanouchi N, Ito R, Saito K, Nakazawa H (2005) Stir bar sorptive extraction and trace analysis of alkylphenols in water samples by thermal desorption with in tube silylation and gas chromatography-mass spectrometry. J Chromatogr A 1(1062):23–29

46. Owens PK, Karlsson L, Lutz ESM, Andersson LI (1999) Molecular imprinting for bio- and pharmaceutical analysis. Trend Anal Chem 18:146–154
47. Yan H, Row KH (2006) Characteristic and synthetic approach of molecularly imprinted polymer. Int J Mol Sci 7:155–178
48. Owens PK, Karlsson L, Lutz ESM, Andersson LI (1999) Molecular imprinting for bio- and pharmaceutical analysis. Trends Anal Chem 18:146
49. Lee HN (2008) Development and applications of novel solvent-minimized techniques in the determination of chemical warfare agents and their degradation products, Ph.D. thesis, National University of Singapore, Singapore
50. Pradhan S, Boopathi M, Kumar O, Baghel A, Pandey P, Mahato TH, Singh B, Vijayaraghavan R (2009) Molecularly imprinted nanopatterns for the recognition of biological warfare agent ricin. Biosens Bioelectron 25:592–598
51. Malosse L, Buvat P, Adès D, Siove A (2008) Detection of degradation products of chemical warfare agents by highly porous molecularly imprinted microspheres. Analyst 133:588–595
52. Le Moullec S, Begos A, Pichon V, Bellier B (2006) Selective extraction of organophosphorus nerve agent degradation products by molecularly imprinted solid-phase extraction. J Chromatogr A 1108:7–13
53. Bossée A (2011) Overview on MIPs experiments at DGA MNRBC. In: Report of the Sixteenth Session of the Scientific Advisory Board, SAB-16/1. Organisation for the Prohibition of Chemical Weapons, The Hague
54. Terzic O (2010) Screening of degradation products, impurities and precursors of chemical warfare agents in water and wet or dry organic liquid samples by in-sorbent tube silylation followed by thermal desorption-gas chromatography-mass spectrometry. J Chromatogr A 1217:4987–4995
55. Terzic O, Swahn I, Cretu G, Palit M, Mallard G (2012) Gas chromatography-full scan mass spectrometry determination of traces of chemical warfare agents and their impurities in air samples by inlet based thermal desorption of sorbent tubes. J Chromatogr A 1225:182–192
56. Alfeeli B, Taylor LT, Agah M (2010) Evaluation of Tenax TA thin films as adsorbent material for micro preconcentration applications. Microchem J 2(95):259–267
57. Alfeeli B, Jain V, Johnson RK, Beyer FL, Hefli JR, Agah M (2011) Characterization of poly (2,6-diphenyl-p-phenylene oxide) films as adsorbent for microfabricated preconcentrators. Microchem J 2(98):240–245
58. Rautio M (ed) (1987) Air monitoring as a means for the verification of chemical disarmament, Part III. Further development and testing of methods. Ministry for Foreign Affairs of Finland, Helsinki
59. Kaipainen A, Kostiainen O, Riekkola M (1992) Identification of chemical warfare agents in air samples using capillary column gas chromatography with three simultaneous detectors. J Microcol Sep 4:245
60. Steinhanses J, Schoene K (1990) Thermal desorption-gas chromatography of some organophosphates and s-mustard after trapping on tenax. J Chromatogr 514:273
61. Hancock JR, Peters GR (1991) Retention index monitoring of compounds of chemical defence interest using thermal desorption gas chromatography. J Chromatogr 538:249
62. Carrick WA, Cooper DB, Muir B (2001) Retrospective identification of chemical warfare agents by high-temperature automatic thermal desorption-gas chromatography-mass spectrometry. J Chromatogr A 925:241
63. Terzic O, Bartenbach S, de Voogt P (2013) Determination of Lewisites and their hydrolysis products in aqueous and multiphase samples by in-sorbent tube butyl thiolation followed by thermal desorption-gas chromatography-full scan mass spectrometry. J Chromatogr A 1304:34–41

Chapter 5
An Overview of Matrix-Assisted Laser Desorption/Ionization (MALDI) Mass Spectrometry and Some of Its Applications

Mark W. Duncan, David Gibson, Ryan Walsh, Afshan Masood, and Hicham Benabdelkamel

Abstract Matrix-assisted laser desorption/ionization time-of-flight (TOF) mass spectrometry (MS) is a powerful tool for both the qualitative and quantitative analysis of wide array of analytes. MALDI-TOF MS is simple, both in design and practice, and offers the advantages that analysis times are very short, the mass range is virtually unlimited, and sensitivity is high. This chapter describes the theory and practice of MALDI and illustrates the utility of approach by reviewing its common applications.

Keywords MALDI • Mass spectrometry • Microbiology • Clinical applications • Imaging

5.1 Introduction

5.1.1 History

The term matrix-assisted laser desorption ionization (MALDI) was coined in 1985 by Franz Hillenkamp and Michael Karas [1, 2]. These investigators found that the amino acid alanine could be ionized more easily if it was mixed with the amino acid

M.W. Duncan (✉) • A. Masood • H. Benabdelkamel
Proteomics Resource, Obesity Research Center, College of Medicine,
King Saud University, Riyadh, Saudi Arabia
e-mail: mark.duncan@ucdenver.edu

D. Gibson
Northern Ireland Centre for Stratified Medicine, University of Ulster,
Altnagelvin Hospital Campus, Londonderry, UK

R. Walsh
Division of Endocrinology, Metabolism & Diabetes, Department of Medicine,
University of Colorado Anschutz Medical Campus, Aurora, CO, USA

J. Banoub (ed.), *Detection of Chemical, Biological, Radiological and Nuclear Agents for the Prevention of Terrorism*, NATO Science for Peace and Security Series A: Chemistry and Biology, DOI 10.1007/978-94-017-9238-7_5

tryptophan and irradiated with a pulsed 266 nm laser. The availability of small and relatively inexpensive nitrogen lasers operating at a wavelength of 337 nm, and the introduction of the first commercial instruments in the early 1990s, lead to widespread adoption of MALDI by researchers.

5.1.2 *Definitions*

MALDI is the commonly used abbreviation for matrix assisted laser desorption/ionization. MALDI is frequently coupled with a time-of-flight mass analyzer (TOF) and the abbreviation therefore becomes MALDI-TOF. Other types of analyzers can be used and are commercially available, but these alternative configurations constitute a small portion of the market because, in practice, the marriage of MALDI ionization with a TOF analyzer is close to ideal.

5.1.3 *An Overview of MALDI*

MALDI is a widely used and powerful "soft" ionization method suitable for almost all analytes that can accept (positive ion mode) or lose (negative ion mode) a proton. Because of its versatility, MALDI has found wide application in the analysis of large and/or labile molecules, lipids, peptides, proteins, oligonucleotides, and synthetic polymers. Special applications of increasing importance include: protein characterization, clinical chemistry, clinical microbiology, single cell analysis, metabolomics and imaging. (These specific applications are discussed in greater detail later.)

For MALDI, the sample is intimately mixed with matrix and dried on a target (or MALDI plate). The target is then loaded into the mass spectrometer where the sample/matrix mixture is irradiated by the laser. This process transfers both energy and charge to the sample molecules: *i.e.*, in positive ion mode the analyte molecules are transferred to the gas phase and protonated. Ions are accelerated out of the source region by an electric field (20–30 kV) and fly down the flight tube which is typically about 1 m long. At the end of the flight tube a detector records both the intensity of the ion current and the time-of-flight.

5.1.4 *The Ionization Process*

In positive ion MALDI the matrix absorbs the laser energy and transfers a proton to the analyte molecule. Together, the analyte and proton form a quasimolecular ion: *e.g.*, $[M+H]^+$. Other possibilities include $[M+Na]^+$ in the case of an added sodium ion, or $[M-H]^-$ in the case of a removed proton. The ions generated are all even-electron species although radical cations (photoionized molecules) can be observed, e.g., in case of matrix molecules or some organic molecules.

5.1.5 The MALDI Mass Spectrum of a Pure Component

In a typical MALDI mass spectrum of a pure compound the major peak corresponds to $[M+H]^{1+}$. In some instances, however, molecules adduct cations rather than a proton. For example, acidic peptides can bind sodium and potassium ions, arising from glass vessels as well as buffers, and these acidic peptides show up at 22 and 38 mass units higher than expected (*i.e.,* higher than the anticipated M+H ion.) In addition, MALDI spectra are also frequently complicated by both multiply charged ions (*i.e.,* $[M+nH]^{n+}$) and singly charged, multimeric forms (*i.e.,* $[nM+H]^{+}$). The propensity to form these more complex species is a function of the matrix, the laser intensity and the analyte (most especially its mass). Molecule-molecule and molecule-matrix adducts are also evident in some spectra.

5.1.6 The Laser

First-generation instruments employed UV lasers such as nitrogen lasers (337 nm), but frequency-tripled or quadrupled Nd:YAG lasers (355 nm and 266 nm respectively) are now more commonly employed. Infrared lasers have sometimes been used because they give a softer mode of ionization, less low-mass interferences, and are compatible with other matrix-free laser desorption mass spectrometry methods.

5.1.7 Time-of-Flight Analyzers

The operating principle of a TOF analyzer is that a population of ions having a distribution of masses and moving in the same direction with (more-or-less) constant kinetic energy will have a distribution of velocities inversely proportional to their mass. Their arrival times at a target plane (parallel to the plane of origin) will be distributed according to the square root of *m/z*. That is to say, for a given energy (E) and distance (d), the mass is proportional to the square of the flight time of the ion: *i.e.,* $m=(2E/d^2)t^2$.

5.1.8 The Matrix and Its Properties

The matrix is central to the ionization process and it is typically a low mass conjugated aromatic molecule. The matrix must be: (a) able to isolate and entrap analyte molecules (*e.g.,* by co-crystallization), (b) able to transfer or accept protons from nonvolatile analytes, (c) chemically inert (*i.e.,* no reactivity with analytes), (d) stable under vacuum in the MALDI source and (e) both soluble in solvents and compatible with analytes and samples. Commonly used matrices and their properties are included in Table 5.1.

Table 5.1 Commonly used MALDI matrices and their features

Matrix	Wavelength	Typical applications
2,5-Dihydroxybenzoic acid (2,5-DHB)	337/355 nm	Proteins and oligosaccharides
Sinapinic acid (SA)	337/355 nm	Proteins
α-Cyano-4-hydroxycinnamic acid (HCCA)	337/355 nm	Peptides and proteins
3-Hydroxycinnamic acid (3-HPA)	337/355 nm	Oligonucleic acids
Picolinic acid (PA)	266 nm	Nucleic acids
2,4,6-Trihydroxyacetophenone (2,4,6-THAP)	337/355 nm	Oligonucleic acids and acidic oligosaccharides
6-Aza-2-thiothymine	266, 337, 355 nm	Oligonucleic acids and acidic oligosaccharides
2-(4′-Hydroxybenzeneazo) benzoic acid (HABA)	337/355 nm	Proteins and carbohydrates
2,6-Dihydroxyacetophenone (2,6-DHAP)	337/355 nm	Oligonucleic acids
3-Aminoquinoline	337 nm	Oligosaccharides
3-Hydroxy picolinic acid	337 nm	Nucleic acids
Nicotinic acids	266 nm	Proteins, peptides and adduct formation
Thiourea	266 nm	Large protein

5.1.9 Sampling from the Target

Typically the laser is fired multiple times at a fixed location on the target and spectra are averaged or summed with the objective of maximizing the observable signal-to-noise ratio. If the amount of analyte is low, it may be necessary to irradiate additional regions on the target and sum/average many of these to yield an acceptable spectrum (Fig. 5.1). For example, when a sample is irradiated 10, 100, 1,000 & 11,000 times and the signal is averaged, there is a marked increase in signal-to-noise in the observed mass spectrum.

5.1.10 Characteristics of the MALDI Spectrum of a Mixture

Ideally, each component in a complex mixture would yield a single peak in the mass spectrum and relative peak heights would reflect relative amounts of each component of the mixture. The reality, however, is that the spectrum of a mixture is (very) complex. Each component often gives rise to multiple peaks and because components differ in how readily they ionize (*i.e.*, their ionization efficiencies), relative peak heights do not reflect relative amounts. Interpretation of the MALDI mass spectrum of a mixture is therefore far from trivial.

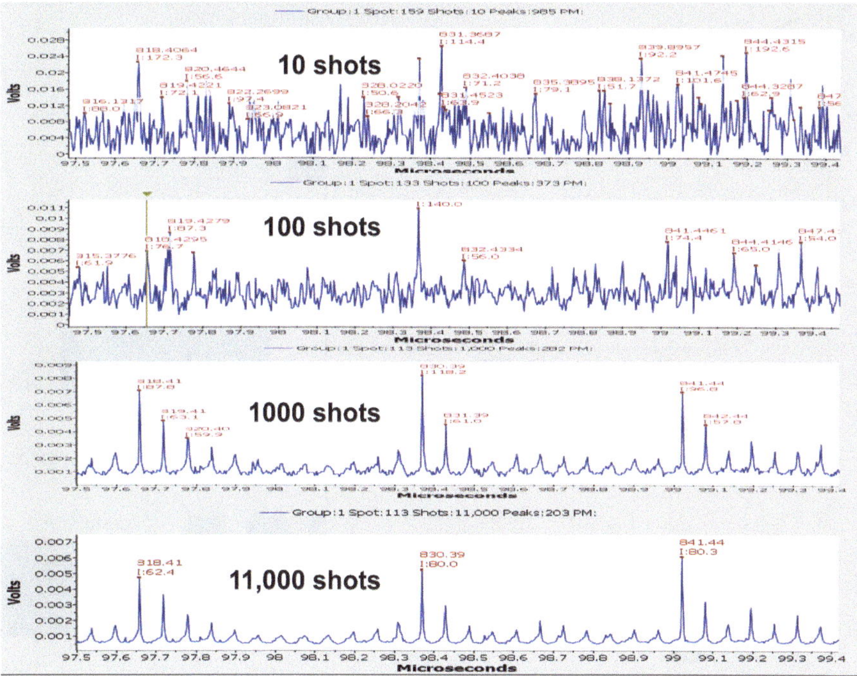

Marvin Vestal, SimulTOF Systems

Fig. 5.1 The impact of increasing the number of laser shots on the signal-to-noise ratio

5.1.11 Instrument Refinements

There have been several refinements to the basic MALDI-TOF configuration aimed at improving both the resolution and information content of spectra. The most important of these are outlined below.

Delayed extraction: In a conventional continuous ion extraction MALDI-TOF mass spectrometer, the achievable resolution is limited by the spread in the initial velocities of ions of identical *m*/*z*. In delayed extraction, a short delay is introduced between the laser pulse and the application of the accelerating voltage to compensate for this. Delayed extraction improves mass resolution [3].

The Reflectron: A reflectron can be located at the end of the flight tube and is comprised of a series of electrodes that form a linear field opposite in direction to the initial accelerating voltage. The reflectron is used to reverse the direction of travel of the ions. It significantly improves mass resolution by ensuring that ions of the same *m*/*z*, but different translational energies, arrive at the detector at the same time [4].

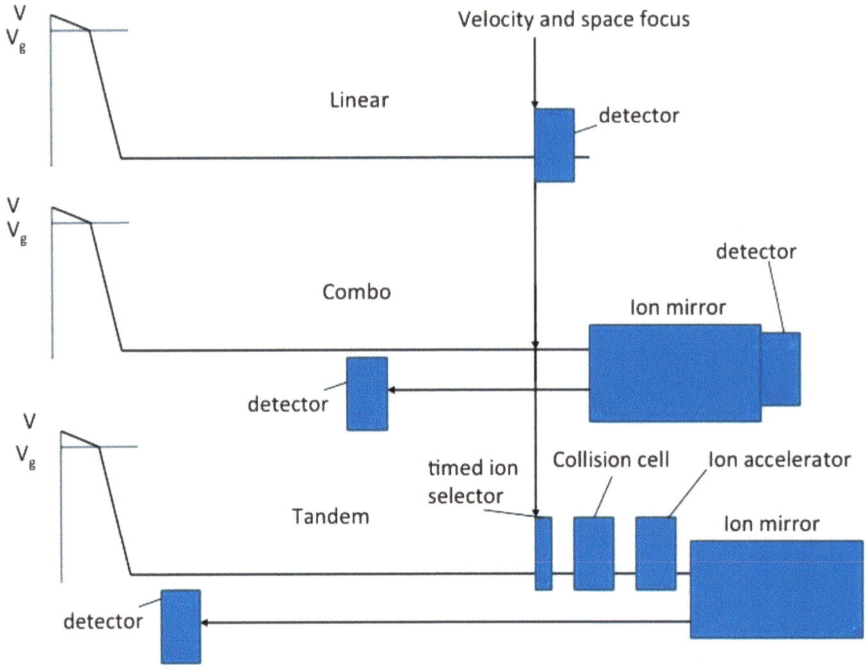

Fig. 5.2 Commonly adopted instrument configurations for MALDI-TOF instruments. From top to bottom, this figure shows the configuration of a simple linear MALDI-TOF system incorporating delayed extraction. The center panels show the incorporation of a reflectron (*i.e.*, ion mirror) to enhance resolution. In the third panel the TOF/TOF design incorporates a collision cell and a second stage of (product) ion separation to give fragmentation and structural information not otherwise accessible by MALDI

Post Source Decay (PSD) and TOF/TOF: There is little to no fragmentation in a typical MALDI mass spectrum. Several approaches have therefore been developed to overcome this limitation and to allow investigators to generate product ion spectra. The first of these was PSD [5]. Although powerful, the approach is limited because it is not possible to select a specific precursor and fragment it. More recently, a high energy collision cell has been introduced between two time-of-flight analyzers to allow selection of a precursor ion of interest, collisional dissociation to induce fragmentation, and then generation of a product ion spectrum [6] (Fig. 5.2).

5.1.12 Major Applications of MALDI (MALDI-TOF)

Some of the most common areas of application of MALDI-TOF mass spectrometry are outlined in the sections that follow.

5.1.12.1 Applications in Organic Chemistry: Characterization of Synthetic Materials and Natural Products

MALDI can be used to aid in the characterization of natural products and synthetic materials. Contrary to common views, MALDI can be applied to very low mass analytes and provides accurate MW assignments with errors in the very low ppm range. This makes it possible to assign elemental compositions [7]. Synthetic polymer characterization is a challenging exercise, but MALDI offers a powerful solution and its applications have been comprehensively reviewed [8].

5.1.12.2 Applications to Proteomics: Peptide Mass Fingerprints (PMF) and MS/MS

It is possible to identify a purified protein by an approach known as peptide mass fingerprints. This is a powerful strategy that is simple and fast to perform that gives valuable information on a "purified" protein and its contaminants. The process involves four steps: (1) protein isolation (PMF requires that a single protein or just a few proteins are present); (2) protein digestion; (3) MALDI-TOF analysis to generate MWs of proteolytic peptides and (4) application of database search algorithms to determine the best match. The algorithm performs a theoretical digest of each entry in the protein database and calculates the mass of all peptides derived from each; the algorithm then compares observed (*i.e.,* experimental) peptide masses with those calculated *in silico* from the database entries (*i.e.,* theoretical). The best matches are determined and scores assigned to each that can be used to assess the certainty of the match. The process generally works well, but factors including the number of masses observed, the mass tolerance/accuracy (set/measured; ppm or Da), protein modifications, and the attributes of the relevant database(s) can all influence the outcome.

Protein identification is feasible by PMF because instead of searching all conceivable sequences, the search space is restricted to consider only the small subset of known proteins that are represented in the database. However, depending on the specific objectives, it may prudent to generate additional data in support of a protein assignment made by PMF. This could include generating one or more peptide MS/MS spectra (*e.g.,* generated by MALDI TOF/TOF) and matching these against the theoretical MS/MS spectra of putative sequence for the peptide.

Notably, these approaches can only be used to assign an ID to proteins that have already been identified and where their sequence resides in, or can be derived from, an accessible database. In some instances, however, matches to closely homologous entries can provide important clues to protein identity, even if the target is not included in the database(s), but for obvious reasons, the exact splice variant, mutant protein and structure homologue may be difficult, if not impossible, to identify. Similarly, complications can arise if proteolysis is incomplete, there are non-specific cleavages or if multiple proteins are present in the sample (*i.e.,* because it becomes computationally intensive to account for all possibilities).

Numerous reviews highlight other applications of MALDI-TOF to peptide and protein characterization [9–14].

Applications in Microbiology

The identification of microbial pathogens by MALDI-TOF mass spectrometry is an important and rapidly growing application. MALDI-TOF is particularly attractive in the field of clinical microbiology because it provides accurate results at markedly reduced costs relative to the more conventional approaches that have been adopted for decades. Further, the rugged, reproducible nature of the technique means fewer replicates are required to confirm pathogen identifications. In addition, training in the MALDI technique typically takes a week; contrastingly, training in the more conventional approaches may take years.

With MALDI-based identification one basic strategy can be used to confidently identify a range of microbial pathogens including bacteria, yeast, molds, fungi and mycobacteria [15]. The approach involves taking a small aliquot of a clinical sample (*e.g.,* urine or blood) and incubating this on a Petri dish with standard agar growth media to amplify the number of pathogens and to generate "pure" colonies. Culture time remains the main rate-limiting step, even with MALDI, because it is necessary to have adequate biomass (*i.e.,* about 10^5 cells/ml) to generate mass spectra of sufficient quality for confident pathogen typing. (In the case of *Mycobacterium tuberculosis*, it can take up to 6–8 weeks to culture the sample.)

After culturing, a representative colony is sampled from the Petri dish with an inoculation loop, smeared onto the MALDI target, matrix is added (*i.e.,* sinapinic acid) and the sample is air-dried. A MALDI mass spectrum of the intact peptides and proteins is then generated. Custom software is used to match the spectral peaks derived from the unknown pathogen peptides and proteins against those in a microbe spectral library. The software assigns a score based on the quality of the match between the sample and the best match in the library. The pathogen peptides/proteins are not identified as a part of this process, but for the most part they are ribosomal proteins. Their sequences, and therefore their masses, are sufficiently constant and unique to allow robust characterization [16]. The approach is reported reproducible from lab-to-lab and is largely independent of culture conditions (*e.g.,* time and temperature) and culture medium [17].

There are, however, several limitations to the technique. For example, atypical organisms or complex samples comprised of multiple pathogens may still require nucleic acid sequencing. Although MALDI approaches to antibiotic susceptibility testing are under development, they are not yet commercially available. As anticipated, species identification is highly dependent on the spectral library, how complete it is, how carefully it is curated, and how well the software matches/discriminates between mass fingerprints.

Despite current limitations, MALDI-based approaches are now routinely adopted, and they allow the reliable characterization of pathogens that would not have been identified by traditional workflows just a few years ago. For example,

MALDI bacterial typing can identify life-threatening infections such as multi-drug resistant bacteria many hours faster than more conventional diagnostic strategies [18]. This is because blood borne bacteria can be detected in culture media after only 6–8 h incubation whereas more conventional approaches require overnight culturing on selective media. Streamlining the process translates directly to better outcomes because early and appropriate therapy is the most important factor in preventing mortality from potentially fatal infections such as septicemia [19].

Rapid change in this area continues. Commercial spectral libraries are regularly updated with new species, workflows are being optimized to incorporate new developments and research into new applications is ongoing. For example, MALDI-based approaches to identify viral pathogens and bacterial endotoxins have also been reported [20–22]. Also, methods have been reported that allow for the direct identification of urine and blood infections without prior culturing [23]. However, when multiple pathogens infect the sample, identification can be problematic because the matching algorithms can fail if the MALDI mass spectrum is a composite derived from several distinct organisms. The problem is compounded because without isolation and enrichment through culturing, high abundance components – not all necessarily derived from the same pathogen – can contribute a strong signal that overwhelms and masks others. Under these conditions it can be difficult to get a high quality match, and less abundant pathogens may be overlooked completely.

Applications of MALDI to Clinical Diagnostics

The development of diagnostics centered round MALDI-TOF mass spectrometry is an area of very active research. A wide variety of endogenous analytes including nucleic acids, glycans, lipids, exogenous small molecules and drug metabolites have been measured in these studies, but peptides and proteins have been the focus of the majority of clinical biomarker studies. Typically these studies involve some approach to profiling (or fingerprinting) multiple components simultaneously, either on the crude sample, or following affinity enrichment, depletion or pre-fractionation strategies. A wide variety of tissue and fluids have been explored including biopsy samples, blood products (*e.g.,* serum and plasma), urine, tears, saliva, and cerebral spinal fluid (CSF). MALDI-TOF has been employed across all phases of biomarker development – from discovery, through verification and validation, to routine testing. In recent years several clinically validated and FDA approved assays have been developed on a MALDI platform and these are now becoming commercially available. In this section some of the strategies adopted, and recent developments from research and commercial sectors are highlighted.

There are many instances of where MALDI has been applied to biomarker discovery. For example, changes in the glycosylation pattern/profile of cellular and circulating proteins have been explored in a variety of cancers and inflammatory disorders. In a case-control study aimed at identifying new glycan biomarkers of epithelial ovarian cancer (EOC), N-glycans were enzymatically cleaved from affinity enriched serum glycoproteins, permethylated to stabilize them and characterized by MALDI [24]. A diagnostic algorithm was then constructed from the peak areas of mannose and fucose containing glycan motifs. This approach reportedly delivered 97 % sensitivity and 98.4 % specificity in the diagnosis of EOC. (This compares with a 97 % sensitivity and 88.9 % specificity for the common tumor antigen CA-125.) In a similar manner, van Swelm et al. [25] describe a MALDI-TOF case-control study of urine collected from psoriatic patients at risk of hepatic fibrosis, an adverse reaction associated with long-term immunosuppression with methotrexate (MTX) therapy. They identified candidate urinary protein biomarkers that may be useful in monitoring MTX-induced hepatic fibrosis. Tear analysis has also been explored as a minimally invasive approach to the diagnosis of breast cancer. Bohm et al. employed MALDI-TOF-TOF to identify over 20 proteins that appeared to be differentially abundant in 25 patients with primary invasive breast carcinoma relative to 25 age-matched healthy controls [26].

Numerous companies now offer bespoke assays, affinity reagents, standards and software solutions based on the MALDI-TOF platform. Examples include the MASStermind® cardiac assay (Pronata, www.pronota.be), Veristrat® lung cancer drug response assay (Biodesix Inc., www.biodesix.com), SISCAPA reagents, affinity depletion frets for MALDI mass spectrometric immunoassay (MSIA®; Thermoscientific, www.thermoscientific.com), MassARRAY® nucleic acid analysis (Sequenom, www.sequenom.com) and MALDI spectral analysis software products from Premier Biosoft (www.premierbiosoft.com) and Bioinformatic Solutions (www.bioinfor.com). The Veristrat® and MassARRAY® tests are prescient examples that showcase the technical capability and clinical potential of MALDI based products.

Sequenom's MassARRAY® system combines primer extension chemistry with MALDI-TOF mass spectrometry for quantification of nucleic acids and detection of genetic variability. The MassARRAY® system provides multiple nucleic acid testing options including genotyping, somatic mutation and methylation profiling, copy number variation and quantitative gene expression. Single nucleotide polymorphism (SNP) genotyping is commonly used for fine mapping of gene variance associated with specific clinical phenotypes (GWAS studies). Sequenom SNP genotyping combines two core elements: first the generation of an allele-specific product from template cDNA and second, extension primers are designed to anneal directly adjacent to every SNP position to be assayed [27]. During subsequent thermocycling, a thermosequence enzyme adds a nucleotide that naturally differs in mass from the other three available. Samples are then cleaned up by ion exchange to remove potential interferents. For these applications 3-hydroxy picolinic acid is used as a matrix to ionize and dissociate the primer-template duplex. The allele specific SNP products can then be discriminated by m/z differences as low as 16 within the mass range m/z 4,000–9,000. The technology was commercialized around the development of a dedicated MALDI-TOF analyzer and specific reagents and has been cited in over 2,000 publications to date.

Veristrat® is a clinically validated protein diagnostic test, released in 2009, that aims to personalize drug treatments for non-small cell lung cancer (NSCLC) patients. Instead of SNP genotyping, Veristrat uses MALDI-TOF to examine a panel of eight spectral features (or specific *m/z* values) in the patient blood sample. A bespoke algorithm uses these features to differentiate between those subjects that are likely to respond to a specific drug type (*i.e.*, a protein kinase inhibitor) and those who likely will not. Biodesix does not define the specific proteins that underpin the Veristrat® signature, but in a subsequent study by Milan et al. [28], serum amyloid A protein 1 (SAA1), together with its truncated forms, are reported to be over-expressed in the plasma of patients classified as Veristrat-poor. These authors report that SAA1 and its variants contribute four out of the eight classifiers composing the Veristrat® signature. To date, Veristrat® has been used in over 5,000 patients and over 80 clinical trials. Ackerly et al. report that 90 % of post-test treatment recommendations positively correlated with Veristrat® test results [29]. The test also has the potential to serve as a breast and colorectal cancer outcome prediction tool.

Applications of MALDI to Imaging

In imaging mass spectrometry a sample of tissue is immobilized on the MALDI target and coated with matrix. Imaging mass spectrometry retains the spatial fidelity of the sample with the objective of locating specific molecules within the sample. Generating an image using MALDI MS is accomplished by rastering the laser across the tissue in order to collect molecular information from a regularly spaced array of positions (or pixels). The molecular information encoded at each location is extracted and plotted to create ion images that can be directly correlated with the location of specific biological molecules.

Typically in MALDI imaging a fresh frozen tissue sample is cut and mounted on a conductive target. Matrix is then applied in an ordered array across the tissue section and mass spectra (or MS/MS spectra) are generated at each x, y coordinate. The distribution of one or more components is visualized within the tissue sample by using a color scale to represent the relative intensity of the components.

5.2 Conclusions

MALDI ionization, especially when combined with time-of-flight ion separation, is a cost-effective, rapid, sensitive, reproducible and exceedingly versatile approach to both qualitative and quantitative analysis. MALDI efficiently ionizes a wide range of compounds including high mass biopolymers and thermally labile compounds and it is a valuable complement to electrospray ionization. Because of these attributes, MALDI is increasingly being applied to issues relating to bioterrorism including the rapid and sensitive identification of potent toxins [30–32] and dangerous pathogens [33–35] that might be intentionally introduced into the environment.

Acknowledgements We thank Dr Marvin Vestal and his colleagues at SimulTOF Systems, Sudbury, MA, for supplying the two figures that are included in this manuscript.

References

1. Karas M, Bachmann D, Hillenkamp F (1985) Influence of the wavelength in high-irradiance ultraviolet laser desorption mass spectrometry of organic molecules. Anal Chem 57(14):2935–2939
2. Karas M et al (1987) Matrix-assisted ultraviolet laser desorption of non-volatile compounds. Int J Mass Spectrom Ion Process 78:53–68
3. Vestal ML, Campbell JM (2005) Tandem time-of-flight mass spectrometry. Methods Enzymol 402:79–108
4. Mamyrin BA et al (1973) The mass-reflectron, a new nonmagnetic time-of-flight mass spectrometer with high resolution. Sov Phys JETP 37:45
5. Kaufmann R et al (1996) Post-source decay and delayed extraction in matrix-assisted laser desorption/ionization-reflectron time-of-flight mass spectrometry. Are there trade-offs? Rapid Commun Mass Spectrom 10(10):1199–1208
6. Medzihradszky KF et al (2000) The characteristics of peptide collision-induced dissociation using a high-performance MALDI-TOF/TOF tandem mass spectrometer. Anal Chem 72(3):552–558
7. Bucknall M, Fung KYC, Duncan MW (2002) Practical quantitative biomedical applications of MALDI-TOF mass spectrometry. J Am Soc Mass Spectrom 13(9):1015–1027
8. Weidner SM, Trimpin S (2010) Mass spectrometry of synthetic polymers. Anal Chem 82(12):4811–4829
9. Yang H, Liu N, Liu S (2013) Determination of peptide and protein disulfide linkages by MALDI mass spectrometry. Top Curr Chem 331:79–116
10. Stuhler K, Meyer HE (2004) MALDI: more than peptide mass fingerprints. Curr Opin Mol Ther 6(3):239–248
11. Hamdan M, Galvani M, Righetti PG (2001) Monitoring 2-D gel-induced modifications of proteins by MALDI-TOF mass spectrometry. Mass Spectrom Rev 20(3):121–141
12. Bonk T, Humeny A (2001) MALDI-TOF-MS analysis of protein and DNA. Neuroscientist 7(1):6–12
13. Roepstorff P (2000) MALDI-TOF mass spectrometry in protein chemistry. EXS 88:81–97
14. Chaurand P, Luetzenkirchen F, Spengler B (1999) Peptide and protein identification by matrix-assisted laser desorption ionization (MALDI) and MALDI-post-source decay time-of-flight mass spectrometry. J Am Soc Mass Spectrom 10(2):91–103
15. Croxatto A, Prod'hom G, Greub G (2012) Applications of MALDI-TOF mass spectrometry in clinical diagnostic microbiology. FEMS Microbiol Rev 36(2):380–407
16. Intelicato-Young J, Fox A (2013) Mass spectrometry and tandem mass spectrometry characterization of protein patterns, protein markers and whole proteomes for pathogenic bacteria. J Microbiol Methods 92(3):381–386
17. Welker M (2011) Proteomics for routine identification of microorganisms. Proteomics 11(15):3143–3153
18. Berrazeg M et al (2013) Biotyping of multidrug-resistant Klebsiella pneumoniae clinical isolates from France and Algeria using MALDI-TOF MS. PLoS One 8(4):e61428
19. Schneiderhan W et al (2013) Work flow analysis of around-the-clock processing of blood culture samples and integrated MALDI-TOF mass spectrometry analysis for the diagnosis of bloodstream Infections. Clin Chem 59(11):1649–1656
20. Gao X et al (2013) Screening of influenza mutations using base-specific cleavage and MALDI mass spectrometry. Clin Chim Acta 420:89–93

21. Peng J et al (2013) Sensitive and rapid detection of viruses associated with hand foot and mouth disease using multiplexed MALDI-TOF analysis. J Clin Virol 56(2):170–174
22. Kilar A, Dornyei A, Kocsis B (2013) Structural characterization of bacterial lipopolysaccharides with mass spectrometry and on- and off-line separation techniques. Mass Spectrom Rev 32(2):90–117
23. Ferreira L et al (2011) Rapid method for direct identification of bacteria in urine and blood culture samples by matrix-assisted laser desorption ionization time-of-flight mass spectrometry: intact cell vs. extraction method. Clin Microbiol Infect 17(7):1007–1012
24. Biskup K et al (2013) Serum glycome profiling: a biomarker for diagnosis of ovarian cancer. J Proteome Res 12(9):4056–4063
25. van Swelm RP et al (2013) Biomarkers for methotrexate-induced liver injury: urinary protein profiling of psoriasis patients. Toxicol Lett 221(3):219–224
26. Bohm D et al (2012) Comparison of tear protein levels in breast cancer patients and healthy controls using a de novo proteomic approach. Oncol Rep 28(2):429–438
27. Tost J, Gut IG (2005) Genotyping single nucleotide polymorphisms by MALDI mass spectrometry in clinical applications. Clin Biochem 38(4):335–350
28. Milan E et al (2012) SAA1 is over-expressed in plasma of non small cell lung cancer patients with poor outcome after treatment with epidermal growth factor receptor tyrosine-kinase inhibitors. J Proteomics 76:91–101
29. Akerley WL et al (2013) The impact of a serum based proteomic mass spectrometry test on treatment recommendations in advanced non-small-cell lung cancer. Curr Med Res Opin 29(5):517–525
30. Ramos Catharino R et al (2005) Aflatoxin screening by MALDI-TOF mass spectrometry. Anal Chem 77(24):8155–8157
31. McGrath SC et al (2011) Detection and quantification of ricin in beverages using isotope dilution tandem mass spectrometry. Anal Chem 83(8):2897–2905
32. Alam SI, Kumar B, Kamboj DV (2012) Multiplex detection of protein toxins using MALDI-TOF-TOF tandem mass spectrometry: application in unambiguous toxin detection from bioaerosol. Anal Chem 84(23):10500–10507
33. Pierce CY et al (2007) Strain and phase identification of the U.S. category B agent Coxiella burnetii by matrix assisted laser desorption/ionization time-of-flight mass spectrometry and multivariate pattern recognition. Anal Chim Acta 583(1):23–31
34. Jiang J et al (2007) An immunoaffinity tandem mass spectrometry (iMALDI) assay for detection of Francisella tularensis. Anal Chim Acta 605(1):70–79
35. Ayyadurai S et al (2010) Rapid identification and typing of Yersinia pestis and other Yersinia species by matrix-assisted laser desorption/ionization time-of-flight (MALDI-TOF) mass spectrometry. BMC Microbiol 10:285

Chapter 6
Field Portable Mass Spectrometry

Stephen A. Lammert

Abstract The emergence of microelectromechanical systems (MEMS) fabrication techniques is prevalent in modern electronics, personal communications systems and many other everyday devices that continually become smaller and more capable. Mass spectrometry (MS) (Fox J, Saini R, Tsui K, Verbeck G, Rev Sci Instrum 80(9):93302–93306, 2009), too has benefited from these fabrication techniques and as a result, there has been an increased focus on instrument miniaturization and field portability. Much of the effort in this area has been in the miniaturization of the mass analyzer where even micron-sized analyzers have been reported. However, the miniaturization of the mass analyzer is not the only barrier to system size reduction. Much of the support hardware (pumping systems, ionization sources, detectors, etc.) has not scaled proportionally, either in size or operational capability. Often the batteries are the largest/heaviest components in a miniature system and, despite improvements in battery technologies, power is still a major limitation for field portability. In addition to MS hardware, the methods for sample acquisition, processing and introduction must be reduced with respect to complexity, size and power requirements while still maintaining sufficient analytical efficiency for adequate detection specifications. Finally, the systems must be easy to use by non-expert operators. All of these analytical figures of merit interplay in such a way that significant tradeoffs must be made when designing a field-portable instrument for a particular application.

The three talks presented at the 2013 NATO Advanced Studies Institute in Sienna Italy and summarized in this chapter cover the progress and obstacles in miniaturization of MS components as well as the interdependencies of the instrumental figures of merit, analytical performance and field applications as they pertain to field-portable miniature MS. The talks included examples from the speakers' past and current fieldable instrument development projects.

Keywords Miniature mass spectrometry analyzers • Field portable GC/MS • Field portable mass spectrometry

---

S.A. Lammert (✉)
Torion Technologies Inc., American Fork, UT, USA
e-mail: steve.lammert@torion.com

J. Banoub (ed.), *Detection of Chemical, Biological, Radiological and Nuclear Agents for the Prevention of Terrorism*, NATO Science for Peace and Security Series A: Chemistry and Biology, DOI 10.1007/978-94-017-9238-7_6

6.1 Introduction

The recent progress in miniaturizing almost every technical gadget in our lives has also included the miniaturizing of chemical analysis instruments, and specifically mass spectrometry [1]. This is part of an increasing drive to take chemical analysis systems out of the laboratory and into the field. While "field-portable" MS systems [2] are not new (NASA launched two portable gas chromatograph-mass spectrometers (GC/MS) systems [3] into the 'field' of Mars in 1975), the past decade has witnessed a large increase in interest and developmental activity in the area of field-portable chemical detection systems, especially those based on MS. This is an expected outgrowth of MS's position as the gold-standard analytical tool for chemical detection and quantitation as well as unknown identification. MS is broadly applicable to a very wide range of chemical classes and compounds and has the added quality of possessing both high sensitivity (for reducing the rate of false negative detection) and high selectivity (for reducing the rate of false positive detection). The applications drivers are many for this new interest in direct field analysis as opposed to field sampling followed by laboratory analysis. The detection and quantitative accuracy of samples that are especially volatile, thermally labile, reactive, or air-sensitive is facilitated by reducing the time between sampling and analysis. In the case where the sample presence or concentration is highly temporal, having the analysis done on-site is beneficial, especially if the actionable levels are important to the decision (such as on-site emissions monitoring for public safety). It also finds advantageous use in the case where many samples must be acquired in order to direct some mitigation activity (e.g., contaminated site remediation or contamination localization). In this case, the ability to rapidly screen multiple samples from the perimeter of a potential hazardous site is greatly enhanced by on-site analysis. Finally, when a rapid time frame for decision-making is important (e.g., as in the case of the detection of chemical agents in the battlefield, or portal security at airports or other installations), on-site analysis is an absolute requirement. When all considerations of analysis time, high sensitivity and high selectivity are taken, as a whole, field-portable MS is the clear frontrunner as a desired on-site analysis tool.

In addition to the analytical examples above, there are other instrument-centered reasons that drive the miniaturization of MS systems. Smaller instruments mean less structural material, and except for a few instrumental components, this can lead to lower materials costs. A few critical components, such as the mass analyzer, may likely have higher costs (especially if traditionally machined) due to increased tolerance requirements. MEMS fabricated components, however, may actually be less expensive to fabricate. To date, any cost advantage has yet to be realized in full since the number of portable instruments sold is only a small fraction of the total MS market. As more application areas adopt the use of field-portable MS, the economies of scale with larger numbers of instrument purchases will begin to appear. And with smaller analytical components, there are other advantages (lower power, smaller thermal zones, and lessened vacuum requirements) that accompany miniaturization.

An area of confusion in this growing field is that of definitions. Field portable and miniature are not necessarily synonymous and are, to some degree, relative. A Siemens 200-lb, 14 ft^3 FT-ICR [4] could certainly be considered miniature (in a relative sense) as well as a significant technology advancement, but it could hardly be moved around sufficiently easy to be generally considered person-portable (especially when one considers the additional size and weight of the gasoline-powered generator). Other systems that have been designed to be considerably more rugged and resistant to external environmental conditions, while retaining similar size and weight to their laboratory counterparts, could be considered fieldable, but again, not miniature or easily portable. Even the term 'field-portable' is complex. Portability refers to the size and weight of the system and the complexity of moving the system around. A transportable system can be carried in a vehicle or on a cart and moved for short distances by a small number (2–3) people. A person-portable system can be moved by one person for a short distance, while a hand-portable system can be easily carried by one person for considerable time or distances. Conversely, fieldability is more a description of the ruggedness (mechanical and environmental) and complexity of the entire sampling and analysis application. An application that requires a chemical work-up procedure on a laboratory scale, prior to analysis on a fieldable instrument does not have the same level of fieldability as one where the sampling and analysis is simple and performed on an instrumental system that can withstand the environmental variability and rough handling that accompanies the simple act of moving an analytical instrument from the lab to the 'field'. Finally, the terms 'miniature' and 'micro' typically refer to the relative scale differences of the mass analyzer or other critical components of the system as they relate to traditional laboratory instruments. Miniature simply means smaller while typically employing fabrication technologies such as traditional machining. Micro is used to imply even smaller-scale device features and the fabrication of these components commonly utilizes micro-electro mechanical systems (MEMS) techniques [1] such as lithography.

6.2 Background of Field Portable Mass Spectrometers

As mentioned above, in 1975 NASA launched two exploration probes each with a portable gas chromatograph-mass spectrometer (GC-MS) destined to study the surface of Mars and, in probably the most visible way, opened the door for taking traditional laboratory instruments out of the laboratory and into the 'field'. Over the ensuing years, advancement in field-portable MS was slow but steady; mostly in the areas of system transportability such as the Viking Spectratrak [5] and the INFICON Hapsite [6] (both are field portable GC/MS systems), ruggedization such as the early vehicle mounted MM-1 chemical weapons MS detector [7], and fieldable, direct sampling techniques combined with a ruggedized ion trap MS as in the Direct Sampling Ion Trap Mass Spectrometer (DSITMS) [8]. And while the overall instrument size and weight did not significantly decrease, ruggedization and field portability opened the door for new applications and users (most notably the military and environmental site characterization). Today, the military (as well as homeland security) is still a primary funding driver for much of the developmental research. NASA continues to employ MS on extraterrestrial probes to study planetary atmospheres, surface composition and the search for life [9]. But new industrial applications such as fugitive emissions monitoring, food safety and quality, and petroleum exploration have begun to grow. Applications have expanded to even harsher analytical environments to include even volcanology [10] and oceanography [11]. As a result of this growing field, the Harsh Environment Mass Spectrometry [12] interest group was formed in 1999 and continues to sponsor a biannual workshop. From these meetings and various other dedicated symposiums in analytical conferences, the progress of this field is gaining wider attention, further driving new requirements for field portable MS applications.

6.3 Miniature Mass Analyzers: Challenges and Approaches

With the rapid rise in MEMS fabrication approaches, an increasing number of researchers have begun to focus their efforts on miniaturizing the mass analyzer. This is an obvious starting point since the challenges of overcoming the ion physics and engineering limitations are important to maintaining the quality of the analytical performance (sensitivity, resolution, selectivity in complex environments, etc.) that makes mass spectrometry so desirable as an analytical technique. It also happens to be somewhat fortuitous in that the mass analyzer is one of the easier components to scale to smaller size. There is another good reason to start with the analyzer since many of the operational conditions and requirements (e.g. required operating voltages, power and vacuum) scale accordingly with reduced analyzer dimensions. Smaller confining optics for ions leads to shorter overall ion path lengths from 'birth' to detection. The mean free path required for analytical effectiveness is thus smaller, allowing higher-pressure operation and smaller vacuum pumps. Smaller mass analyzers may also lead to smaller ionizers and detectors (although not necessarily proportionally) and the collective analyzer components can now be housed in smaller vacuum chambers, with a concomitant reduction in volume and mass. Smaller mass analyzer dimensions also lead to lower ion optic fields, lowering the voltage requirements which also aids in reducing the complexity of the supporting electronics. Ultimately, this power reduction path leads to improved prospects for battery operation; a major crux point in the tradeoffs of converting a smaller MS to a portable MS. However, there are also corresponding new challenges that arise which broadly separate into two categories of limitations – analytical and physical. The analytical limitations refer to performance metrics such as, sensitivity, selectivity, resolution, etc. For example, smaller ion sources or mass analyzers will support fewer ions, thus limiting the detection limits of the overall system. From a physical perspective, the quality of the ion optic components with respect to alignment, spacing dimensional precision or electrode shape often suffers with miniaturization, primarily due to machining limitations. These geometric or alignment errors (or even surface roughness for that matter) can ultimately degrade the analytical performance and lead to sensitivity loss or reduction in analytical selectivity resulting from reduced mass resolution.

Despite these obstacles, significant progress has been demonstrated in miniaturizing all current forms of mass analyzers, [13, 14] including time-of-flight and sector instruments, as well as 2D [15] and 3D [16] quadrupole devices. Even FT-ICR systems have been made smaller as described in the introduction, although the stringent high-vacuum and magnetic field requirements for FT-ICR instruments have limited this progress to the transportable stage. There are specific advantages and disadvantages for miniaturization and field portability when considering each type of mass analyzer. Time-of-flight MS, for example has the simplest mass analyzer for miniaturization (a field-free flight tube), but requires a low operating pressure and therefore larger and heavier vacuum pumps.

Ion trap mass analyzers are an especially attractive choice for miniaturization. They have a simple electrode structure that does not require especially high machining precision in the larger, laboratory-based systems. They operate at, and actually require, a higher operating pressure than other forms of mass analyzer, lessening the requirements for the supporting vacuum hardware. Ion traps are capable of 'tandem-in-time' [17] multiple stages of MS (MS^n) and, therefore, do not require separate mass analyzers for each stage of MS as do 'tandem-in-space' instruments, such as sector or 2D quadrupole instruments for example. The geometry of ion trap mass analyzers lends itself nicely to miniaturization and especially MEMS fabrication. The simple, three-electrode design assembles conveniently into 3 'layers': top and bottom electrodes with an interposed ring electrode. Finally, the fundamental physical equations that control ion trap performance scale favorably with analyzer size. A 2× reduction in size leads to a 4× reduction in operating voltage and a 16× reduction in power. However, ion traps face two challenges when considering miniaturization: (1) maintaining sufficient ion capacity for detection [18] and (2) maintaining the trapping field quality and, therefore, good mass resolution and sensitivity performance, with decreasing dimensions due to the inability to maintain the required geometry, alignment and surface quality. The first affects all attempts at ion trap miniaturization (traditionally machined 'miniature' versions as well as MEMS-fabricated 'micro' versions). The second becomes increasingly important in 'micro' versions as the fabrication dimensions are reduced [18].

Ion Capacity The first fundamental challenge becomes obvious when viewing an ion trap mass analyzer as an 'ion bottle'. There is a limited capacity of the device to 'hold' ions due to ion-ion repulsion forces (known as space charge). Space charge affects the dimensions of the trapped ions within the electrostatic field. If the size of the ion cloud exceeds some optimum size, mass resolution (and even sensitivity) suffers. Therefore, a smaller 'bottle' can hold even fewer ions before space charge becomes debilitating. There have been two approaches to addressing the loss of ion capacity that accompanies trap miniaturization. The first is to assemble many of these smaller traps into an array. This approach maximizes the voltage scaling

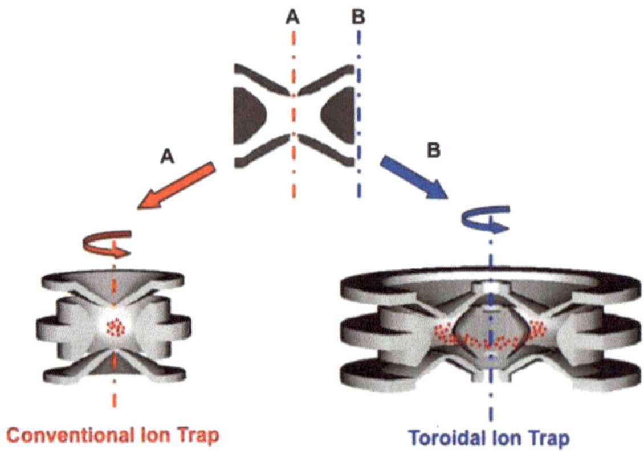

Fig. 6.1 Comparison of a traditional 3D Paul ion trap (**a**) to a toroidal ion trap (**b**). (Modified with permission from Ref. [20])

advantage, but comes with the additional challenge of designing an efficient coupling of the ionization and detection ion optics elements to ensure equal illumination or filling as well as equal sampling to the detector for each trap element. In addition, the increased capacitance of the trap array typically requires higher operating frequencies and power. Many examples of this approach are available [19]. The second approach to restoring ion capacity is to address the 'dimensionality' of the mass analyzer. Recently, linear 2D quadrupole mass filters have been employed as ion traps by including a gate on each end of the filter, thus creating a DC potential barrier to the ion path out the ends. The advantage of these linear ion traps (LITs) is that the fundamental operating conditions (rf power, etc.) is controlled by the cross sectional dimensions, but the ion capacity is controlled by the length along the field-free dimension. LITs have become common ion optical components in laboratory instruments and now are also gaining interest in the miniature mass analyzer arena. The limit as to how long one can practically extend the trapping dimension is governed primarily by the geometry of the trap/detector ion optics (again addressing the efficiency of detecting all ions ejected from the trap). In addition, because of the DC end gates, the trapping field is not uniform from end-to-end. An alternate approach to this second 'dimensionality' avenue of restoring ion capacity is to change the 'geometric dimensionality'. This can be seen conceptually in Fig. 6.1. The root geometry is the cross section for a conventional, 3D ion trap. It is this geometry (and it's internal dimensions) that governs the operating characteristics for the analyzer. The traditional ion trap device is produced by rotating this cross section around a center axis (A), which results in, to first order, a cylinder geometry. If, however, the rotation axis is offset to position (B), the resulting geometry is toroidal in nature. This "Toroidal Ion Trap" device [20] has the same cross-sectional dimension and, therefore, the same operating parameters as the traditional 3D ion trap, but possesses a greatly enhanced trapping volume for that same dimension. The spacing and shape of the electrode curvature must be redesigned in order to correct the trapping fields and account for this added dimension. However, when accomplished, the mass analysis performance of this extended capacity ion trap analyzer is comparable to the conventional 3D ion traps. For miniaturization, however, this increased

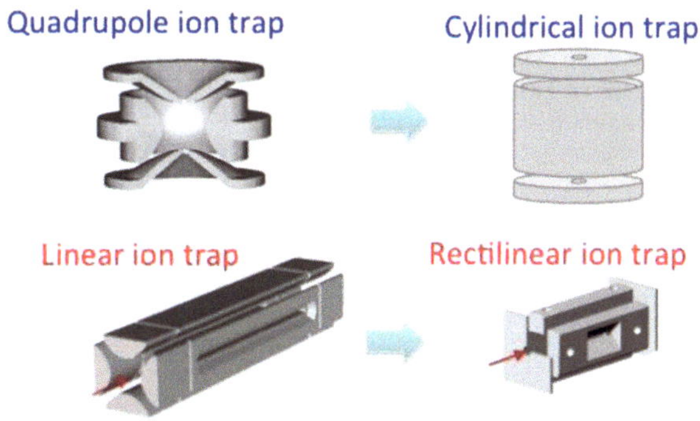

Fig. 6.2 A 3D Paul ion trap (*top left*) and a 2D quadrupole mass filter (*bottom left*) and their respective simplified geometries (Modified with permission from Ref. [22])

volume can be 'traded back' by reducing the cross-section dimension with the resulting reduction in operating voltage. So, for example, if the cross-section dimension is reduced by a factor of 5, the operating voltage is reduced by a factor of 25 and the power by a factor of 125. This miniature toroidal ion trap [21] is the mass analyzer engine of the Torion Technologies (American Fork, UT) TRIDION®-9 field portable GC/MS system described below.

Field Integrity and Electrode Dimension Tolerances As the dimensions of the mass analyzer decrease, a point will eventually be reached whereby the ability to conventionally machine the ion optic components will no longer be feasible. This limitation stems from three separate and sequentially encountered constraints as the analyzer dimension is reduced. The first constraint is reached when complex geometries make the conventional machining process overly difficult or impossible. At this point, the required curvatures (e.g., the hyperbolic curved surfaces of ion trap and quadrupole electrodes) must be compromised for more simple geometries. As shown in Fig. 6.2, the complex surface characteristics of both the 2D and 3D quadrupole devices can be replaced by more planar versions that are considerably easier to machine, while still maintaining required alignment and spacing tolerances. The curved poles of the linear ion trap can be replaced with flat-surfaced rods as in the case of the Rectilinear Ion Trap [22]. The hyperbolic electrodes of the quadrupole ion trap can be replaced with a cylinder and two flat endcaps. In each case, the resulting electrostatic trapping fields for each of the simplified versions are <u>similar</u> to those of the original devices, but typically are compromised to some degree. These compromises can result in reduced performance (mass resolution, etc.) or more complex ion motion; however, modeling approaches can minimize the magnitude of these effects. This planar-surface approach has been optionally employed in some <u>miniature</u> analyzer development efforts; however, it becomes required in <u>micro</u> dimensioned analyzers as the analyzer dimensions reach the sub-millimeter scale.

The second constraint on fabrication of mass analyzer components is reached when either the component feature size is below the machining cutting tool size, or when the required component size or alignment tolerance falls below the traditional machining capability (typically around 0.0001″). At this point, MEMS techniques

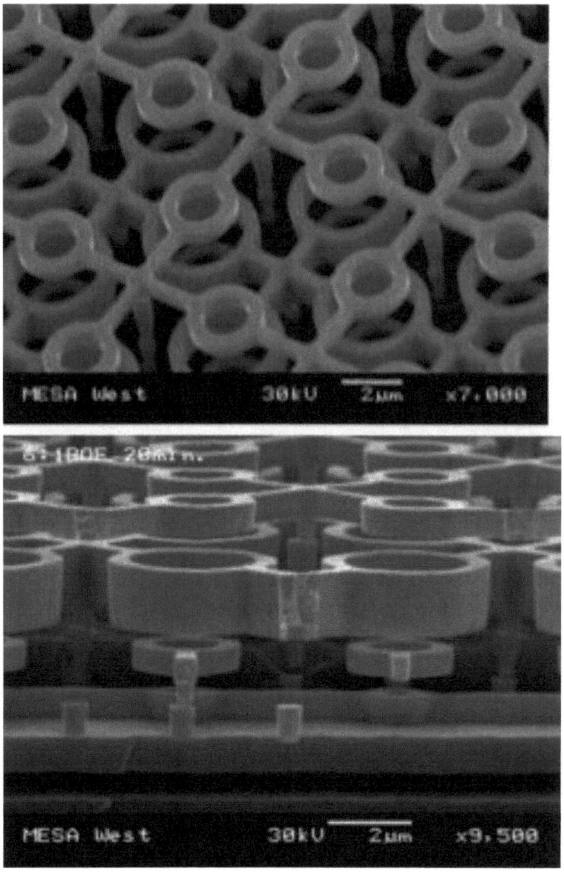

Fig. 6.3 An array of micro-scale cylindrical ion traps (Reprinted with permission from Ref. [23])

take over and enable components to be fabricated whose dimensions are on the scale of the tolerance limits of conventional machining. Figure 6.3 shows one example of a cylindrical ion trap array from Sandia National Lab [23] where the radius of the trapping cylinder is on the order of 1 μm. These results have been achieved in numerous laboratories with various MEMS approaches on many different types of mass analyzers and ion optics. In most of these cases, the devices are three-dimensional in nature and, allowing for the planar surface approximations, truly scaled-down versions of their laboratory precursors. As such, they are subject to a new set of conditions that can lead to performance loss. Now, at this point, a third constraint comes into play, and misalignment errors (either spatial, concentric, or parallelness) can affect performance. Furthermore, because these devices rely on the signal from a summed array to maintain ion capacity, performance degradation of the full array assembly can result if all of the array elements are not identical to each other.

An emerging approach attempts to mitigate some of the alignment and equivalence problems of micro arrays as well as allow a prescriptive adjustment of the

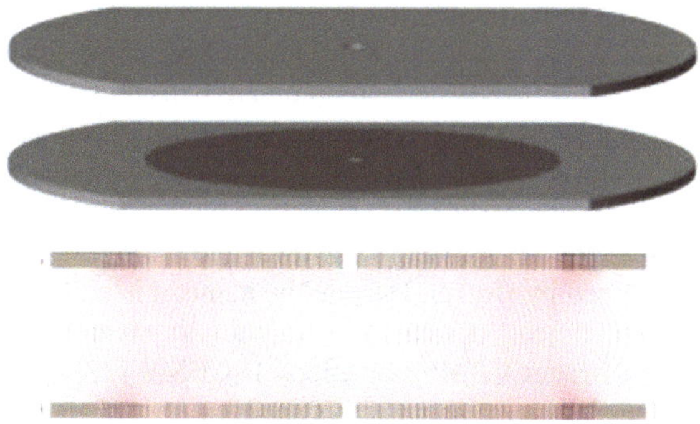

Fig. 6.4 A planar 3D Paul trap. The trapping fields are generated from surface electrodes on two planar, parallel plates (Reprinted with permission from Ref. [24])

field quality. In this approach, the field is further approximated as the number of required electrodes is reduced so that the actual trapping field is produced between two parallel plates and without a third dimension electrode. These planar devices [24] consist of two optically flat planar substrates on which multiple electrode traces are lithographically deposited in specifically designed patterns. Each trace can be then be independently biased in such a way that the collection of potentials produces an optimized field that approximates its 3-D analog. An example of this can be seen in Fig. 6.4, the planar Paul trap. In this case, the 3D quadrupole ion trap field is approximated using two planar electrode substrates with surface deposited concentric electrode rings, each biased in such a way to create a trapping field between the two plates that is very similar to that of the 3D quadrupole ion trap. This device shows mass resolution performance comparable to that of its 3D predecessor. The approach has been employed to produce other ion trap mass analyzers including a "halo" trap [25] (a two-plate version of the toroidal ion trap), a planar linear ion trap [26], and a "coaxial" trap, where a toroidal ion trap field surrounds a cylindrical ion trap field, all using only two electrodes. Because of the high resolution and accuracy of the lithography methods employed to fabricate the electrode traces, this planar device strategy offers a promising novel approach to smaller, high performance miniature mass analyzers.

6.4 Field-Portable Mass Spectrometry Systems

Over the past decade or two, the field of portable MS systems has focused more on fieldability (i.e., ruggedness, simplicity of operation, etc.) than on portability (miniaturization). Because of this, the analytical engine of these instruments showed only small progress in miniaturization. During this period, conventional scale quadrupole-based devices (2D, 3D) have been the most widely used mass analyzers employed in fieldable MS systems. These devices have simple ion optics and provide the nominal mass resolution required for most field portable MS applications. Initially, 2D linear quadrupole mass filter-based systems led the way. Three systems in particular, the US Army's 'Fly-Away Laboratory' based on an Agilent GC/MS system, and the above mentioned Viking Spectrotrak and INFICON Hapsite portable GC/MS system predominated. In addition, Bruker-Daltonics had developed and employed a vehicle-mounted, chemical weapons reconnaissance MS capability in the form of the MM-1, a rugged, but very heavy (~300 lb) quadrupole-based detector. The early 1990s saw the development of the first ion trap-based fieldable system, the US military's chemical and biological mass spectrometer (CBMS), also developed by Bruker. Despite the name, this early instrument focused primarily on the 'biological' aspects of the military threat arena. Later, the follow-on instrument to the Bruker CBMS, the Block II-CBMS [27, 28] began its development at Oak Ridge National Laboratory (ORNL) with the aim of providing both chemical and biological threat detection. While all of these systems use nominal-dimensioned, conventionally machined quadrupole filters or ion traps, the instrumental development focus was on portability, power-reduction and ruggedness. Still, with the exception of the Hapsite, these instruments were large (2-drawer file cabinet size) and heavy (>100 lb).

The remarkable examples of milli- and micro-meter scale mass analyzers described in Sect. 6.3 have emboldened many researchers in this field to predict that cellphone size MS systems are just around the corner. However, the mass analyzer (and even the associated ion source or detector) is not necessarily the limiting factor in achieving this lofty goal. The required support technologies (e.g., vacuum pumps, electronics, high pressure detectors and high capacity, low weight batteries, etc.) have not kept pace in this area. And as discussed below, other factors such as broad sample applicability, highly reliable analytical results, and ease-of-use for non-expert operators all work against the near prospect of such a small system size.

6.4.1 *Tradeoffs and Constraints*

The miniaturization of the mass analyzer is not the only barrier to the overall MS system size reduction. Much of the support hardware technology (pumping systems, ionization sources, detectors, etc.) has not scaled proportionally, either in size or operability. While concomitant progress in these technologies may be lagging due to low market demands, there are also significant technical challenges to overcome. Reliable, rugged, and smaller vacuum pumps as well as sensitive ion detectors capable of operation at higher pressure still limit the development of smaller systems. Analytical 'add-ons' such as chromatography or MS^n, which provide additional selectivity and identification confidence, also require an increase in space, power or complexity. These may need to be eliminated if the instrument size/weight/power requirements are to be reduced to such small dimensions. From a physics perspective, there are tradeoffs between system size or weight and heat dissipation as field portable systems may have to function in the desert or artic, complicating the electronics cooling design. Furthermore, systems that must operate in hazardous environments (e.g., chemical agent background) must be completely sealed to the

Fig. 6.5 Tradeoffs in miniaturizing mass spectrometers

outside environment, further complicating the heat dissipation problem. For every effort to make the system more rugged in order to handle the shock and vibration that accompany field analysis, weight must be added. Finally, smaller analyzer dimensions lead to increased likelihood of arcing, especially for high voltage components whose operation voltage doesn't necessarily scale with size (e.g., CDEM detectors). The graphic in Fig. 6.5 is meant to show the inter-relationship between the various tradeoffs and goals of miniaturization. The figure can be imagined as a plumbing diagram with flexible bellows in each arm. Any attempt to reduce one of the parameters feeds back into the system at another point. For example, 'squeezing' the *False Positive Rate* bellows will likely result in an increase in either the *Size/Weight* arm (due to adding selectivity in the form of an orthogonal technique such as gas chromatography) or the *Complexity* arm (in the case of adding an additional stage of MS, or MS/MS). Reducing the *Size/Weight* component often comes at the expense of increased *Susceptibility to Shock/Vibration* and so on.

One example of a system whose fieldable features reside near the middle of the interdependent tradeoffs summarized in Fig. 6.5 is the TRIDION-9 field portable GC/MS system (Fig. 6.6). There are several other systems (both commercial and under development in universities or government research laboratories) that are either smaller and lighter with some analytical or operational tradeoffs, (e.g., Purdue University [29], Analytical Instrumentation Research Institute [30], First Detect [31], or 908 Devices [32]) or heavier and more rugged but with

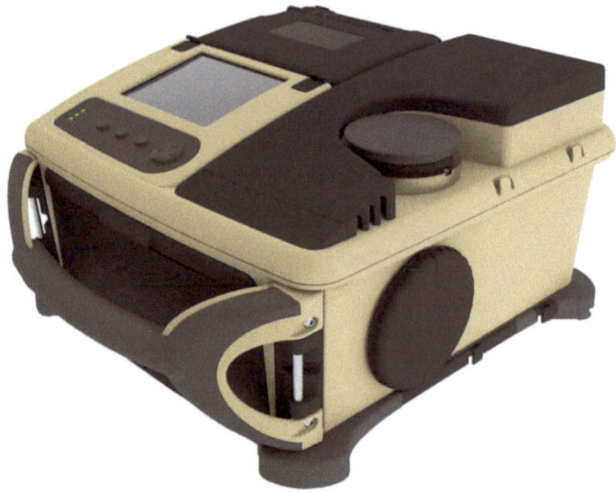

Fig. 6.6 The TRIDION-9 field-portable GC/MS from Torion Technologies

added capabilities (e.g., FLIR 424 [33] or Bruker [34] E²M). The TRIDION-9 is one attempt at providing a light-weight, field-portable GC/MS with analytical capability as close as possible to that obtained by laboratory-based GC/MS systems. The salient features of this system are that it is small (38 cm × 39 cm × 43 cm) and easily person-portable (13 kg). These metrics include all utilities (an integrated, ~100–200 sample capacity helium cylinder) and battery power for 3 h of operation (power draw is 120 W peak during startup, then drops to 60 W average). The analytical engine is the above-mentioned 'gold standard' GC/MS as these orthogonal analytical techniques provide both high sensitivity and high selectivity. The GC is a low-thermal mass, resistively-heated, 5 m, 100 μm capillary GC column which is temperature programmable at up to 1 degree/s, providing very fast, high resolution chromatographic separations (2 min analysis, <5 min cycle time). Sample introduction is via either a solid phase microextraction (SPME) fiber or by a sorbent-bed based needle trap. The MS is a toroidal ion trap as described above with electron ionization (EI) and unit mass resolution (<0.5 da., FWHM) across the entire 43–500 da. mass range. The ion trap analyzer is heated (which allows >1,000 analyses before any required cleaning or maintenance). The entire chassis is completely sealed so that it can be operated in chemically contaminated areas and decontaminated afterwards owing to the chemical agent resistant coating and materials. In order to accomplish this, the chassis itself is designed to be an integrated heat sink for the enclosed electronics without the need for outside air exchange. This allows it to be operated between 5 and 40 °C without opening the sealed exchange ports and up to 45 °C by opening the exchange ports or applying cold packs to strategically-placed heat exchangers on the outside of the chassis. It can be operated across a full, non-condensing humidity range. The chassis is designed to protect the system from vibration and shock (drops up to 16″). The time from cold startup to system ready is <5 min and <15 min to first sample analysis to system (after blank and system validation/calibration).

6.5 Fieldable Mass Spectrometry Sampling Support

The final requirement in the transition of laboratory MS applications to the field lies in the area of sampling and data processing/interpretation. Even if the mass analyzer is miniature or micro, and the overall MS-based system possesses the size, ruggedness and performance characteristics required of the application, the sample collection, handling, processing and transfer to the instrument must also be proportionally miniaturized and simplified if one is to avoid having a hand-portable MS, but a shipping crate full of sampling accessories. Many of the parameters in sample handling and introduction are direct tradeoffs with other required parameters by the MS. For example, small turbo pump systems cannot accommodate the injection of the typical liquid quantities (~1 μl) of volatile solvents. Furthermore, since most of these fieldable MS systems end up in the hands of non-expert operators, the sampling and preparation steps must be reduced to as automated a process as possible, often to non-contact, direct analysis methods such as the emerging DART [35] or DESI [36] techniques. Many similar approaches, collected under the general 'Ambient Ionization' [37] category are being developed in research laboratories for a variety of applications, however are complicated to use outside of the laboratory. Even techniques in which the sample requires some additional processing steps (e.g., extraction, derivatization, transfer to instrumental interfaces, etc.) require automated, fieldable (i.e., battery operated) and rapid processing equipment in order to fit the fieldable philosophy. An example of one of these units is the Torion SPS®-3 sample preparation station (Fig. 6.7, right). This battery-powered unit can take vapor samples collected via a portable pump onto conventional sorbent tubes, and transfer them automatically, along with a quantitative aliquot of internal standard, to a 'needle trap' sample introduction unit (compatible with the injector of the

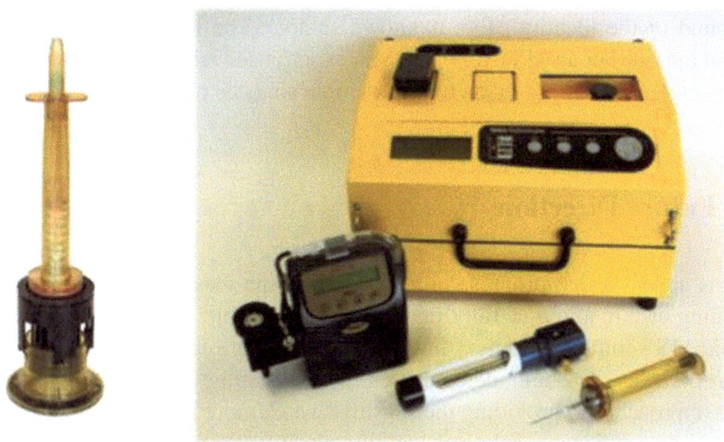

Fig. 6.7 The ADS surface sampling SPME interface (*left*) and the SPS-3 field-portable vapor and liquid sample preparation station (*right*)

TRIDION-9). The system can also acquire target samples from liquid-phase samples (heated or unheated) either by headspace or direct immersion using SPME or vapor/air samples using the needle trap, or solid samples by headspace (heated or unheated). Sampling temperature control and timing is all automated for the user. Sample adsorbed on surfaces can be efficiently and quickly acquired onto an SPME fiber using the ADS® sampler (Fig. 6.7, left). The ADS creates a vacuum over the surface by retracting a lockable 'syringe-like' plunger, which significantly enhances the sampling efficiency for semivolatile materials adsorbed to surfaces. The SPME sampler pierces a septum on the back of the plunger and can be left for long periods of time to collect low-level samples.

6.5.1 A New Generation of 'Mass Spectrometrists'

A final, important consideration in the development of fieldable mass spectrometer systems is 'who is the final user?' As mass spectrometry detection becomes more portable and lower cost, new application areas for its utility will open. As it is unlikely that the numbers of traditional, university-trained analytical chemists will keep pace, a new generation of 'operators' will fill the requirement. Consider one obvious example in the US military, where these instruments are sufficiently commonplace that ground soldiers are often the ones using the equipment. In a similar vein, TSA agents typically use explosives detectors (typically IMS-based systems) in airport security checkpoints. These new 'analysts' rely heavily on results reported by the instrument, which, obviously, depends on the calibration, operational validity, and service status of the system. These applications require simple (or better yet, no) sample handling or preparation and fully automated instrument operation, including automatic diagnostics, calibration and validation capabilities. In addition, the instrument must accurately report the result – either the detection or absence of a target compound or the identity of an unknown sample. The instruments must reliably perform the roles of an MS expert, (i.e., an experienced laboratory mass spectrometrist) from instrument power-on to final sample analysis report.

6.6 Future Directions

It seems apparent that progress in miniaturizing the mass analyzer has out-paced everything else required to fully realize a hand-portable, broadly applicable field-portable MS solution. The next few years will require significant progress in the support hardware (vacuum, detectors, etc.) and sampling devices and strategies, as well as approaches to providing the sensitivity and, more importantly, selectivity to allow highly confident, analytical identifications in complex environments. If the goal of an iPhone-size system is ever to be achieved or become generally useful, there is still plenty of work to do on the horizon.

References

1. Fox J, Saini R, Tsui K, Verbeck G (2009) Microelectromechanical system assembled ion optics: an advance to miniaturization and assembly of electron and ion optics. Rev Sci Instrum 80(9):93302–93306
2. Lopez-Avila V, Hill HH (1997) Field analytical chemistry. Anal Chem 69(12):289R–305R
3. Rushneck DR, Diaz AV, Howarth DW, Rampacek J, Olson KW, Dencker WD, Smith P, McDavid L, Tomassian A, Harris M, Bulota K, Biemann K, LaFleur AL, Biller JE, Owen T (1978) Viking gas chromatograph-mass spectrometer. Rev Sci Instrum 49:817–834
4. Jones JJ, Wilkins CL (2004) Bacterial pyrolysis products analyzed using a portable high resolution 1 Tesla Fourier transform mass spectrometer. In: Proceedings of the 52nd ASMS conference on mass spectrometry and allied topics, Nashville, 23–27 May 2004
5. Enfield W, Bender SF (1997) Keenen MR, Thronberg SM, Hightower MM, Environmental technology verification report. Field portable gas chromatograph/mass spectrometer. Viking Instruments Corporation SpectraTrak™ 672; Sandia National Laboratories, Albuquerque, December 1997, 103 pp
6. Smith PA (2012) Person-portable gas chromatography: rapid temperature program operation through resistive heating of columns with inherently low thermal mass properties. J Chromatogr A 1261:37–45
7. http://www.gulflink.osd.mil/foxnbc/
8. Wise MB, Thompson CV, Buchanan MV, Merriweather R, Guerin MR (1993) Direct sampling ion trap mass spectrometry. Spectroscopy 8:14–22
9. Stalport F, Glavin DP, Eigenbrode JL, Bish D, Blake D, Coll P, Szopa C, Buch A, McAdam A, Dworkin JP, Mahaffy PR (2012) The influence of mineralogy on recovering organic acids from Mars analogue materials using the "one-pot" derivatization experiment on the Sample Analysis at Mars (SAM) instrument suite. Planet Space Sci 67(1):1–13
10. Diaz JA, Pieri D, Arkin R, Gore E, Griffin TP, Fladeland M, Bland G, Soto C, Madrigal Y, Castillo D, Rojas E, Achi S (2010) Utilization of in situ airborne MS-based instrumentation for the study of gaseous emissions at active volcanoes. Int J Mass Spectrom 295(3):105–112
11. Short RT, Fries DP, Toler SK, Lembke CE, Byrne RH (1999) Development of an underwater mass-spectrometry system for *in situ* chemical analysis. Meas Sci Technol 10:1195–1201
12. http://www.hems-workshop.org
13. Austin DE, Lammert SA, Mass analyzer miniaturization; Encyclopedia of mass spectrometry, vol 7. Elsevier, Amsterdam (in Press)
14. Badman ER, Cooks RG (2000) Miniature mass analyzers. J Mass Spectrom 35:659–671
15. Dawson PH (1976) Quadrupole mass spectrometry and its applications. Elsevier, Amsterdam. Republished by American Institute of Physics, Woodbury (1995)
16. March RE, Hughes RJ (1989) Quadrupole storage mass spectrometry. Wiley Interscience, New York
17. Johnson JV, Yost RA, Kelley PE, Bradford DC (1990) Tandem-in-space and tandem-in-time mass spectrometry: triple quadrupoles and quadrupole ion traps. Anal Chem 62(20):2162–2172
18. Tian Y, Higgs J, Li A, Barney B, Austin DE (2014) How far can ion trap miniaturization go? Parameter scaling and space-charge limits for very small cylindrical ion traps. J Mass Spectrom 49:233–240
19. Ouyang Z, Gao L, Fico M, Chappell WJ, Noll RJ, Cooks RG (2007) Quadrupole ion traps and trap arrays: geometry, material, scale, performance. Eur J Mass Spectrom 13(1):13–18
20. Lammert SA, Plass WR, Thompson CV, Wise MB (2001) Design, optimization and initial performance of a toroidal RF ion trap mass spectrometer. Int J Mass Spectrom 212:25–40
21. Lammert SA, Rockwood AA, Wang M, Lee ML, Lee ED, Tolley SE, Oliphant JR, Jones JL, White RW (2006) Miniature toroidal radio frequency ion trap mass analyzer. J Am Soc Mass Spectrom 17:916–922

22. Ouyang Z, Wu G, Song Y, Li H, Plass WR, Cooks RG (2004) Rectilinear ion trap: concepts, calculations, and analytical performance of a new mass analyzer. Anal Chem 77:4595–4605
23. Blain MG, Riter LS, Cruz D, Austin DE, Wu G, Plass WR, Cooks RG (2004) Towards the hand-held mass spectrometer: design considerations, simulation, and fabrication of micrometer-scaled cylindrical ion traps. Int J Mass Spectrom 236:91–104
24. Zhang Z, Peng P, Hansen BJ, Miller IW, Wang M, Lee ML, Hawkins AR, Austin DE (2009) Paul trap mass analyzer consisting of opposing microfabricated electrode plates. Anal Chem 81(13):5241–5248
25. Wang M, Quist HE, Hansen BJ, Peng Y, Zhang Z, Hawkins AR, Rockwood AL, Austin DE, Lee ML (2011) Performance of a halo ion trap mass analyzer with exit slits for axial ejection. J Am Soc Mass Spectrom 22:369–378
26. Hansen BJ, Niemi RJ, Hawkins AR, Lammert SA, Austin DE (2013) A lithographically patterned discrete planar electrode linear ion trap mass spectrometer. J Microelectromech Syst 22:875–883
27. Hart KJ, Wise MB, Griest WH, Lammert SA (2000) Design, development and performance of a fieldable chemical and biological agent detector. Field Anal Chem Tech 4:93–110
28. Griest WH, Wise MB, Hart MB, Lammert SA, Thompson CV, Vass AA (2001) Biological agent detection and identification by the Block II chemical biological mass spectrometer. Field Anal Chem Technol 5(4):177–184
29. Gao L, Song Q, Patterson GE, Cooks RG, Ouyang Z (2006) Handheld rectilinear ion trap mass spectrometer. Anal Chem 78:5994–6002
30. Yang M, Kim T-Y, Hwang H-C, Yi S-K, Kim D-H (2008) Development of a palm portable mass spectrometer. J Am Soc Mass Spectrom 19(10):1442–1448
31. http://www.1stdetect.com
32. http://www.908devices.com
33. http://gs.flir.com/detection/chemical/mass-spec
34. http://www.bruker.com/products/cbrne-detection/ms/e2m/overview.html
35. Gross J (2014) Direct analysis in real time-a critical review on DART-MS. Anal Bioanal Chem 406(1):63–80
36. Badu-Tawiah AK, Eberlin LS, Ouyang Z, Cooks RG (2013) Chemical aspects of the extractive methods of ambient ionization mass spectrometry. Annu Rev Phys Chem 64:481–505
37. Harris GA, Galhena AS, Fernandez FM (2011) Ambient sampling/ionization mass spectrometry: applications and current trends. Anal Chem 83(12):4508–4538

Chapter 7
MALDI Imaging Mass Spectrometry

Erin H. Seeley and Richard M. Caprioli

Abstract Matrix assisted laser desorption/ionization imaging mass spectrometry (MALDI IMS) is a powerful tool for the *in situ* analysis of endogenous biomolecules and exogenous pharmaceutical compounds in thin tissue sections. Hundreds to thousands of analytes can be detected and mapped in a single imaging experiment without *a priori* knowledge of the exact molecules present in the section. Advances in sample preparation and in instrumentation have allowed for MALDI IMS to become a high throughput technique allowing for statistical analysis and diagnostic classification of clinically important specimens. This review discusses the current state of the technology with regards to advances in sample preparation techniques and instrumentation. These include the expansion of the technique to the analysis of formalin-fixed, paraffin embedded tissues as well as the ability to acquire IMS data at the single cell level. Applications to skin cancer, infectious disease, diabetes, and age-related macular degeneration are highlighted.

Keywords Imaging Mass Spectrometry • MALDI • Melanoma • Infectious disease • Diabetes • Age-related macular degeneration

7.1 Introduction

Mass spectrometry (MS) has become an essential analytical tool for the investigation of molecular processes in tissues and cells. Specifically, matrix-assisted laser desorption/ionization (MALDI) imaging mass spectrometry (IMS) provides the capability of mapping the spatial distribution of biomolecules directly in tissue sections, including proteins, peptides, lipids and other metabolites. Such mapping often leads to a new understanding of the underlying biology since the spatial localization of essential biomolecules is not disturbed during the analysis. IMS does not require a target-specific reagent such as an antibody or any prior knowledge of

E.H. Seeley • R.M. Caprioli (✉)
Mass Spectrometry Research Center and the Department of Biochemistry,
Vanderbilt University School of Medicine, Nashville, TN, USA
e-mail: richard.m.caprioli@vanderbilt.edu

J. Banoub (ed.), *Detection of Chemical, Biological, Radiological and Nuclear Agents for the Prevention of Terrorism*, NATO Science for Peace and Security Series A: Chemistry and Biology, DOI 10.1007/978-94-017-9238-7_7

which biomolecule may have changed in expression level; therefore, IMS is an excellent discovery tool. IMS can also map post-translational modifications and protease clips and truncations in tissues.

Typically, imaging mass spectrometers operate in the microprobe mode, that is, each pixel in an ordered array is collected one at a time to cover the area of interest within a given field of the sample. Each individual pixel measurement, typically of a diameter of 1–150 μm, ablates an area of the sample to give rise to a spectrum of ionized compounds at high sensitivity and high-throughput analysis over the entire field of the sample. Spectra from thousands of ablated spots make up the dataset which can then be mined to provide molecular maps of the sample. Thus, any signal in the mass spectrum can be plotted as an intensity of a specific *m/z* value over the entire array, producing a molecular map of the compound at that *m/z* value. The approach is illustrated in Fig. 7.1 showing both the overall process and the resulting maps generated from the data set.

Nearly every ionization process and every mass analyzer may be used for imaging purposes. Each has advantages and disadvantages in terms of molecular weight range, molecular types amenable for imaging, sensitivity, routinely achievable spatial resolution, etc. One of the earliest MS technologies used for imaging employed secondary ionization mass spectrometry (SIMS) [1, 2]. SIMS is capable of producing extremely high resolution images (~100 nm) for the spatial analysis of elements and small molecules ($m/z < 1{,}000$ Da). However, it is not effective for mapping large peptides and proteins. Direct laser ablation also covers a similar mass range and has spatial resolution in the low micrometer range [3]. Liquid based spray ionization methods such as desorption electrospray ionization (DESI) have been used for imaging and have the advantage of being ambient methods and so samples not amenable to a vacuum lock introduction system can be analyzed [4]. DESI can achieve a spatial resolution of 200–500 μm and generally is most effective for the measurement of small molecules. MALDI MS has the advantages of high spatial resolution (<5 μm) and is amenable to a large suite of biomolecules ranging in MW from 100 to over 100,000 [5, 6], but has the added preparation step of requiring an energy absorbing matrix to be applied to the sample prior to analysis.

In a given analysis, the choice of spatial resolution, mass range, sensitivity required, data set size, and time of analysis all play into the decision in how best to set up the experiment. In general, many investigators want the highest spatial resolution possible, but this is not always the best choice. In going to high spatial resolutions, the number, diameter, and pitch of the pixels become important considerations since these all will lead to large data file sizes and long data acquisition times. Most importantly, the ‘sensitivity’ of the analysis will also fall dramatically as resolution increases since the area ablated in a given pixel decreases as the square of the radius of the pixel. The interrelationship of acquisition parameters is shown diagrammatically in Fig. 7.2. For most analyses, it is best to choose the lowest resolution needed to answer the analytical/biological question at hand. This will minimize the data file size and give the shortest acquisition time while still providing the sensitivity needed to answer the question at hand.

The field of imaging mass spectrometry is quite broad and in fact the technology is rapidly developing and its application to biological and medical research issues

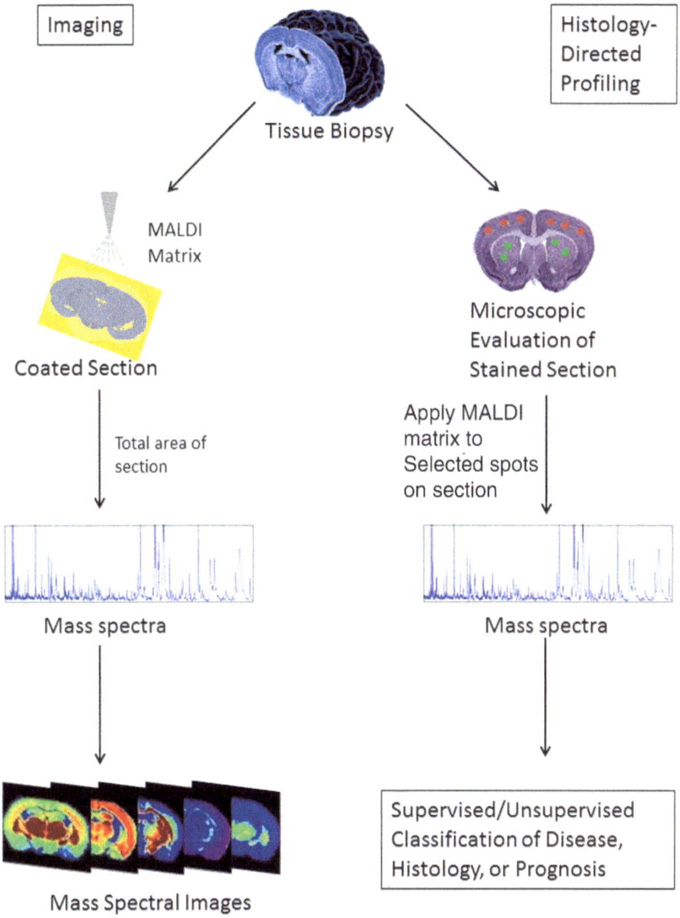

Fig. 7.1 Imaging and Histology Directed Profiling Mass Spectrometry. *Left* – In the imaging approach a tissue section is collected onto a target and the entire section is coated with matrix. Thousands of mass spectra are collected in an ordered array from the entire section. Each peak from the collected spectra can be displayed as a function of its spatial localization and relative intensity. Hundreds to thousands of images can be generated from one single imaging experiment. *Right* – In the histology-directed profiling approach, only a few spectra are collected from each tissue section in specific areas of biological interest as indicated by the colored dots on the stained section that is used to guide the experiment. Typically 10–20 spectra are collected per sample. Hundreds of samples may be analyzed as part of a clinical experiment. Statistical analyses are carried out on the spectra to classify them according to their mass spectral diagnosis

are increasing as well. This review will focus on MALDI Imaging MS because it has the greatest utility in performance for common applications in medical and biological research. It is useful for both qualitative and quantitative analysis of tissue and for assessing spatial and temporal changes that occur in biological systems. For a more in depth discussion of other MS imaging technologies and comparisons, the reader is referred to more comprehensive reviews of the field [7–9].

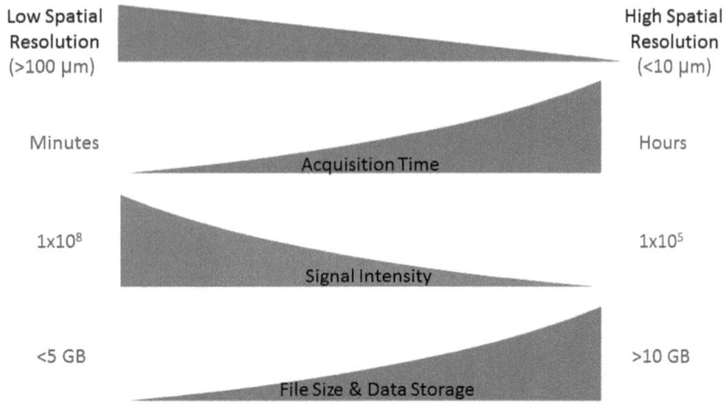

Fig. 7.2 The Trade-off between Spatial Resolution and Time, Signal Intensity, and File Size. While higher spatial resolution will provide finer detail in the generated image, other factors must be considered. As the spatial resolution increases, the acquisition time and the overall file size increase exponentially. At the same time, the signal intensity or sensitivity is decreased as the amount of analyte below a sampled area is reduced. The user must consider the ultimate goal of the imaging experiment and use the appropriate spatial resolution to meet that goal

7.1.1 Technology

MALDI IMS can map the specific location of molecules such as drugs, lipids, small organic metabolites, peptides, and proteins directly from fresh frozen tissue and formalin fixed paraffin embedded tissue sections with lateral image resolution down to 5 µm. Thin frozen sections (5–10 µm thick) are cut, thaw-mounted on target plates and then coated with a thin layer of a matrix molecule, often this is a low MW crystalline organic compound such as dihydroxybenzoic acid, 3,5-dimethoxy-4-hydroxy-cinnamic acid (sinapinic acid) and other like compounds. Tissue samples are irradiated by a laser resulting in the ablation of ionic molecular species that are measured as their mass-to-charge (*m/z*) values. A single pixel then is associated with a full mass spectrum containing thousands of signals. The intensity of any of these *m/z* signals can be plotted over the entire array of pixels to generate a map of that ion over the analyzed area of the sample (Fig. 7.1). From one raster of the tissue, many hundreds to thousands of images can be generated.

The analyzer most commonly used for the direct analysis of biomolecules in tissues for the mapping of peptides and proteins utilizes time-of-flight (TOF) analyzer technology. Modern high speed lasers have repetition rates of 1–5 kHz and the duty cycle of a modern TOF analyzer is an excellent match for the pulsed laser process. The TOF analyzer also has the advantage of a theoretically unlimited mass range, high ion transmission efficiency, multiplex detection capability and simplicity in instrument design and maintenance [10]. Other analyzers can bring special capabilities. High resolving power instruments such as the Fourier transform ion cyclotron resonance (FTICR) instruments, orbital trapping instruments, and orthogonal TOF instruments provide exact mass determinations. Ion mobility instruments utilize a gas phase separation process that may be important to separate out two or more nominally isobaric species. Ion traps [11, 12] are relatively simple instruments that provide MS/MS capabilities for imaging, along with other analyzers such as TOF-TOF [13], orthogonal TOF and orthogonal quadrupole-TOF (Q-TOF) [14, 15], and FTICR [16, 17] analyzers. These several analyzers provide capabilities for protein identification, high mass resolution acquisition, and the ability to detect small molecules such as drugs and metabolites. The reader is referred several publications for more thorough descriptions [11, 13–16, 18] of analyzer capabilities.

The mapping of intact proteins in tissue is usually performed on a linear TOF instrument so as to achieve the highest sensitivity. Ions formed and desorbed during the laser pulse are extracted and accelerated into the field free region of the TOF analyzer. Ions are usually detected by a multi-channel plate detector and the time of flight of the various ions is inversely proportional to their *m/z* values. This time measurement is then converted to *m/z* through appropriate calibration procedures. For the analysis of low molecular weight species, an ion mirror or reflector can be used in the ion flight path to compensate for the initial velocity/energy distribution and improve mass resolution [19].

7.1.2 Matrix Application

MALDI IMS utilizes an energy absorbing matrix to affect desorption and ionization of compounds in the sample. Some of the commonly used matrices are small organic molecules that co-crystallize with analytes present in the tissue. These include 3,5-dimethoxy-4-hydroxy-cinnamic acid (SA), 2,5-dihydodroxybenzoic acid (DHB) and α-cyano-4-hydroxycinnamic acid (HCCA). The particular solvent or solvent combination and specific matrix utilized may vary depending on the specific biomolecules and tissue being analyzed [20]. A 50:50 (v/v) acetonitrile/water or ethanol/water mixture is a common solvent system used for matrix application. SA is preferred for the analysis of high MW proteins [20] while DHB and HCCA are generally used in the analysis of peptides and lower MW analytes. The MALDI matrix need not be a crystalline compound; liquid matrices have been employed in many published reports [21, 22] as well as the use of metal particles [23] and nanoparticles [24].

The matrix is deposited onto the surface of the tissue in such a way as to minimize the lateral dispersion of the molecules of interest. This can be accomplished by applying matrix solution to the tissue either as a high density spotted array or a homogenous spray coating [20]. A continuous and homogenous spray coating is best for high spatial resolution imaging. A spotted array is perferred when only a small area is to be imaged and generally this approach gives high reproducibility and high quality mass spectra. Several robotic spotting devices are commercially available and utilize acoustic [25], piezoelectric [13], inkjet printer [26], and capillary deposition processes [27]. Several robotic spray coating devices are also commercially available and utilize a mist/nebulizing method [28] or a thermally-assisted spray deposition [29]. From these arrays of matrix coated samples, each micro spot or pixel coordinate is then automatically analyzed by MALDI MS.

7.1.3 Profiling/Histology Directed Imaging

In some cases, the complete molecular image of a tissue is not required but rather one or more relatively small areas within the tissue are targeted for analysis. Such cell-type specific molecular imaging, termed profiling or histology directed analysis, minimizes data acquisition time, file size, and costs. The molecular data obtained from these discrete regions are then mapped back to the optical scan of the sample performed by microscopy to provide placement of the MS data precisely onto the tissue section. Any number of spots may be chosen to analyze with the size of the spot (user variable) depending on the information content needed from the biological tissue analyzed. This approach is designed to obtain comparative measurements from different regions within a single tissue specimen or to compare multiple pieces of tissue. The profiling approach generally does not make use of high spatial resolution but rather maps molecular measurements back to the histology obtained from the tissue (Fig. 7.1). Statistical confidence is obtained by analyzing a sufficient number of spots on the tissue and this will vary depending on the specific needs and focus of the analytical question. Samples may be fresh frozen or formalin fixed, with each having a somewhat different tissue preparation protocol.

For profiling frozen tissue, the sample preparation protocols are relatively simple. Biopsies or other tissue should be frozen in liquid nitrogen immediately after removal to preserve the morphology and prevent any significant proteolysis. Cryostat sections 6–12 μm thick are thaw-mounted onto target plates that can be directly inserted into the mass spectrometer. Gold-coated or stainless steel metal plates and glass slides that have a conductive coating are typically used, the conductive character being necessary for instruments having high voltage ion sources. For protein mapping, gentle rinsing of the tissue with ethanol serves as a fixation process and helps remove lipids and salts that can interfere with ionization of the proteins. When microscopic examination is needed to establish the areas to be analyzed, stained sections on glass slides are used and areas of interest registered with respect to several fiducial marks placed on the slide (often a magic marker dot in several places on the slide will suffice). MALDI matrix is then applied and the sample is subsequently analyzed. For mapping of small molecules such as metabolites and drugs and for lipids, organic solvent washing of the tissue is not recommended because it may redistribute these analytes or even remove them from the tissue.

Since there are large tissue banks of stored biopsies, formalin fixed paraffin embedded (FFPE) tissue often must be analyzed. FFPE tissue requires a trypsin digestion step following the normal antigen retrieval process used in histology since most of the proteins are cross-linked from the formalin fixation. Trypsin will hydrolyze and release peptides from those protein domains that are not cross-linked. This can be accomplished by micro-spotting trypsin solution directly on areas of interest on the section. In this process, arginyl tryptic peptides are favored since the guanido group of the arginine side chain does not cross-link as efficiently as does the side chain ε-amino group of lysine. Identification of the sequence of the tryptic peptides by MS/MS directly from the tissue and a query of the protein database usually identifies the nominal protein that the fragment was released from. It is important to remember that this process simply identifies the nominal or base protein, that is, it does not necessarily identify the active post-translationally modified form of the protein present in the original tissue.

7.1.4 Application 1: Tissue Profiling and Diagnosis of Melanocytic Skin Tumors

As a clinical tool for diagnosis of melanocytic tumors, MALDI IMS has major advantages over most other tissue genomic/proteomic techniques. MALDI IMS can be used to study FFPE tissue and can access stored biopsies in tissue banks, determine the exact location of peptide expression patterns compared to corresponding histopathologic attributes on a matched H&E slide, utilizes very little tissue, and does not require laborious preparation steps (such as laser capture microdissection, DNA/RNA/protein extraction, gel electrophoresis, etc.). The assay is performed *in-situ* and a large number of different biomolecules are measured both in targeted and non-targeted modes to provide native distributions in the tissues.

A test study of over 100 patient biopsies of Spitz nevi (SN) and Spitzoid malignant melanoma (SMM) was performed in collaboration with pathologist at Yale University Medical School and Harvard Medical School. Initially, whole slide imaging was used to examine melanocytic skin tumors: benign dermal nevus, primary malignant melanoma, and metastatic malignant melanoma, all in male patients (age range: 41–64 years), and all from the head or scalp. Because these were all FFPE tissue specimens, trypsin was applied to the sections prior to matrix deposition in order to digest pieces of the proteins that were not involved in crosslinks. Candidate peptides having specific *m/z* values were successfully identified that were preferentially over-expressed in both cases of primary and metastatic melanomas, but not in the dermal nevus. Conversely, several peptides were identified that are only preferentially over-expressed in the benign dermal nevus and not in the two malignant lesions (Fig. 7.3).

In a larger clinical study, IMS data were used to discriminate benign from malignant Spitzoid lesions using robotic matrix deposition techniques with a 200 μm spatial resolution. The differentiation of SN from SMM can be challenging

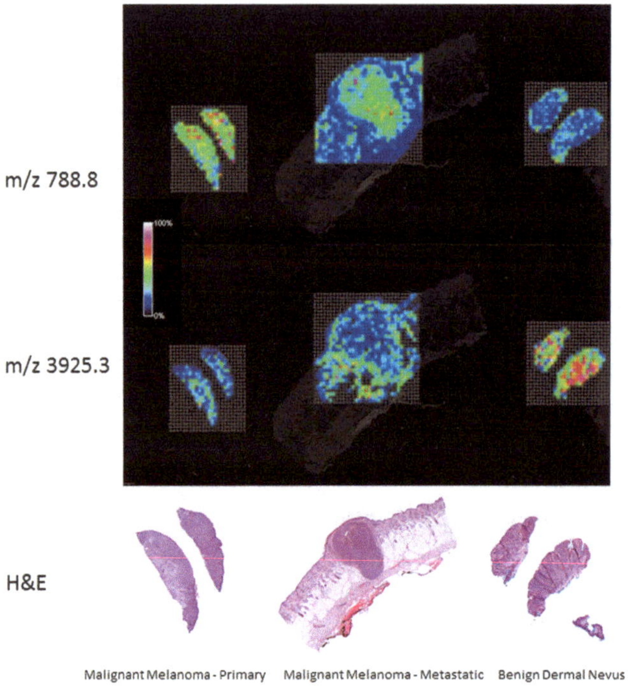

Fig. 7.3 Imaging Mass Spectrometry of Melanocytic Skin Lesions. On-tissue tryptic digestion and peptide imaging were carried out on representative malignant melanoma, metastatic melanoma, and a benign dermal nevus. Examples are shown of peptides with higher expression in the malignant tissue (*m/z* 788.8) and higher expression in the benign tissue (*m/z* 3,925.3)

in dermatopathology. Normally, the diagnosis of SN and SMM is based on histopathologic criteria, clinical features, and follow-up data, confirming that none of the lesions diagnosed as SN recurred or metastasized. In our studies, melanocytic tumor and tumor microenvironment (dermis) from 114 cases of SN and SMM (Yale Spitzoid Neoplasm Repository) were analyzed and from the mass spectra from each sample, classification models were built using a training set of biopsies from 26 SN and 25 SMM separately for tumor and dermis. The classification algorithms developed on the training data set were validated on a second different set of 30 SN and 29 SMM. Protein expression differences between the melanocytic areas of SN and SMM biopsies were analyzed and a classification model consisting of five peptides that showed a high confidence differentiation the two types of lesions was generated. The results of this biocomputational analysis showed that 29 of 30 SN and 26 of 29 SMM biopsies were correctly categorized based on the histology data in the validation set. At this point, SN were correctly classified with 97 % sensitivity and 90 % specificity in the validation set. This work clearly demonstrated that MALDI IMS can routinely and reliably differentiate SN from SMM in FFPE tissue based on proteomic differences [30].

7.1.5 Application 2: Imaging the Immune Response to Infection in Whole-Animal Sections

The aim of this study is to investigate the invasion of host tissue by *Staphlococcus aureus* that often leads to the formation of pus-filled abscesses characterized by the extensive participation of host neutrophils. The specific neutrophil factors that are active to combat bacterial outgrowth within the abscess have yet to be fully described. We have used IMS to identify and localize bacterial and host proteins that are expressed in a *S. aureus* infected murine abscess as a result of infection [31, 32]. The *S. aureus* abscess shows macroscopic differences from healthy tissue and lends itself to investigation by IMS. Specifically, 6–8 week old female C57BL/6 mice were inoculated via the retro-orbital route with a sublethal dose of *S. aureus* in order to analyze protein distributions in infected tissue. Tissue from uninfected and infected animals was sectioned on a cryostat, coated with matrix (sinapinic acid), and analyzed by MALDI MS. The molecular maps produced from the imaged tissue reveal numerous proteins with specific distribution patterns for each of the tissue types. For example, one protein at *m/z* 10,165 localized specifically to the bacterial abscesses (Fig. 7.4). Using tryptic hydrolysis, peptide sequencing by MS/MS and protein database queries, this protein was identified as S100A8, a component of the calprotectin (CP) S100A8/S100A9 heterodimer. CP is a member of the S100 family of EF-hand Ca binding proteins, which are abundantly recruited to sites of inflammation [33]. CP is largely considered a neutrophil-specific protein and represents approximately 50 % of the protein complement of neutrophil cytoplasm. More recently, CP was reported to be a major component of neutrophil extracellular traps or NETs [34], and furthermore, it is not found in *S. aureus* lesions from neutropenic mice [35]. From these and other studies, it is concluded that CP is recruited to *S. aureus* infected murine tissue in a neutrophil-dependent manner.

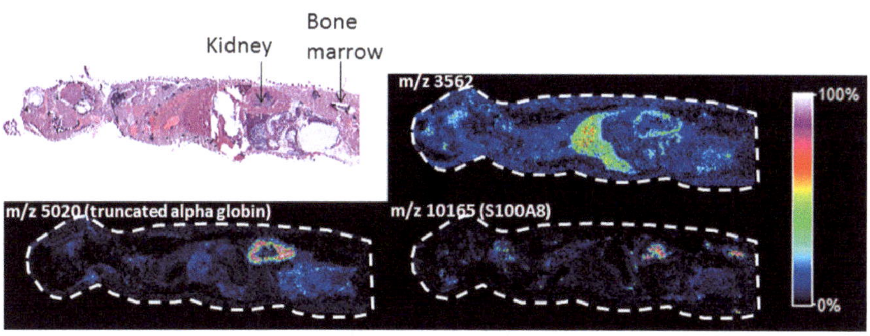

Fig. 7.4 Whole Body Imaging of a *S. aureus* Infected Mouse. Sagittal sections were collected of an infected animal that showed *S. aureus* containing abscesses. A protein as *m/z* 3,562 shows localization to the liver, while a protein at *m/z* 5,020, a truncated form of alpha globin, is localized to the cortex of the kidney. S100A8 (*m/z* 10,165), part of the host innate immune response, is present within the bacterial abscesses and in the bone marrow where it is produced

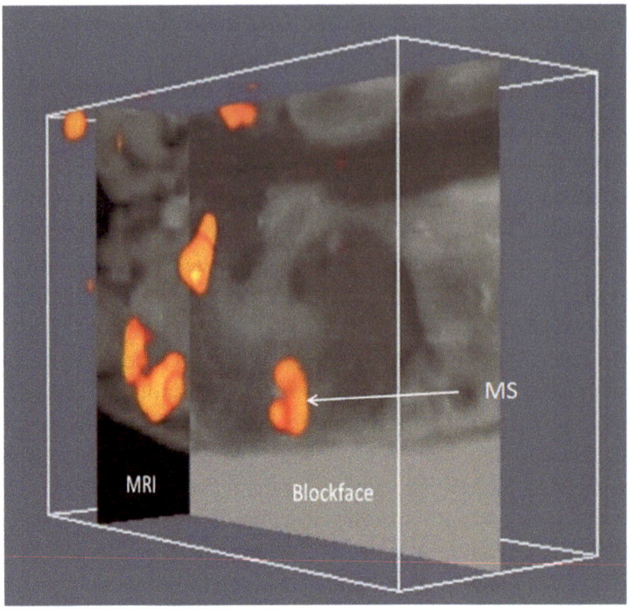

Fig. 7.5 3D Imaging Mass Spectrometry of a Mouse Kidney with Co-registration to Blockface and Magnetic Resonance Imaging. *In vivo* MRI measurements were registered to *ex vivo* blockface and mass spectrometry measurements in the same three dimensional space. A single plane of each of the MRI (*black pane*) and blockface (*gray pane*) data are shown perpendicular to each other with the MS data (*orange*, *m/z* 10,165) shown as a 3D volume. S100A8 localizes to the *S. aureus* abscesses within the kidney

A three dimensional approach to IMS has also been accomplished in a murine model of *S. aureus* infection [36]. In this approach mice were retro-orbitally infected as before and the mice were subjected to magnetic resonance imaging measurements prior to sacrifice to allow for *in vivo* visualization of the infection process. Animals were frozen into ice blocks to allow for rigid sectioning and sections containing the infected kidneys were collected every 210 μm throughout the kidney region. Additionally, digital images of the blockface were collected every 30 μm to facilitate registration. Protein IMS was carried out on a total of 45 tissue sections. These data were rendered into a three-dimensional image volume and then co-registered to the blockface and MRI. All three images modalities could be visualized in one 3D volume space. Figure 7.5 shows the co-registration of all three imaging modalities with the mass spectral volume of S100A8 displayed.

These results (i) demonstrate that calprotectin is recruited to infected murine tissue, (ii) identify nutrient metal chelation as a potent host defense against bacterial infection, and (iii) establish IMS as a powerful new tool in the field of bacterial pathogenesis which will be exploited in the proposed studies to identify the spatial distribution of proteins and elements that are critically important to the host-pathogen interaction. Notably, IMS can be expanded to whole animal sections to determine the impact of infection on global protein and metal distribution across an entire infected animal.

7.1.6 Application 3: Cellular Mechanisms of Renal Glomerular Diseases

The major aim of this study is to determine the molecular changes that occur in the glomerulus in diabetic nephropathy (DN), a complication of long term diabetes. Improvements in IMS spatial resolution have provided the means to obtain molecular information from distinct cell types within the kidney. In addition, instrumental and procedural advances have greatly improved detection levels of components in tissues and have allowed the mapping of disease related molecular modifications. Recent work in our lab has employed a gas-phase ion enrichment technique termed CASI (continuous accumulation of selected ions) using a Fourier transform ion cyclotron resonance (FTICR) mass spectrometer for direct application to the analysis of tissues [37]. The major advantage of CASI is that it allows the dynamic range of a selected mass window to be greatly increased. Briefly, ions generated by MALDI are selected using a mass selective quadrupole. These ions are then stored in a linear hexapole ion trap and the accumulation cycle is repeated at the repetition rate of the laser, typically 1 kHz or higher. The enriched ion population is then transferred to the ICR cell for detection. The use of CASI can increase signal intensities up to over 1,000-fold and thereby represents a significant dynamic range improvement over common mass spectral profiling methods while maintaining the multiplex advantage of mass spectrometry. In the study of DN, CASI FTICR IMS was employed to analyze phosphatidylethanolamine (PE) glycation in the kidney directly from a thin tissue section (Fig. 7.6). The identification of the several PE species and their modifications were assessed with tandem MS. The data showed that many lipid species were differentially expressed in the wild-type and eNOS-/- db/db (DN mouse model) kidneys, most notably a series of PEs containing an Amadori rearrangement moiety from either glucose adduct to the lipid amino group or other activated glycated intermediate (termed collectively as Amadori products). Amadori-PEs were localized to the cortex of a kidney from a eNOS-/- db/db mouse but were not detected in the wild-type kidneys. A decrease in these species was observed in the kidneys of diabetic mice treated with pyradoxamine (Fig. 7.6). It was found that altered lipids were specific to different morphological regions of the kidney. For example, the signal at *m/z* 1,151.7055 corresponding to ganglioside GM3, was present (at our limit of detection) only in glomeruli whereas the sulfatide SM3 signal at *m/z* 1,042.6712 was localized to the tubulointerstitial region of the kidney. Similarly, the signal at *m/z* 909.5497 (phosphatydylinositol (PI) 40:6) was mapped to the cortex whereas the signal at *m/z* 906.6345 (sulfatide SM4) was mapped to the medulla.

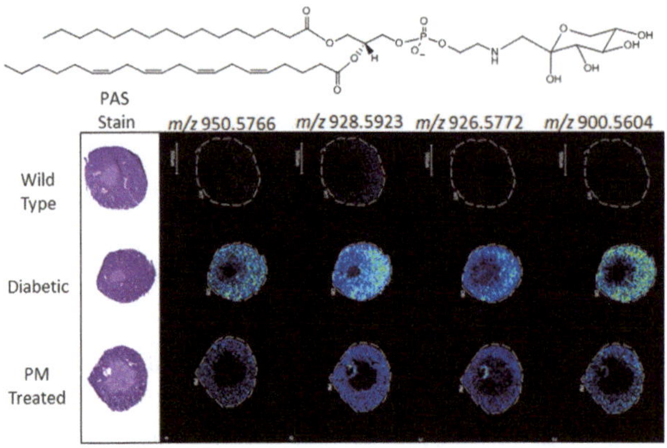

FTMS Analysis of Glycated Lipids

m/z	Identification	ppm
902.5769	Glucose-PE(36:3)	-0.59
904.5925	Glucose-PE(36:2)	-0.55
910.5831	Glucose-PE(P-38:5)	-1.86
912.5983	Glucose-PE(P-38:4)	-1.32
924.5618	Glucose-PE(38:6)	-1.22
926.5772	Glucose-PE(38:5)	-0.82
928.5923	Glucose-PE(38:4)	-0.37
948.5621	Glucose-PE(40:8)	-1.52
950.5766	Glucose-PE(40:7)	-0.26
952.5925	Glucose-PE(40:6)	-0.52

Fig. 7.6 FTICR Imaging of Glycated Lipids. *Top left* – An example of a lipid modified with a glucose attached to the headgroup. *Bottom left* – Images of glucose modified lipids. These lipids show localization to the cortex of the kidney of the diabetic mouse, but are nearly absent in the kidney of the wild type mouse. These modified lipids show decreased expression in the kidney of the mouse treated with pyrodoxamine (PM). *Right* – Summary of glucose modified lipids observed from the imaging experiment

The MALDI MS technology has advanced to where tissues can be routinely imaged at 10 μm spatial resolution. Imaging at high spatial resolution is necessary in order to determine the molecular changes that occur within small structures such as the glomerulus. MALDI MS was performed at 10 μm to generate lipid images with modified commercial instrumentation. At these resolutions the substructure of the glomeruli are easily differentiated within the kidney cortex. The imaged section was then washed to remove the MALDI matrix and stained with periodic acid Schiff (PAS) to correlate the IMS data with the histology from the same glomeruli. Using this approach, signals were detected that were specific to the glomeruli of all experimental groups, as well as those that were specific to disease.

7.1.7 *Application 4: Retinoid Metabolite Imaging in Age-Related Macular Degeneration (AMD)*

The long term goal of this project involves the imaging of retinoid metabolites to identify those metabolites that make up lipofuscin and to map their spatial distributions and abundances in retina tissue. The identification of novel retinoids and bis-retinoid metabolites and the collection of their spatial abundance profiles in the retinal pigment epithelium (RPE) at high spatial resolution was performed using IMS. Imaging of retina tissue in cross section has the advantage of allowing visualization of distinct cell layers and changes in analyte concentrations within these layers in disease. Sample preparation methods for imaging retina tissue in cross section currently involve embedding the entire eye in carboxymethylcellulose prior to sectioning. A key issue for this tissue is managing to keep the different cell layers intact during

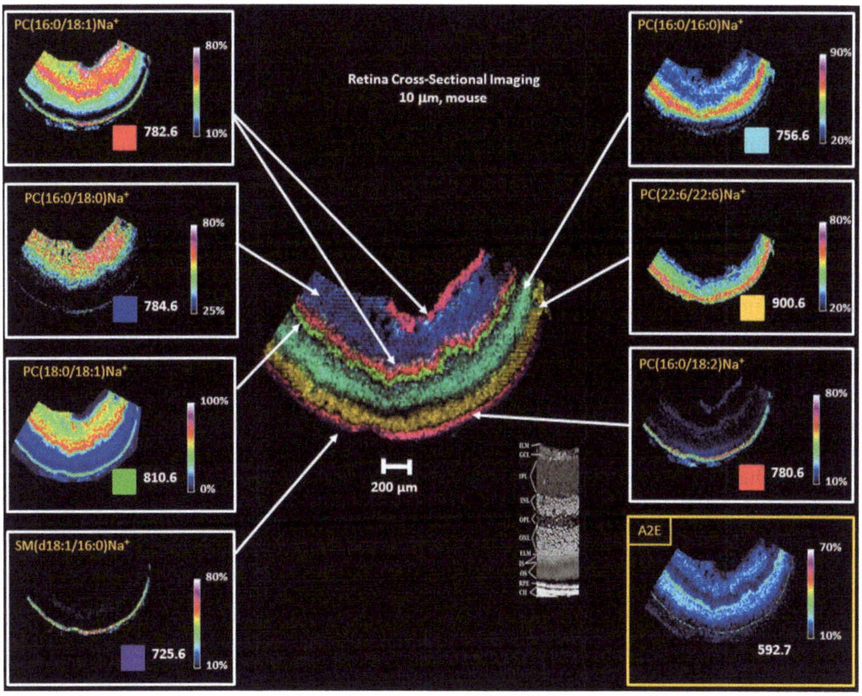

Fig. 7.7 High Resolution Imaging Mass Spectrometry of a Mouse Retina. *Center* – MALDI image of a mouse retina in cross section indicating different cell layers by color. Each color represents a different measured ion *m/z* value. Individual ion images are shown in the periphery with molecular assignments. Reference image (*center* panel insert) of cell layers in retina cross section [38]

and following sectioning. The current protocol has been successfully applied to mouse and rat retina and in early experiments, to human retina. Figure 7.7 presents several individual ion images from mouse retina showing cell layer specificity in their localization. These data were acquired from an ABCR-/- mouse retina as a model of Stargardt's disease where retinoid metabolites accumulate similar to AMD. These data were acquired with 10 μm spatial resolution. Molecular assignments were made using both accurate mass measurements and tandem mass spectrometry on a 9.4 T FTICR instrument. Note that the signal for the expected major retinoid metabolite, A2E at *m/z* 592.7, is shown in the lower right panel to be present in the single cell layer of the RPE. The other signal in the inner retina layers corresponds to an isobaric lipid as determined by tandem mass spectrometry.

7.2 Conclusion

MALDI IMS has become a very powerful tool with which to map biomolecules is tissue sections. It has major advantages in that it is an excellent discovery tool because it does not require a target specific reagent such as an antibody. The product of the analyses is an image obtained from a direct molecular map of the biomolecules in tissues that can change their expression levels as a result of a biological perturbation or disease. These images can be easily correlated with other image modalities such as microscopy or MRI [36]. IMS technology provides a multiplex image process in that many hundreds to thousands of images can be generated from a single raster of a tissue section. In leaving biomolecules in place in tissues in their native microenvironment, new biological insights can be obtained and correlations made that would otherwise be missed using methods that require grinding and extraction of tissues and therefore lose positional information and only provide averaged measurements over the entire sample.

Acknowledgements The authors thank Rossitza Lazova (Yale University Medical School) and Alireza Sepehr (Harvard University Medical School) for contribution and histological evaluation of melanoma samples and Eric Skaar (Vanderbilt University School of Medicine) for collaborative efforts with the *S.aureus* studies. The work on diabetic nephropathy was carried out by Kerri Grove and Billy Hudson (Vanderbilt University School of Medicine). Studies of AMD were performed by Kevin Schey and David Anderson (Vanderbilt University School of Medicine) in collaboration with Rosalie Crouch (Medical University of South Carolina). Funding from NIH/NCRR 8P41 GM103391 and NIH/NIGMS 5R01 GM58008 are gratefully acknowledged.

References

1. Fletcher JS, Vickerman JC, Winograd N (2011) Curr Opin Chem Biol 15:733–740
2. Guerquin-Kern J-L, Coppey M, Carrez D, Brunet A-C, Nguyen CH, Rivalle C, Slodzian G, Croisy A (1997) Microsc Res Tech 36:287–295
3. Vaeck L, Struyf H, van Roy W, Adams F (1994) Mass Spectrom Rev 13:189
4. Takats Z, Wiseman JM, Gologan B, Cooks RG (2004) Science 306:471–473
5. Caprioli RM, Farmer TB, Gile J (1997) Anal Chem 69:4751–4760
6. Chaurand P, Hayn G, Matter U, Caprioli RM (2004) Zhipu Xuebao 25(205–206):216
7. Norris JL, Caprioli RM (2013) Chem Rev 113:2309–2342
8. McDonnell LA, Heeren RMA (2007) Mass Spectrom Rev 26:606–643
9. Seeley EH, Schwamborn K, Caprioli RM (2011) J Biol Chem 286:25459–25466
10. Dass C (2001) Principles and practice of biological mass spectrometry. Wiley, Hoboken
11. Verhaert PD, Conaway MCP, Pekar TM, Miller K (2007) Int J Mass Spectrom 260:177–184
12. DeKeyser SS, Kutz-Naber KK, Schmidt JJ, Barrett-Wilt GA, Li LJ (2007) J Proteome Res 6:1782–1791
13. Groseclose MR, Andersson M, Hardesty WM, Caprioli RM (2007) J Mass Spectrom 42:254–262
14. Khatib-Shahidi S, Andersson M, Herman JL, Gillespie TA, Caprioli RM (2006) Anal Chem 78:6448–6456
15. Li Y, Shrestha B, Vertes A (2007) Anal Chem 79:523–532
16. Taban IM, Altelaar AFM, Van der Burgt YEM, McDonnell LA, Heeren RMA, Fuchser J, Baykut G (2007) J Am Soc Mass Spectrom 18:145–151
17. Altelaar AFM, Taban IM, McDonnell LA, Verhaert P, de Lange RPJ, Adan RAH, Mooi WJ, Heeren RMA, Piersma SR (2007) Int J Mass Spectrom 260:203–211
18. McLean JA, Ridenour WB, Caprioli RM (2007) J Mass Spectrom 42:1099–1105
19. Mamyrin BA (1994) Int J Mass Spectrom 131:1–19
20. Schwartz SA, Reyzer ML, Caprioli RM (2003) J Mass Spectrom 38:699–708

21. Shrivas K, Hayasaka T, Goto-Inoue N, Sugiura Y, Zaima N, Setou M (2010) Anal Chem 82:8800–8806
22. Meriaux C, Franck J, Wisztorski M, Salzet M, Fournier I (2010) J Proteomic 73:1204–1218
23. Hayasaka T, Goto-Inoue N, Zaima N, Shrivas K, Kashiwagi Y, Yamamoto M, Nakamoto M, Setou M (2010) J Am Soc Mass Spectrom 21:1446–1454
24. Bhat AR, Wu HF (2010) Rapid Commun Mass Spectrom 24:3547–3552
25. Aerni HR, Cornett DS, Caprioli RM (2006) Anal Chem 78:827–834
26. Baluya DL, Garrett TJ, Yost RA (2007) Anal Chem 79:6862–6867
27. Guo J, Colgan TJ, DeSouza LV, Rodrigues MJ, Romaschin AD, Siu KW (2005) Rapid Commun Mass Spectrum 19:2762–2766
28. Schuerenberg M, Luebbert C, Deininger S-O, Ketterlinus R, Suckau D (2007) Nat Methods 4:iii
29. Yamada Y, Hidefumi K, Shion H, Oshikata M, Haramaki Y (2001) Rapid Commun Mass Spectrum 25:1600–1608
30. Lazova R, Seeley EH, Keenan M, Gueorguieva R, Caprioli RM (2012) Am J Dermatopathol 34:82–90
31. Stoeckli M, Chaurand P, Hallahan DE, Caprioli RM (2001) Nat Med 7:493–496
32. Reyzer ML, Caprioli RM (2007) Curr Opin Chem Biol 11:29–35
33. Striz I, Trebichavsky I (2004) Physiol Res 53:245–253
34. Urban CF, Ermert D, Schmid M, Abu-Abed U, Goosmann C, Nacken W, Brinkmann V, Jungblut PR, Zychlinsky A (2009) PLoS Pathog 5:e1000639
35. Corbin BD, Seeley EH, Raab A, Feldmann J, Miller MR, Torres VJ, Anderson KL, Dattilo BM, Dunman PM, Gerads R, Caprioli RM, Nacken W, Chazin WJ, Skaar EP (2008) Science 319:962–965
36. Attia AS, Schroeder KA, Seeley EH, Wilson KJ, Hammer ND, Colvin DC, Manier ML, Nicklay JJ, Rose KL, Gore JC, Caprioli RM, Skaar EP (2012) Cell Host Microbe 11:664–673
37. Fuchser J, Cornett DS, Becker M (2008) Application note # FTMS-37 high resoluting molecular imaging of pharmaceuticals at therapeutic levels. Bruker Daltonics, Billerica
38. Mullins RF, Skeie JM (2010) Essentials of retinal pathology. In: Pang IH, Clark AF (eds) Animal models of retinal diseases, vol 46. Humana Press, New York, pp 1–11

Chapter 8
Bacterial Identification by Mass Spectrometry

Christopher R. Cox and Kent J. Voorhees

Abstract Mass spectrometry has become an important tool for the clinical microbiologist for bacterial identification. Because matrix assisted laser desorption ionization-time of flight mass spectrometry (MALDI-TOF-MS) has been shown to provide reliable protein profiling data to identify many bacteria, three commercially available instruments are available. Phage amplification (PA) has added new dimensions including sensitivity, specificity, and reduced culturing time to bacterial characterization by MALDI-TOF-MS-based protein identification. In this procedure, a sample suspected of containing a target bacterium is infected with a species-specific phage and allowed to incubate for two to three hours. If an increase in phage protein is observed in the MALDI spectrum, amplification has occurred and is indicative of the presence of the target bacteria. As an extension of this, a rapid method of determining antibiotic resistance using PA with MALDI-MS has also been developed. Other non-protein-based mass spectrometry techniques for bacterial identification have also been explored. As an example, fatty acid profiling employing MALDI with CaO as a saponification catalyst has been shown to provide highly specific identification to the strain level. A comparison of metal oxide FA profiles to protein profiles obtained from a commercial instrument is presented.

Keywords Bacteria • Identification • Protein • Profiling • MALDI MS • Phage amplification • Antibiotic resistance • Matrix free • CaO • Fatty acids

8.1 Introduction

Since the first applications in the mid-1960s, mass spectrometry techniques for bacterial profiling and identification have evolved to offer a new horizon for clinical microbiologists. Pyrolysis-gas chromatography (PyGC) was the first analytical technique widely applied to bacteria profiling [1–6]. Reiner demonstrated that pyrolysis of cellular components could be used to successfully profile bacteria [1].

C.R. Cox • K.J. Voorhees (✉)
Department of Chemistry, Colorado School of Mines, Golden, CO 80401, USA
e-mail: kvoorhee@mines.edu

J. Banoub (ed.), *Detection of Chemical, Biological, Radiological and Nuclear Agents for the Prevention of Terrorism*, NATO Science for Peace and Security Series A: Chemistry and Biology, DOI 10.1007/978-94-017-9238-7_8

In a series of papers [1–6], Reiner et al. used PyGC to differentiate bacteria to the genera, species, and subspecies level. Early work by Able et al. demonstrated the feasibility of bacterial classification using lipid extraction and transesterification followed by gas chromatographic analysis [7]. Moss extended the method to include a saponification and methylation step and showed that profiles of trimethylsilyl derivatives from whole-cell lysates could be used to classify *Clostridia* to the species level [8]. Microbial Identification Inc. (MIDI) introduced the first commercial unit in the early 1980s employing fatty acid (FA) profiling to identify bacteria [9]. The MIDI technique utilized an *ex situ* saponification/methylation procedure similar to the Moss technique [8] and gas chromatography applied to bacteria that had been cultured under strict conditions. The major drawback of the MIDI approach was the need for tightly controlled culture of organisms (typically up to 24 h) and sample preparation and analysis, which took an additional 2–3 h. The sample preparation time was improved by using an *in situ* reaction involving heat and tetramethyl ammonium hydroxide (TMAH) as a replacement for the saponification/methylation step [10, 11]. A comparison of MIDI and *in situ* TMAH results showed that the two methods were equivalent in identifying bacteria [2, 12].

In the 1970s, Muezelaar developed a bacterial profiling procedure based on pyrolysis and mass spectrometry (PyMS) [13, 14]. Mass spectrometry was incorporated to eliminate problems inherent to gas chromatography [13]. Their PyMS instrument consisted of a Curie-point pyrolyzer interfaced to a computer-controlled, low-energy electron ionization quadrupole mass spectrometer. Reproducibility of the technique was shown to be satisfactory for profiling of diverse samples [15]. Data from three different bacteria showed that significant differences in PyMS profiles existed between strains. Meuzelaar and his coworkers were the first to use pattern recognition methods on PyMS data [16], the addition of which expanded its usefulness and led to development of other MS bacterial identification techniques. Up to that point, data analysis had only been visual.

The U.S. Army recognized the potential of PyMS and incorporated the technique as the basis of the chemical and biological mass spectrometer (CBMS). The first generation instrument (Block I) consisted of a micro furnace connected to an ion trap MS designed to generate pyrolysis biomarkers for bacterial identification [17]. Bruker currently manufactures this instrument for bio-security applications. A second generation CBMS (Block II) was configured to utilize TMAH with a detection protocol focused on fatty acids [18]. Fatty acid methyl ester libraries were developed; however, the instrument was never commercially manufactured.

Because of its relatively simple sample preparation in comparison to other methods, matrix assisted laser desorption ionization time-of-flight mass spectrometry (MALDI-TOF-MS) is becoming an increasingly popular tool for protein-based bacterial identification. Reviews describing diverse MALDI-TOF-MS bacterial classification methods have been published [19, 20]. Holland et al. first showed that MALDI-TOF-MS protein profiles from whole cell bacteria could be used to differentiate various species [21]. In the mass range of *m/z* 5,000–20,000, Gram-positive and negative bacteria produced unique protein profiles characteristic of each phylotype. Seminal work by Bizzini et al. [2] changed the landscape for clinical

MALDI-TOF MS protein analysis for bacterial ID [22]. In this study, 22 bacterial isolates previously identified by traditional methods were successfully classified to the species level with 92.3 % accuracy, and it was concluded by the authors that MALDI-TOF MS protein profiling was a fast and reasonably reliable technique. Two FDA-approved commercial MALDI-TOF MS instruments, the Bruker Microflex Biotyper and the BioMerieux VITEK MS are currently available for bacterial ID using protein profiles [23–25]. Both support substantial databases for ID of unknown profiles, and are accurate to within 92–98 % [22]. This approach does have several drawbacks including sensitivity, need for exact overnight culturing, and mixture analysis, all of which can be overcome by using phage amplification [26].

Several other MS-based detection platforms have been used for detection of bacteria. Among these are electrospray mass spectrometry (ESI-MS) [27], polymerase chain reaction (PCR)-ESI-MS [28], direct analysis in real time–MS (DART-MS) [29], and desorption electrospray ionization MS (DESI-MS) [30]. These will not be discussed in this chapter.

8.2 Colorado School of Mines (CSM) Research

Work conducted during the last 4 years at the CSM Advanced Biodetection Laboratory has included:

- Evaluation of phage amplification-MALDI-MS for rapid bacterial detection and identification [26, 31, 32].
- Development of phage amplification for determination of antibiotic resistance [33].
- Development of new MALDI-TOF-MS sample pretreatment methods for enhanced resolution of viral capsid protein structures [4, 34].
- Development of metal oxides as replacements for traditional MALDI matrices [35].
- Development of bacterial fatty acid profiling using CaO with MALDI-MS [36].

8.3 MALDI-TOF-MS

MALDI-TOF MS was developed in 1988 [37] and has traditionally been used for analysis and identification of high molecular weight analytes such as peptides, proteins, and polymers [38–40]. The basic concept of MALDI-TOF MS focuses on the volatilization of target compounds by laser ablation of samples embedded in a UV-absorbing matrix [40]. Typical MALDI-TOF MS sample preparation involves mixing a sample (biological fluids such as serum, urine, tissue extracts, or whole bacterial or viral preparations) with an organic matrix solution that co-crystallizes a target compound with matrix on a chargeable target grid. Samples under vacuum are subjected to a laser pulse resulting in the liberation of ions from the sample/matrix surface. Liberated ions are then accelerated by an electrical field towards a detector. Smaller molecules are accelerated to higher velocities than heavier molecules and thus reach the detector sooner. The flight time for a given target molecule is then proportional to its relative mass to charge ratio (*m/z*). For singularly charge molecules, the *m/z* value equals the mass of the analyte. The signal of each detection event can be used to generate a spectral profile with a mass to charge ratio along the x-axis, and by a peak-intensity on the y-axis for each sample.

8.4 Phage Amplification (PA) MALDI-TOF-MS

Bacteriophage (phage) amplification (PA) coupled to MALDI-TOF-MS expands both the selectivity and sensitivity of instrument-based bacterial identification. PA MALDI-TOF-MS requires species-specific bacterial infection and allows for bacterial detection and identification in as little as 2 h. This method lowers the limit of detection by several orders of magnitude over conventional biochemical detection methods. Phages are viruses that infect bacteria and multiply by utilizing the host biosynthetic machinery [41]. During natural phage propagation, a bacterial cell is infected with a species-specific phage and progresses through the infection cycle, resulting in the release of progeny phage. With each progeny phage produced, there is an amplification of both protein and nucleic acid content, which can be used as secondary indicators of a specific bacterial host.

Figure 8.1 summarizes phage amplification. Some phages contain a icosahedral capsid (head) composed of approximately 180 copies of one or two proteins, which encapsulates the nucleic acids. The number of progeny phage produced from each infection (burst size) can vary from a few tens up to as high as 10,000 [42]. The burst size multiplied by the number of specific proteins gives the theoretical amplification, which can in turn be exploited as a means of species-specific MALDI-TOF-MS signal enhancement.

The first PA experiments with MALDI-TOF-MS detection were conducted in our laboratory with MS2 coliphage and *Escherichia coli* [43]. The MALDI limit of detection (LOD) for *E. coli* was 10^7 cfu/mL. Following a 40 min amplification with MS2, 5×10^4 cfu/mL were detected. Figure 8.2 shows spectra generated during a PA experiment. Panels A and B, show spectra for *E. coli* and MS2, respectively. Panel C is a spectrum taken at the onset of infection (time = 0) with the phage introduced below the detection limit. After 40 min a peak at 13.7 kD for the MS2 capsid protein was observed, indicating the presence of *E. coli*.

Since these early experiments, several phage-host pairs have been investigated at the Colorado School of Mines for bacterial detection and identification by phage amplification. These are summarized in Table 8.1.

Burkholderia will be presented as an example of CSM research because it represents the complete process from phage isolation and characterization to antibiotic resistance determination [44]. Phage ϕX216 was isolated and characterized in collaboration with Dr. Herbert Schweizer at Colorado State University [45], with an 80 min replication cycle and burst size of 120 pfu per infected cell. Figure 8.3a, b

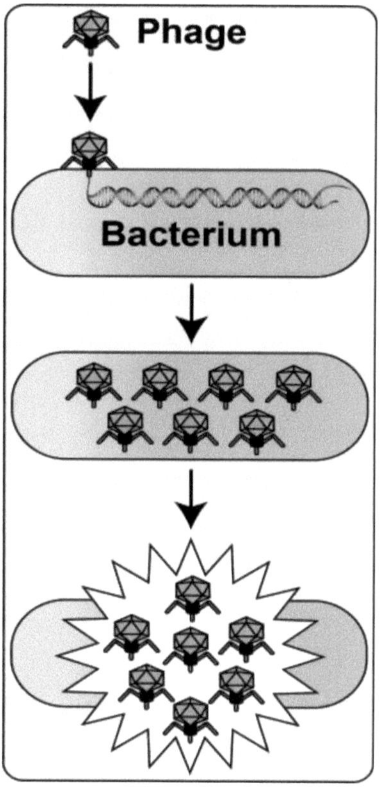

Fig. 8.1 Phage amplification resulting from infection of a single bacterium with a single phage results in the production and release of a large number of detectable progeny phage. This can be exploited as a means of signal amplification

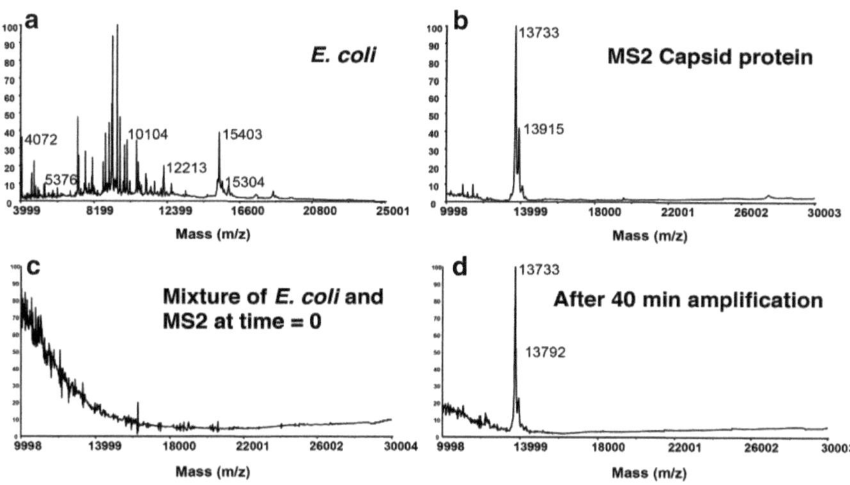

Fig. 8.2 PA MALDI-TOF MS detection of *E. coli*. Mass spectra of (**a**) *E. coli*, (**b**), MS2, (**c**) *E. coli*-MS2 mixture before PA (bacterial and phage concentrations below the MALDI-TOF-MS LOD), and (**d**) the same mixture after 40 min PA

Table 8.1 Bacteria and phages studied at CSM

Phylotype	Phage
Escherichia coli	MS2
	T1-T7
Salmonella spp.	MS2
	MPSS1
Yersinia pestis	ϕA1122
	PKR
	R
	V
	Y
	CC1[a]
Acinetobacter baumannii	AC54
	BS46
Burkholderia pseudomallei/mallei	ϕX216
	ϕ52237
	ϕ1026b
Bacillus anthracis	γ
Listeria monocytogenes	A511
Enterococcus faecalis	1A
	1B
	120
	182
	1/V12
Enterococcus faecium	P3
	P13
	113
Staphylococcus aureus	187

[a]Environmental isolate collected, characterized and sequenced by Dr. Christopher R. Cox

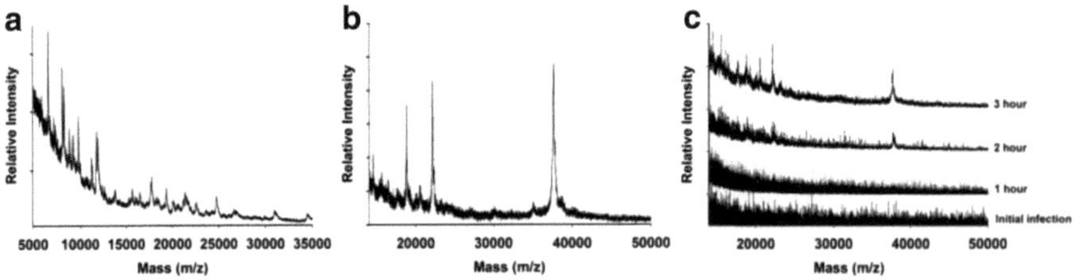

Fig. 8.3 Species-specific detection of *B. pseudomallei* using PA MALDI-TOF MS. (**a**) Mass spectrum of *B. pseudomallei* Bp82, and (**b**) phage ϕX216. (**c**) PA measured over a 3-h timeframe

show MALDI-TOF spectra of *B. pseudomallei* Bp 82 and phage ϕX216, respectively. The prominent peak at 37 kDa is the major capsid protein. The spectra in Fig. 8.3c illustrate ϕX216 amplification following species-specific infection of *B. pseudomallei*. The phage infection was initiated with an input phage concentration below the MALDI-TOF-MS LOD (2.0×10^7 pfu/mL). Phage amplification was observed, as exemplified by the appearance of the 37 kDa peak, 2 h after the initial infection, and served as an indicator of the presence of *Burkholderia*.

The phage capsid is primarily held together by non-covalent bonds which can be partially disassembled by laser ablation [34]. Some phages listed in Table 8.1 contain capsid proteins held together by covalent disulfide bonds that provide additional strength to their overall capsid structure. It was determined that a reducing agent such as ***β***-mercaptoethanol (BME) added sample spots prior to MALDI-TOF-MS analysis contributed significantly to capsid disassembly [4, 34]. Figure 8.4a shows a spectrum of *Yersinia pestis* phage ϕA1122 without BME added. Figure 8.4b shows the same phage refluxed in 90 % acetic acid at 90 °C for an hour, while Fig. 8.4c shows the result of pre-treating ϕA1122 with BME. The capsid protein has a mass at 36.5 kDa, which was observed in Fig. 8.4c only after disulfide bond reduction. Other reducing agents investigated include dithiothreitol (DTT) and tris-(2-carboxyethyl)phosphine (TCEP). Based on the intensity of the capsid peak, BME produced stronger MALDI-TOF-MS peaks than DTT or TCEP.

8.5 Antibiotic Resistance Determined by PA MALDI-TOF-MS

Antibiotic resistance testing using PA with MALDI-TOF-MS was first reported in 2004 [46]. The protocol is an extension of the identification process described in Sect. 8.4. Phage infection only occurs with a viable host (Fig. 8.5). Antibiotics act faster in killing susceptible hosts than PA, allowing for a simple means to determine bacterial response to an antibiotic. Hence, if an antibiotic is added to a bacterial culture followed by addition of phage, a positive test for PA will only be obtained if the bacteria are antibiotic resistant.

Figure 8.6a illustrates the MALDI-TOF spectra obtained from an antibiotic susceptible strain of Bp 82 *B. pseudomallei* grown without the addition of ceftazidime [44]. Figure 8.6b shows the result of PA attempted on the susceptible strain in the presence of 15 μg/mL of ceftazidime. No capsid protein peak was observed. In contrast, the spectrum from an antibiotic resistant isogenic mutant strain shows the capsid protein peak after 2 h of PA (Fig. 8.6c). This same approach for antibiotic resistance using lateral flow immunochromatography as the detector was approved by FDA for clinical application [4, 47].

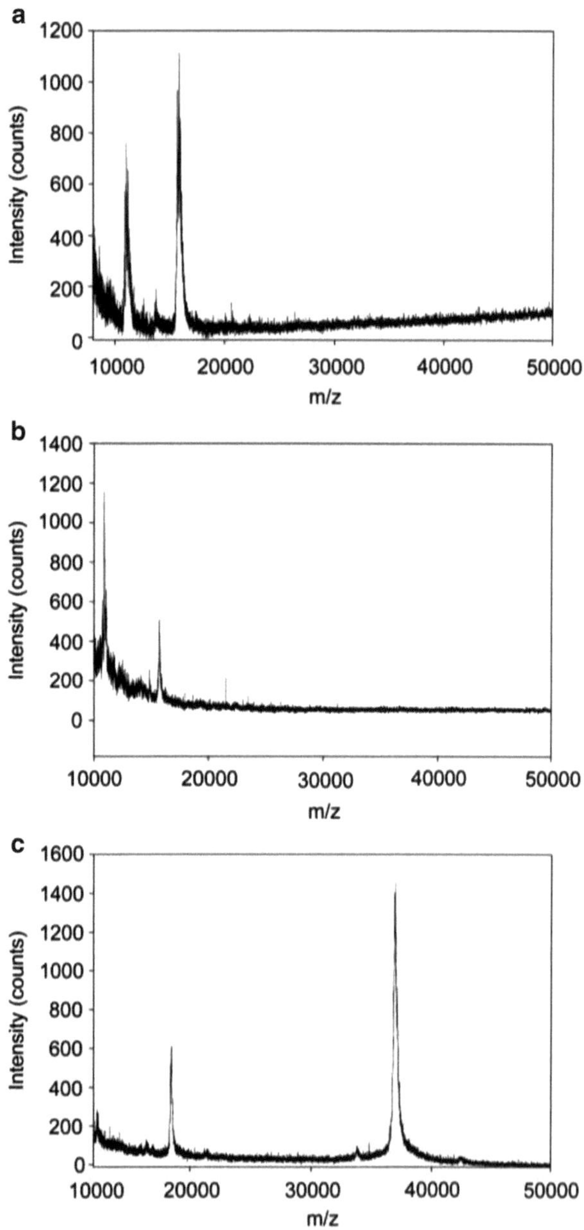

Fig. 8.4 MALDI-TOF-MS spectra of *Y. pestis* phage ϕA1122. (**a**) with conventional sample preparation and analysis, (**b**) refluxed with 90 % acetic acid at 90 °C for 1 h, and (**c**) following pretreatment with BME

8.6 Metal Oxide Laser Ionization-MS (MOLI-MS)

Conventional MALDI matrices produce background peaks in the mass range less than 1,000 Da. For example, a MALDI-TOF spectrum of 2,5-dihydroxy benzoic acid (DHB) has a strong peak at m/z 273 resulting from the dimer of DHB minus

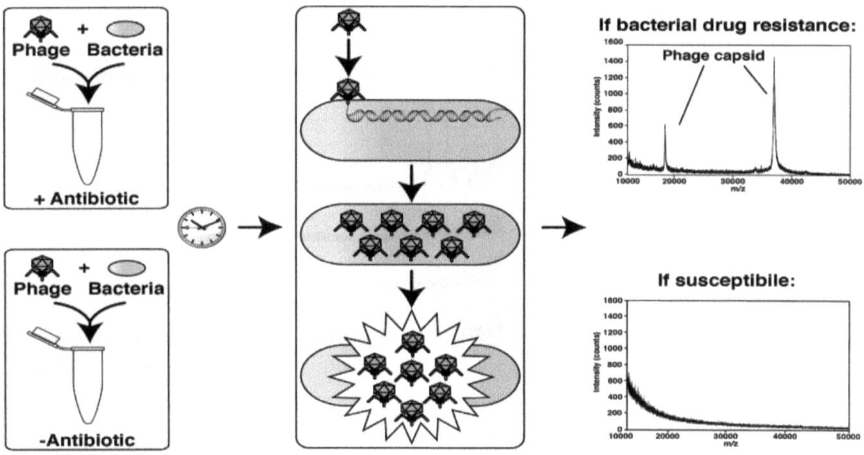

Fig. 8.5 Schematic representation of multiplexed PA-MALDI-TOF-MS for bacterial detection, ID and drug resistance determination

water, which overwhelms and masks any analyte peaks in this region. Several strategies have been explored to minimize this problem including the development of matrix-free systems [48–54], which generally involve interactions between an analyte and a functionalized surface during laser irradiation [54].

Early studies in our laboratory aimed at eliminating background focused on CaO and NiO for the analysis of glycerides [35]. NiO produced mostly intact molecular species for mono-, di-, and tri-glycerides, fatty acids, phospholipids, carbohydrates, and peptides. MOLI-MS using NiO has also been applied to complex mixtures of lipids from vegetable oil shortening and bacterial and algal extracts. In contrast, CaO shows a catalytic effect and efficiently cleaves lipid ester groups into fatty acids and glycerols. A 30 s chloroform/methanol extract from bacteria and algae produced a fatty acid profile that was consistent with previous work [35, 36].

MOLI-MS using CaO in the positive-ion mode produced a calcium/fatty acid adduct and a carboxylate ion in the negative mode [36]. Figure 8.7 shows distinctive negative-ion FA mass spectra of four representative bacterial phylotypes. Positive-ion profiles are shown in Fig. 8.8.

The MOLI data (11 FA peaks in the range of C14:0 to C21:0 from extracts from five Gram positive and five Gram negative bacteria [55] (Table 8.2), when analyzed by PCA (Fig. 8.9a, b) showed that replicates from each organism grouped in unique spaces. Cross validation (CV) of spectral data from either positive or negative ionization modes resulted in greater than 96 % accuracy and differentiation at the species level. When CV results were compared for the two ionization modes, negative-ion data produced higher reproducibility [5, 55]. An interesting aspect of this work is that bacterial protein and fatty acid profiles can be obtained in parallel using a single MALDI-TOF-MS instrument, and there are bacterial groups that cannot be identified by protein profiling that have previously been reported by MIDI as distinguishable [56–59].

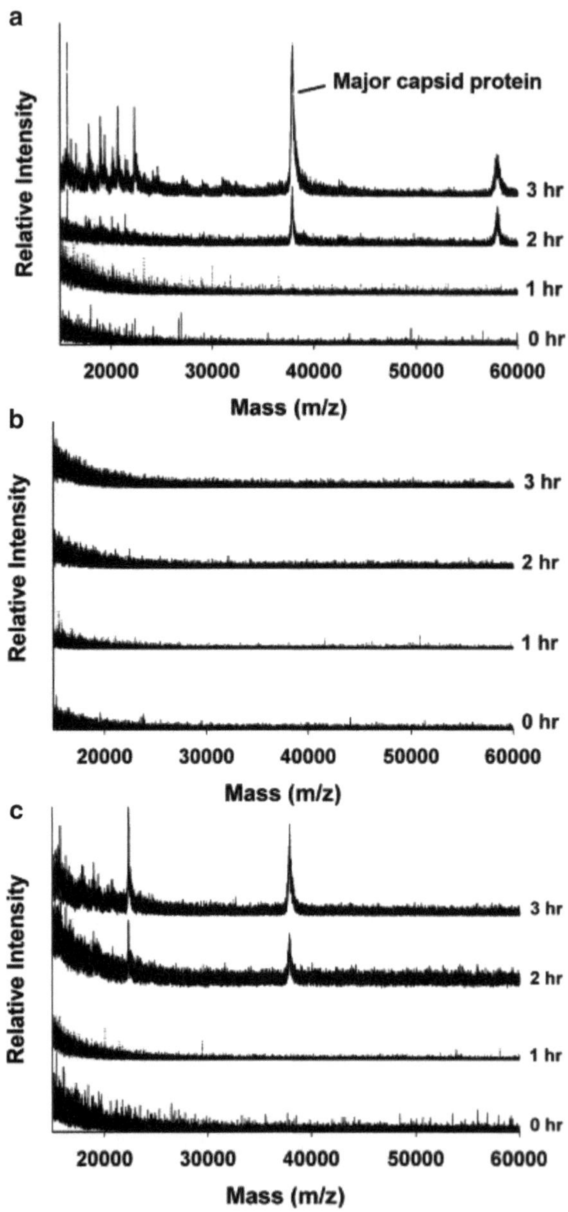

Fig. 8.6 *Burkholderia pseudomallei* ID and drug resistance determination by PA-MALDI-TOF MS. (**a**) PA targeting wild type sensitive strain, no drug. (**b**) Same reaction +15 μg/ml ceftazidime. (**c**) PA targeting drug resistant isogenic mutant +15 μg/ml ceftazidime

8.7 Strain Level Bacterial ID Using MOLI MS

Analysis conducted with the Bruker Biotyper revealed a failure of that technology to differentiate *Shigella* and *Salmonella* from *E. coli*. The strains investigated and the Biotyper results are summarized in Table 8.3 for five individual analyses

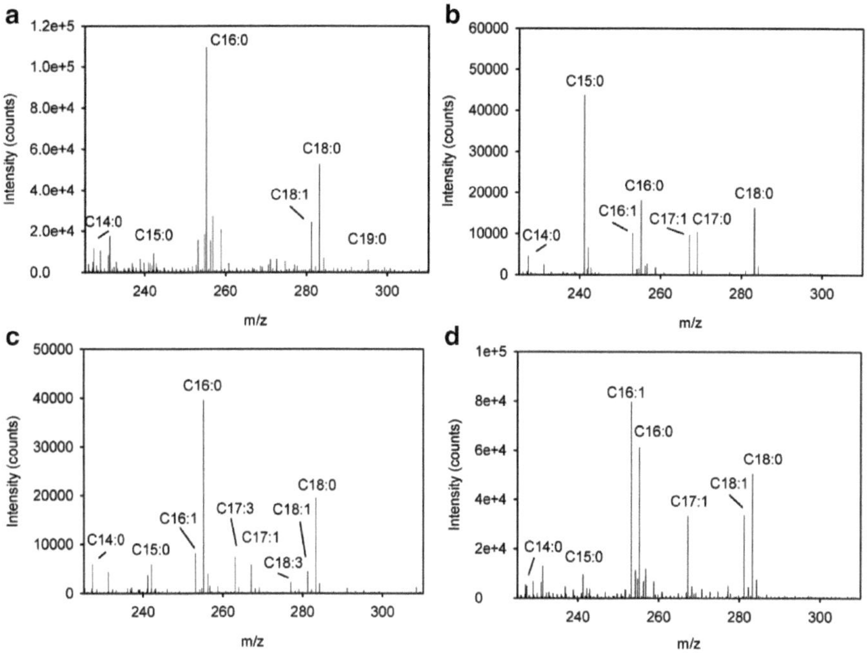

Fig. 8.7 Negative-ion mass spectra of selected bacteria. (**a**) *E. faecalis*, (**b**) *B. anthracis*, (**c**) *E. coli*, and (**d**) *Y. pestis*

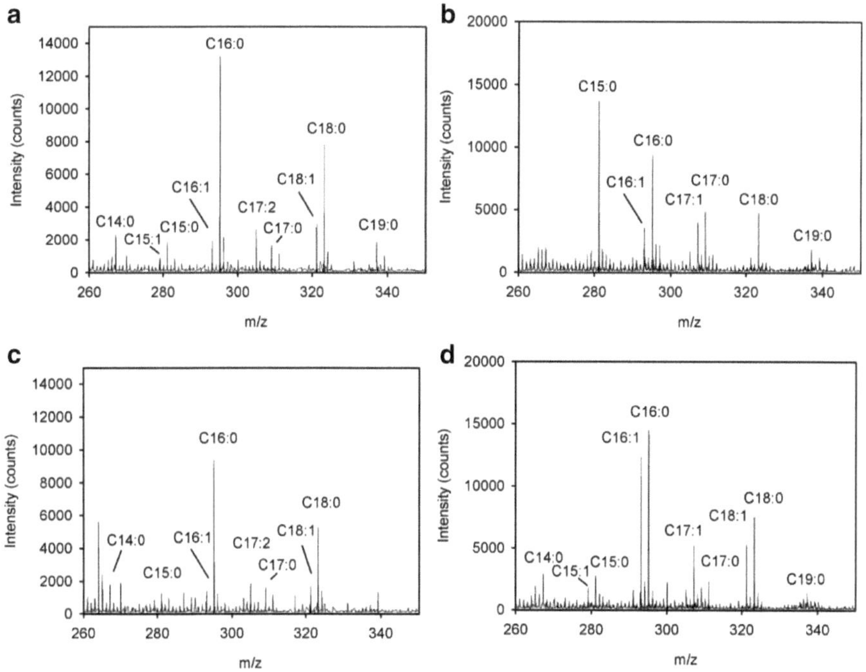

Fig. 8.8 Positive-ion mass spectra of selected bacteria. (**a**) *E. faecalis*, (**b**) *B. anthracis*, (**c**) *E. coli*, and (**d**) *Y. pestis*

Table 8.2 Bacterial phylotypes used

	Strain
Gram positive	
Bacillus anthracis	Sterne
Enterococcus faecalis	V583
Clostridium putrefaciens	ATCC 25786
Listeria monocytogenes	ATCC 19112
Staphylococcus aureus	ATCC 27660
Gram negative	
Acinetobacter baumannii	AC54
Escherichia coli	ATCC 15597
Francisella tularensis	LVS
Salmonella typhimurium	ATCC 13311
Yersinia pestis	A1122

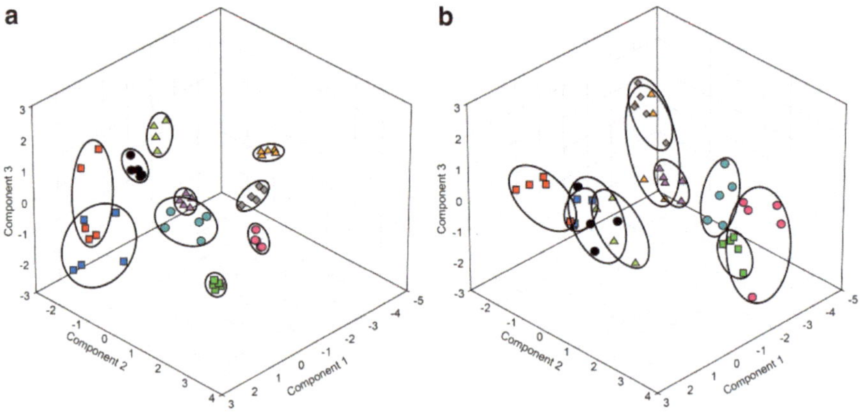

Fig. 8.9 PCA plot of bacterial MOLI MS spectra derived in (**a**) negative- and (**b**) positive-ion modes. *B. anthracis*, *C. putrefaciens*, *E. faecalis*, *L. monocytogenes*, *S. aureus*, *A. baumannii*, *E. coli*, *F. tularensis*, *S. typhimurium*, *Y. pestis*

Table 8.3 Summary of Bruker Biotyper ID trials

Phylotype tested	Biotyper results
E. coli K12	Identified as *E. coli*
E. coli ATCC 15597	Identified as *E. coli*
Salmonella enterica ATCC 14028	Identified at genus level only
Salmonella typhimurium ATCC 19585	Identified at genus level only
Shigella boydii ATCC 9207	Misidentified as *E. coli*
Shigella flexneri PHS-1059	Misidentified as *Enterobacter cloacae*

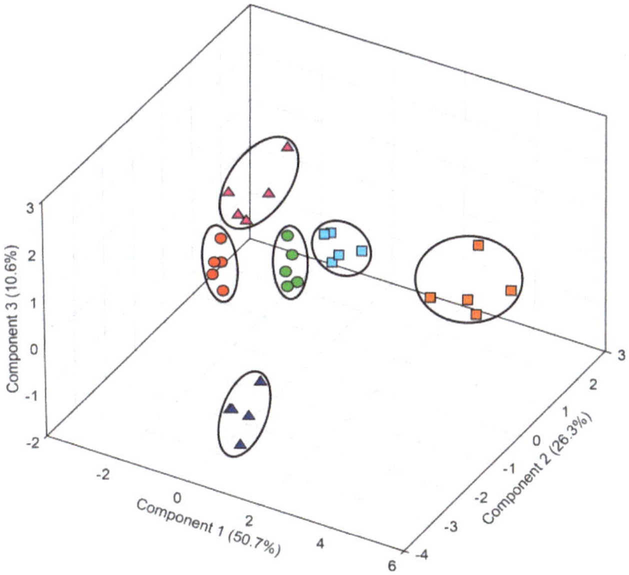

Fig. 8.10 PCA differentiation of enterobacteriaceae using MOLI MS negative ion spectra obtained with nanostructured CaO catalysis. ● *E. coli* K12, ● *E. coli* ATCC 15597, ▲ *Salmonella enterica* ATCC 14028, ▲ *S. typhimurium* ATCC 19585, ■ *Shigella boydii* ATCC 9207, ■ *S. flexneri* PHS-1059

performed on separate but identical colonies of pure cultures grown overnight and aseptically removed from LB agar plates. Two different *E. coli* strains were identified to the species level but incorrectly assigned at the strain level. Two *Salmonella* species were identified to the genus level but could not be characterized to the species level. Importantly, *Shigella boydii* ATCC 9207 was misidentified as *E. coli*, and *S. flexneri* PHS-1059 was misidentified as *Enterobacter cloacae*. This problem led us to compare MOLI-MS in the same capacity.

Figure 8.10 shows the 3-D PCA score plot of nanoscale CaO-catalyzed MOLI MS spectra that clearly differentiate two strains of *E. coli* from two strains of *Salmonella* and *Shigella*. Cross validation of this data provided 100 % correct classification to the strain level. This example serves to further establish MOLI-MS-based bacterial ID, which can be used with existing MALDI spectrometers already in use in a rapidly increasing number of laboratories, and can serve to complement current and widely accepted ID methods.

8.8 Conclusions

Phage amplification has been demonstrated to overcome many of the problems associated with protein profiling. Sensitivity, specificity and total analysis time have all been improved. A procedure using PA has been developed for rapid antibiotic resistance determination, which requires 2–3 h compared to 24–72 h using conventional techniques. MALDI-TOF-MS using CaO as a matrix replacement and catalyst has also been shown to produce fatty acid profiles that allow for bacterial identification to the strain level. A comparison of MALDI-TOF MS protein and fatty acid profiling methods demonstrated that many of the organisms that cannot be differentiated by protein profiling are readily distinguishable by fatty acid-based approaches.

Acknowledgements Portions of the described research were made possible through funding provided by the Defense Threat Reduction Agency (W81XWH-07-C-0061), the National Science Foundation (CHE-1229156), and the National Institutes of Health (5U54 AI065357).

References

1. Reiner E (1965) Identification of bacterial strains by pyrolysis-gas-liquid chromatography. Nature 206:1272–1273
2. Reiner E, Beam RE, Kubica GP (1967) A rapid chemotaxonomic method for distinguishing mycobacterial strains. J Chromatogr 27:495–496
3. Reiner E, Beam RE, Kubica GB (1967) Chemotaxonomic studies of some gram negative bacteria by means of Pyrolysis–Gas–Liquid Chromatography. Nature 217:191–194
4. Reiner E, Hicks JJ, Ball MM, Martin WJ (1972) Rapid characterization of *salmonella* organisms by means of pyrolysis-gas-liquid chromatography. Anal Chem 44:1058–1061
5. Reiner E, Hicks JJ (1972) Differentiation of normal and pathological cells by pyrolysis-gas-liquid chromatography. Chromatographia 5:525–529
6. Haddadin JM, Stirland RM, Preston NW, Collard P (1973) Identification of *Vibrio cholerae* by pyrolysis gas-liquid chromatography. Appl Microbiol 25:40–43
7. Abel K, Deschmertzing H, Peterson JI (1963) Classification of microorganisms by analysis of chemical composition. I. Feasibility of utilizing gas chromatography. J Bacteriol 85:1039–1044
8. Moss CW, Lewis VJ (1967) Characterization of clostridia by gas chromatography. I. Differentiation of species by cellular fatty acids. Appl Microbiol 15:390–397
9. Microbial ID, I (1993) Operating manual, vol. 3.0, Newark, DE
10. Challinor JM (1989) A pyrolysis-derivatization-gas chromatography technique for the structural elucidation of some synthetic polymers. J Anal Appl Pyrol 16:323–333
11. Shadkami F, Helleur R (2010) Recent applications in analytical thermochemolysis. J Anal Appl Pyrol 89:2–16
12. Basile F, Beverly MB, Abbas-Hawks C, Mowry CD, Voorhees KJ, Hadfield TL (1998) Direct mass spectrometric analysis of in situ thermally hydrolyzed and methylated lipids from whole bacterial cells. Anal Chem 70:1555–1562
13. Meuzelaar HL, Kistemaker PG (1973) A technique for fast and reproducible fingerprinting of bacteria by pyrolysis mass spectrometry. Anal Chem 45:587–590
14. Meuzelaar HLC, Posthumus MA, Kistemaker PG, Kistemaker J (1973) Curie point pyrolysis in direct combination with low voltage electron impact ionization mass spectrometry. New method for the analysis of nonvolatile organic materials. J Anal Chem 45:1546–1549
15. Levy RL, Fanter DL, Wolf CJ (1972) Temperature rise time and true pyrolysis temperature in pulse mode pyrolysis gas chromatography. Anal Chem 44:38–42
16. Eshuis W, Kistemaker PG, Meuzelaar HLC (1970) Some numerical aspects of reproducibility and specificity, analytical pyrolysis. In: Jones CER, Cramers CA (eds) Analytical pyrolysis. Elsevier, New York, pp 151–166

17. Department of Defense Chemical and Biological Defense Program, annual report to Congress and Performance Plan (2001)
18. Griest WH, Wise MB, Hart KJ, Lammert SA, Thompson CV, Vass AA (2001) Biological agent detection and identification by the Block II chemical biological mass spectrometer. Field Anal Chem Technol 5:177–184
19. Lay JOJ (2001) MALDI-TOF mass spectrometry of bacteria. Mass Spectrom Rev 20:172–194
20. Sandrin TR, Goldstein JE, Schumaker S (2013) MALDI TOF MS profiling of bacteria at the strain level: a review. Mass Spectrom Rev 32:188–217
21. Holland RD, Wilkes JG, Rafii F, Sutherland JB, Persons CC, Voorhees KJ, Lay JO Jr (1996) Rapid identification of intact whole bacteria based on spectral patterns using matrix-assisted laser desorption/ionization with time-of-flight mass spectrometry. Rapid Commun Mass Spectrom 10:1227–1232
22. Bizzini A, Durussel C, Bille J, Greub G, Prod'hom G (2010) Performance of matrix-assisted laser desorption ionization-time of flight mass spectrometry for identification of bacterial strains routinely isolated in a clinical microbiology laboratory. J Clin Microbiol 48: 1549–1554
23. Eigner U, Holfelder M, Oberdorfer K, Betz-Wild U, Bertsch D, Fahr AM (2009) Performance of a matrix-assisted laser desorption ionization-time-of-flight mass spectrometry system for the identification of bacterial isolates in the clinical routine laboratory. Clin Lab 55:289–296
24. Erhard M, Hipler UC, Burmester A, Brakhage AA, Wostemeyer J (2008) Identification of dermatophyte species causing onychomycosis and tinea pedis by MALDI-TOF mass spectrometry. Exp Dermatol 17:356–361
25. Cox C R, Voorhees KJ (Year) of Conference bacterial identification by mass spectrometry NATO advance study institute: detection of chemical, biological, radiological and nuclear agents for the prevention of terrorism, Siena, Italy
26. Rees JC, Voorhees KJ (2005) Simultaneous detection of two bacterial pathogens using bacteriophage amplification coupled with matrix-assisted laser desorption/ionization time-of-flight mass spectrometry. Rapid Commun Mass Spectrom 19:2757–2761
27. Wunschel D, Fox KF, Fox A, Nagpal ML, Kim K, Stewart GC, Shahgholi M (1997) Quantitative analysis of neutral and acidic sugars in whole bacterial cell hydrolysates using high-performance anion-exchange liquid chromatography-electrospray ionization tandem mass spectrometry. J Chromatogr A 776:205–219
28. Ecker DJ, Sampath R, Massire C, Blyn LB, Hall TA, Eshoo MW, Hofstadler SA (2008) Ibis T5000: a universal biosensor approach for microbiology. Nat Rev Microbiol 6:553–558
29. Pierce CY, Barr JR, Cody RB, Massung RF, Woolfitt AR, Moura H, Thompson HA, Fernandez FM (2007) Ambient generation of fatty acid methyl ester ions from bacterial whole cells by direct analysis in real time (DART) mass spectrometry. Chem Commun (Camb) 8:807–809
30. Kaleta EJ, Clark AE, Johnson DR, Gamage DC, Wysocki VH, Cherkaoui A, Schrenzel J, Wolk DM (2011) Use of PCR coupled with electrospray ionization mass spectrometry for rapid identification of bacterial and yeast bloodstream pathogens from blood culture bottles. J Clin Microbiol 49:345–353
31. Madonna AJ, Basile F, Ferrer I, Meetani MA, Rees JC, Voorhees KJ (2000) On-probe sample pretreatment for the detection of proteins above 15 KDa from whole cell bacteria by matrix-assisted laser desorption/ionization time-of-flight mass spectrometry. Rapid Commun Mass Spectrom 14:2220–2229
32. Cox CR, Rees JC, Voorhees KJ (2012) Modeling bacteriophage amplification as a predictive tool for optimized MALDI-TOF MS-based bacterial detection. J Mass Spectrom 47:1435–1441
33. Cox CR, Kvitko BH, Voohrees KJ, Schweizer HP (2013) Conference translational technologies for rapid *Burkholderia* ID and drug resistance determination NIH NIAID Regional Centers of Excellence for Biodefense and Emerging Infectious Disease Research, Seattle, WA
34. McAlpin C, Cox CR, Matyi S, Voorhees KJ (2010) Enhanced MALDI-TOF MS analysis of bacteriophage major capsid proteins with β-mercaptoethanol pretreatment. Rapid Commun Mass Spectrom 24:11–14

35. McAlpin CR, Voorhees KJ, Corpuz AR, Richards RM (2012) Analysis of lipids: metal oxide laser ionization mass spectrometry. Anal Chem 84:7677–7683
36. Voorhees KJ, McAlpin CR, Cox CR (2012) Lipid profiling using catalytic pyrolysis/metal oxide laser ionization-mass spectrometry. J Anal Appl Pyrol 98:201–206
37. Karas M, Hillenkamp F (1988) Laser desorption ionization of proteins with molecular masses exceeding 10,000 daltons. Anal Chem 60:2299–2301
38. Tang K, Taranenko NI, Allman SL, Chang LY, Chen CH (1994) Detection of 500-nucleotide DNA by laser desorption mass spectrometry. Rapid Commun Mass Spectrom 8:727–730
39. Alomirah HF, Alli I, Konishi Y (2000) Applications of mass spectrometry to food proteins and peptides. J Chromatogr A 893:1–21
40. Hillenkamp F, Karas M, Beavis RC, Chait BT (1991) Matrix-assisted laser desorption/ionization mass spectrometry of biopolymers. Anal Chem 63:1193A–1203A
41. Kutter E, Raya R, Carlson K (2005) Molecular mechanisms of phage infection. In: Kutter E, Sulakvelidze A (eds) Bacteriophages biology and applications. CRC Press, New York, pp 165–222
42. Davis JE, Sinsheimer RL (1963) The replication of bacteriophage MS2. 1. Transfer of parental nucleic acid to progeny phage. J Mol Biol 6:203–207
43. Madonna AJ, Van Cuyk S, Voorhees KJ (2003) Detection of *Escherichia coli* using immunomagnetic separation and bacteriophage amplification coupled with matrix-assisted laser desorption/ionization time-of-flight mass spectrometry. Rapid Commun Mass Spectrom 17:257–263
44. Saichek N R, Cox CR, Schweizer HP, Voorhees KJ (2013) Conference bacteriophage amplification MALDI-TOF-MS as a means of rapid *Burkholderia pseudomallei* diagnostic identification and antibiotic resistance determination 113th General Meeting of the American Society for Microbiology, Denver, CO
45. Kvitko BH, Cox CR, DeShazer D, Johnson SL, Voorhees KJ, Schweizer HP (2012) phiX216, a P2-like bacteriophage with broad Burkholderia pseudomallei and B. mallei strain infectivity. BMC Microbiol 12:289
46. Majors, LK, Doan, LG, Rees JC, Voorhees KJ (2004) A rapid method to determine minimum inhibitory concentration of antibiotics in *Staphylococcus aureus* by bacteriophage amplification MALDI-TOF MS, 52nd American Society of Mass Spectrometry Conference on Mass Spectrometry and Allied Topics, Nashville, TN
47. Jefferson E (2011) FDA clears first test to quickly diagnose and distinguish MRSA and MSSA, FDA News Release
48. Wei J, Buriak JM, Siuzdak G (1999) Desorption-ionization mass spectrometry on porous silicon. Nature 399:243–246
49. Yanes O, Woo HK, Northen TR, Oppenheimer SR, Shriver L, Apon J, Estrada MN, Potchoiba MJ, Steenwyk R, Manchester M, Siuzdak G (2009) Nanostructure initiator mass spectrometry: tissue imaging and direct biofluid analysis. Anal Chem 81:2969–2975
50. McLean JA, Stumpo KA, Russell DH (2005) Size-selected (2–10 nm) gold nanoparticles for matrix assisted laser desorption ionization of peptides. J Am Chem Soc 127:5304–5305
51. Wen X, Dagan S, Wysocki VH (2007) Small-molecule analysis with silicon-nanoparticle-assisted laser desorption/ionization mass spectrometry. Anal Chem 79:434–444
52. Kinumi T, Saisu T, Takayama M, Niwa H (2000) Matrix-assisted laser desorption/ionization time-of-flight mass spectrometry using an inorganic particle matrix for small molecule analysis. J Mass Spectrom 35:417–422
53. Sunner J, Dratz E, Chen YC (1995) Graphite surface-assisted laser desorption/ionization time-of-flight mass spectrometry of peptides and proteins from liquid solutions. Anal Chem 67:4335–4342
54. Chen CT, Chen YC (2004) Molecularly imprinted TiO2-matrix-assisted laser desorption/ionization mass spectrometry for selectively detecting alpha-cyclodextrin. Anal Chem 76: 1453–1457

55. Voorhees KJ, Jensen KR, McAlpin CR, Rees JC, Cody R, Ubukata M, Cox CR (2013) Modified MALDI MS fatty acid profiling for bacterial identification. J Mass Spectrom 48:850–855
56. Saffert RT, Cunningham SA, Ihde SM, Jobe KE, Mandrekar J, Patel R (2011) Comparison of Bruker Biotyper matrix-assisted laser desorption ionization-time of flight mass spectrometer to BD Phoenix automated microbiology system for identification of gram-negative bacilli. J Clin Microbiol 49:887–892
57. Dubois D, Grare M, Prere MF, Segonds C, Marty N, Oswald E (2012) Performances of the Vitek MS matrix-assisted laser desorption ionization-time of flight mass spectrometry system for rapid identification of bacteria in routine clinical microbiology. J Clin Microbiol 50:2568–2576
58. Martiny D, Busson L, Wybo I, El Haj RA, Dediste A, Vandenberg O (2012) Comparison of the Microflex LT and Vitek MS systems for routine identification of bacteria by matrix-assisted laser desorption ionization-time of flight mass spectrometry. J Clin Microbiol 50:1313–1325
59. Schmitt BH, Cunningham SA, Dailey AL, Gustafson DR, Patel R (2013) Identification of anaerobic bacteria by Bruker Biotyper matrix-assisted laser desorption ionization-time of flight mass spectrometry with on-plate formic acid preparation. J Clin Microbiol 51:782–786

Chapter 9
Analysis of Bio-nanoparticles by Means of Nano ES in Combination with DMA and PDMA: Intact Viruses, Virus-Like-Particles and Vaccine Particles

Guenter Allmaier, Victor U. Weiss, Marlene Havlik, Peter Kallinger, Martina Marchetti-Deschmann, and Wladyslaw W. Szymanski

Abstract For characterization of whole viruses, vaccine particles and virus-like-particles (VLPs) besides immunological and functional parameters usually methods as electron microscopy (EM), liquid phase separation or light scattering techniques are applied. The use of nano electrospraying (nano ES) to transfer such bio-nanoparticles (NP) from the liquid phase into the gas phase and ionization is a relative new development, i.e. to bring such kind of nano-objects as intact species into the gas-phase at atmospheric pressure. Now it is possible to generate ions with multiple charges as well as a single charge fixed on such spherical bio-NPs.

Here, we want to present two techniques which open up new avenues of analysis of intact viruses, VLPs and vaccine particles. Namely, a nano electrospray (nano ES) source with a charge manipulation (reduction) device is coupled to a separation device called differential mobility analyzer (DMA) followed by a detection system (either a condensation particle counter or Faraday cup). Two types of separation devices either a single DMA (gas-phase electrophoretic mobility macromolecular analyzer, GEMMA) or two in parallel used DMAs (PDMA) will be described and discussed. The size-separated bio-NPs are transferred in most cases then into the universal detector CPC (condensation particle counter). The other very interesting part of both instruments is the possibility to collect size-separated bio-NPs for subsequent further characterization as image generation by means of atomic force microscopy (AFM) or immunological evaluation by means of DotBlot via a specific antibody. Both devices will be compared and their development will be described. Both systems are allowing the handling of bio-NPs from 2.5 to several hundred nm and are measuring number concentrations.

G. Allmaier (✉) • V.U. Weiss • M. Havlik • M. Marchetti-Deschmann
Institute of Chemical Technologies and Analytics, Vienna University of Technology, Getreidemarkt 9, A-1060 Vienna, Austria
e-mail: guenter.allmaier@tuwien.ac.at

P. Kallinger • W.W. Szymanski
Faculty of Physics, University of Vienna, Boltzmanngasse 5, A-1090 Vienna, Austria

J. Banoub (ed.), *Detection of Chemical, Biological, Radiological and Nuclear Agents for the Prevention of Terrorism*, NATO Science for Peace and Security Series A: Chemistry and Biology, DOI 10.1007/978-94-017-9238-7_9

The characterization by means of nano ES GEMMA with sample collection after DMA separation and nano ES PDMA of intact viruses – human rhinovirus (HRV serotype 2; so-called "common cold" virus), of inactivated viruses – tick-borne encephalitis virus (TBEV) vaccines and of VLPs will be shown.

Keywords Bio-nanoparticles • Viruses • VLP • Vaccines • Size determination • DMA • GEMMA • MacroIMS • PDMA • Particle collections • AFM • EM • Immuno detection

9.1 Introduction

For detailed bioanalytical characterization of intact viruses, vaccine particles and virus-like-particles (VLPs) besides immunological and functional parameters usually methods as transmission electron microscopy (TEM), scanning electron microscopy (SEM) as well as atomic force microscopy (AFM) in the different modes, analytical ultracentrifugation, size exclusion chromatography (SEC), asymmetric flow field-flow fractionation (AF4), multiangle light scattering (MALS) or dynamic light scattering (DLS) are used. The above mentioned types of nano-objects (mostly in a size below 150 nm) are interesting examples of so-called bio-nanoparticles (bio-NP).

A great deal of research over the last decades was dedicated to investigate the technological opportunities as well as health/environmental effects related to airborne bio-NPs (nanoaerosols) both of natural and man-made origin. Bio-NPs have gained particular attention as a result of their role in medicine, biology, nanosciences and nanotechnology, where they can be used in different ways as for example building blocks for nanostructures, whereas their size, surface nature and concentrations dominate their behavior. Bio-NPs such as vaccines, viruses, VLPs, or engineered NPs (drug carriers such as liposomes or dendrimers) are now available in higher qualities as well as large quantities. This, however, poses a new metrological challenge for biosciences in general and biotechnology.

Measuring the electrical particle mobility of such bio-NPs yields a possibility for size classification of airborne NPs under ambient pressure conditions. This method was initially designed to characterize aerosols in the upper sub-micrometer size range. In the recent decade, various designs of mobility analyzers were developed and adapted to extend this method into the single digit nanometer range [1–5]. This technique, combined with an elegant and gentle method for aerosol generation from solution, as for example the nano electrospray (nano ES) technique with charge reduction, was employed for the characterization of e.g. individual proteins or bio-specific complexes and protein aggregates, high-molecular weight polymers as engineered dendrimers, lipoparticles or viruses according to their electrical mobility (EM) equivalent size [6, 7]. This device was termed GEMMA and this acronym stands for gas-phase electrophoretic mobility macromolecular analyzer. However, if completely unknown components appear in the measured size spectrum, the information from a differential mobility analyzer (DMA) with a condensation particle counter (CPC) or Faraday cup detector alone is frequently insufficient for satisfactory data interpretation, giving rise to the need for more comprehensive characterization of bio-NPs. The extension of the DMA technique into the sub-10 nm size range – where it merges with chemical analytical methods such as SEC or mass spectrometry – opens up new fields of applications.

9.2 Experimental

Analysis and characterization of airborne bio-NPs based on the utilization of the EM of a charged particle in a defined electrostatic field appears to be one of the best approaches currently available for non-destructive nanoparticle research. A DMA can be used to measure the particle size distribution of aerosols [3, 8, 9]. Prior to entering the DMA, the particles need to be charge-conditioned (i.e. charge reduced). This can be achieved by means of radioactive sources such as Kr-85, Po-210 or Am-241, soft X-rays, or corona discharge [10]. Particles exiting the DMA can be detected using a condensation particle counter (CPC) or a Faraday cup. Characterization of unknown bio-NP species with the DMA technique requires its calibration with known particles [11]. The actual ambient bio-NP size distribution can be calculated, taking into account the equilibrium charge distribution [12]. The first two GEMMA prototypes, one in operation in the USA and one in Austria (Fig. 9.1), consisted of a nano ES part with a Po-210 source (Fig. 9.2) and a eight

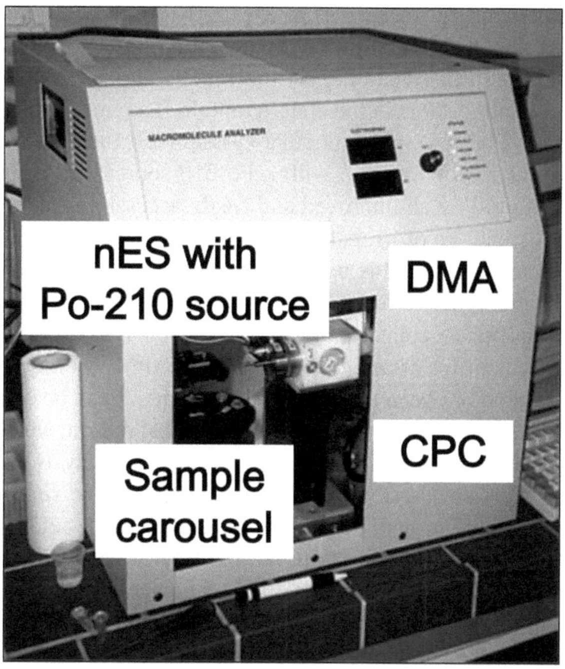

Fig. 9.1 Prototype GEMMA system with a sample introduction carousel (up to eight samples to be place inside), nano ES source with the Po-210 source, the DMA and the detection system CPC

Fig. 9.2 Prototype nano ES source (inlets for CO_2 gas and particle-free air) with an attached Po-210 source to perform the charge reduction. *Left side* is the fused silica spray tip introduced and on the *right side* (black tubing) the single charged bio-NPs are transported into the DMA

sample introduction carousel followed by a nano DMA allowing the separation and detection by means of an appropriate CPC (TSI Inc, Shoreview, MN, USA) of 2.5 nm sized bio-NPs [6]. In the nano ES part special attention has been paid to the design of fused silica spray tip (Fig. 9.3 top) to form a Taylor cone (Fig. 9.3 bottom) operating stable in the cone-jet mode. The first commercial available GEMMA system (Fig. 9.4) was a modular system (TSI Inc, Shoreview, MN, USA) consisting of a nano ES part again with a Po-210 source and a single sample introduction part followed by an improved and easily accessible nano DMA (Fig. 9.5). Because bio-NPs may be of irregular shape (e.g. tomato mosaic virus is a forming long rods or membrane proteins exhibit cubic shapes), the actual particle "size" (in terms of length, width and height) and EM diameter may be quite different. Consequently, it is useful to collect particles of a given EM diameter for further examination by means of for example TEM, SEM or AFM. Moreover, once a monomobile bio-NP fraction has been extracted from a DMA, it can be collected as well as enriched and used for further characterization. For that reason, we developed the parallel differential mobility analyzer (PDMA) to perform analysis and collection simultaneously.

This device (Fig. 9.6) is designed to enable simultaneous recording of the size and distribution of bio-NPs in question and the selection as well as the collection of a very specific size fraction for further investigations [5]. Parallel to the scanning DMA 2 (Analyzer, i.e. a GEMMA device), which delivers the complete size distribution of the investigated bio-NP sample, an identical separation unit DMA 1 (Classifier) operates at one given voltage setting (the so-called extraction voltage)

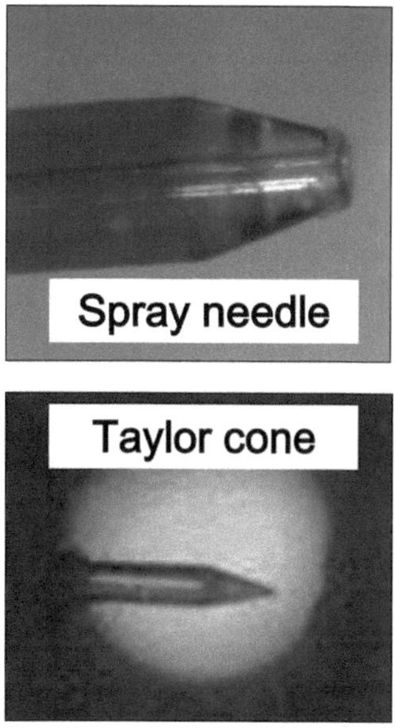

Fig. 9.3 The special shaped fused silica spray tip (*top*) to allow electrospraying in the cone jet mode. Formed Taylor cone of bio-NP solution by means of the shown spray tip in the cone jet mode monitored through the source window prior to charge reduction (*bottom*)

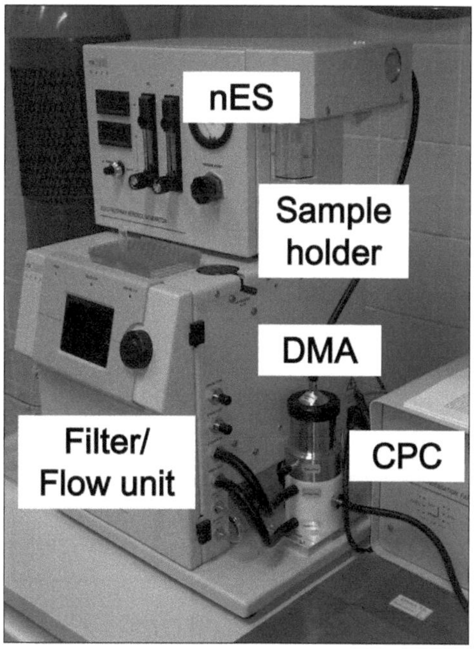

Fig. 9.4 First modular commercial GEMMA system with a nano ES source incorporating the Po-210 source allowing the infusion of one sample. On the left side the filter system/flow control unit is connected to the DMA which by itself is connected to nano ES source. The DMA is hooked up to the detection system (a standard CPC for low-sized NPs)

Fig. 9.5 The NP separation device DMA with the sample introduction from the top. The exit for the excess flow and the inlet sheath flow are shown. The lower right exit of the DMA is connected directly to the CPC (using n-butanol or water)

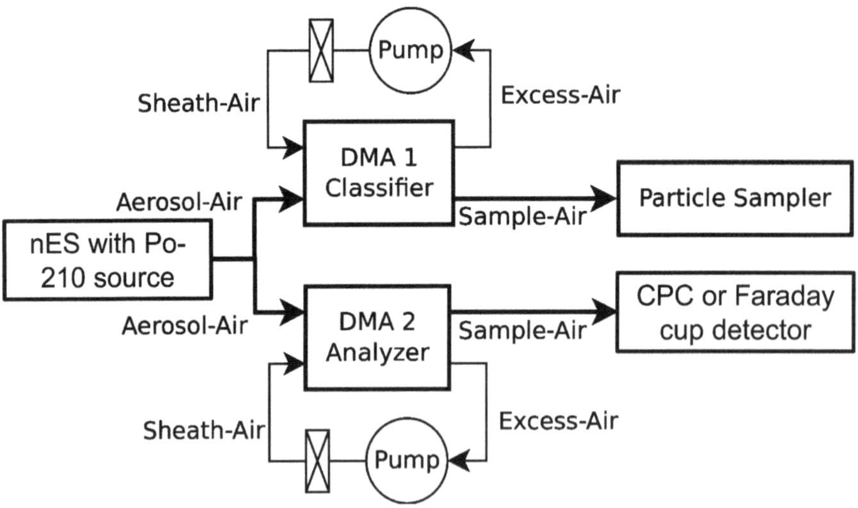

Fig. 9.6 Schematic concept of the two home-built PDMA systems allowing the simultaneous size-monitoring of the separated bio-NPs (DMA 2 Analyzer) and parallel separated (DMA 1 Classifier) bio-NP collection by means of an electrostatic particle collecting device (Particle Sampler)

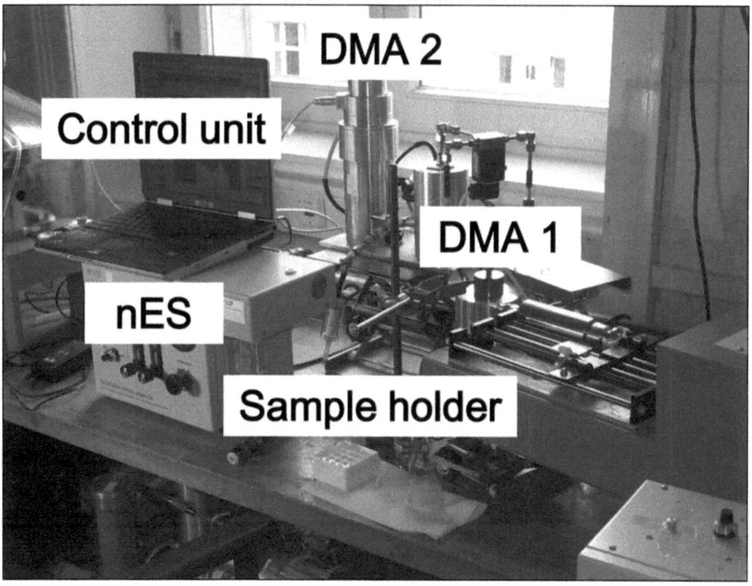

Fig. 9.7 First home-built PDMA system allowing the simultaneous size-monitoring (DMA 2 Analyzer) and parallel separated (DMA 1 Classifier) collection by means of an electrostatic particle collecting device (sampling either on a solid surface or in a liquid). As monitoring detector a Faraday cup was incorporated into the DMA 1

and performs the sampling (i.e. collection) of the selected size class of bio-NPs (Fig. 9.7). For the experimental data presented here, the investigated bio-NP sample was aerosolized and charge-reduced by means of a nano ES with a Po-210 source (TSI Inc, Shoreview, MN, USA). As particle detector, a Faraday cup (Fig. 9.7) or water- (Fig. 9.8) based CPC (TSI Inc, Shoreview, MN, USA) was used. The bio-NPs extracted from DMA 1 can also be injected into further characterization instruments, such as for example a high-molecular mass spectrometer (MS) or a gas-phase fluorescence detector, without affecting the operation of DMA 2. Figure 9.8 shows the details of the second generation PDMA utilized custom-built DMAs (Tapcon & Analysensysteme, Salzburg, Austria) based on a design published elsewhere [2]. The radii of the outer and inner electrodes are 25 mm and 18 mm respectively, and the separation length is 65 mm. The differently designed DMAs are operated at an air flow of up to 50 L/min (Lpm) – much higher than the first generation PDMA. The resulting sizing accuracy and resolution of the measuring system are of much higher quality than the prototype GEMMA, the commercial GEMMA and first generation home-built PDMA system. Figure 9.9 illustrates clearly the enhanced performance of the applied DMAs achieved with human rhinovirus serotype 2 (HRV2) samples. Figure 9.9, top spectrum shows the first successful analysis of HRV2 by the prototype GEMMA system (6). The GEMMA spectrum in Fig. 9.9, middle part shows already an improved resolution whereas the spectrum at the bottom of

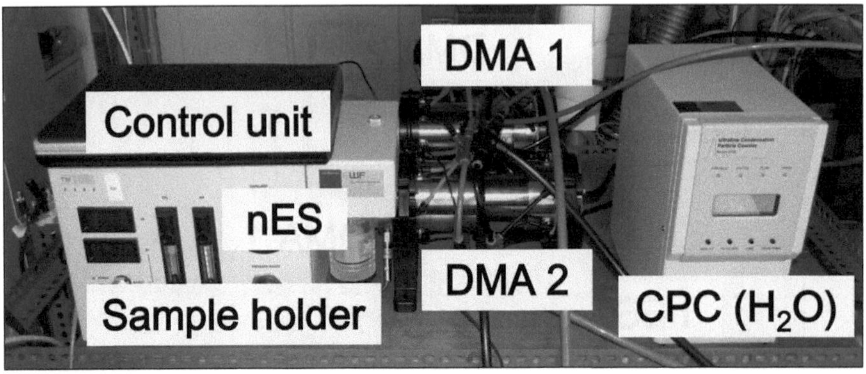

Fig. 9.8 Second custom-built PDMA system with high – up to 50 L/min – flow rate allowing the same experiments as the first home-built PDMA but now having attached a so-called water-based CPC. The resolving power of the DMA (1 and 2 as well as to the commercial DMA) has been significantly improved

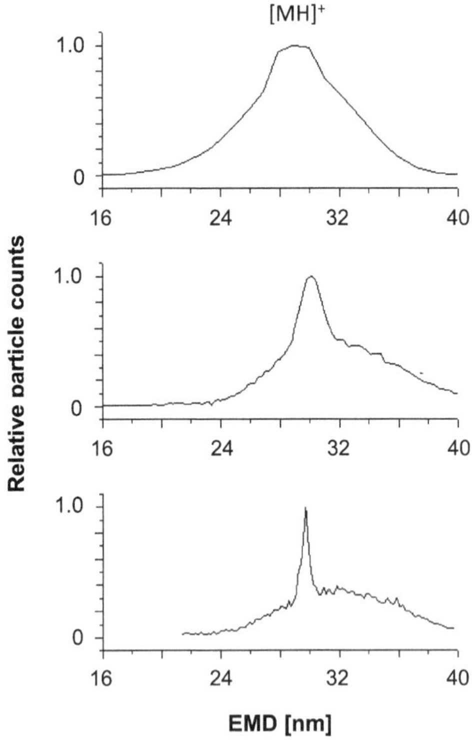

Fig. 9.9 Normalized size (GEMMA) spectra of intact human rhinovirus preparation (in 50 mM ammonium acetate, pH 6.8) obtained by means of the prototype GEMMA system (*top*), the first commercial GEMMA system (*middle*) and the second generation PDMA system with high flow rate (*bottom*). In all cases a CPC was used for detection. The determined bio-NP size was 29.75 nm taken from the bottom spectrum

Fig. 9.9 demonstrates the significantly increased resolving power of the DMA (Vienna type DMA design [2]) in the second generation PDMA. For this latter measurement, the two DMAs were not operated in parallel, but DMA 2 (Analyzer) was run just in the analyzing mode (determination of complete size distribution of the HRV2 sample).

In addition from the commercial GEMMA system the n-butanol CPC was removed and replaced by a commercially available collecting device (Fig. 9.10, left top). In order to collect a specified monomobile fraction leaving the nano DMA, an electrostatic bio-NP sample collector allowing the deposition of the size-selected bio-NPs on a solid surface was used (TSI Inc, Shoreview, MN, USA). The deposition of nanoparticles on various surfaces in this device is driven by the electrostatic field that is applied, and the particles can be sampled directly onto e.g. a TEM grid, an AFM mica platelet (Fig. 9.10, center) or a piece of nitrocellulose membrane (5×5 mm) or other substrates. An AFM image of a collected bio-NP fraction is depicted in Fig. 9.10, right bottom. SEM, TEM, AFM, SEC, AF4, MALS, MS or DLS data of size classified nanoparticles with known EMs enable verification of the calibration of a nano DMA or provide directions for its size classification performance for the case of non-spherical bio-NPs. But the most important point is the

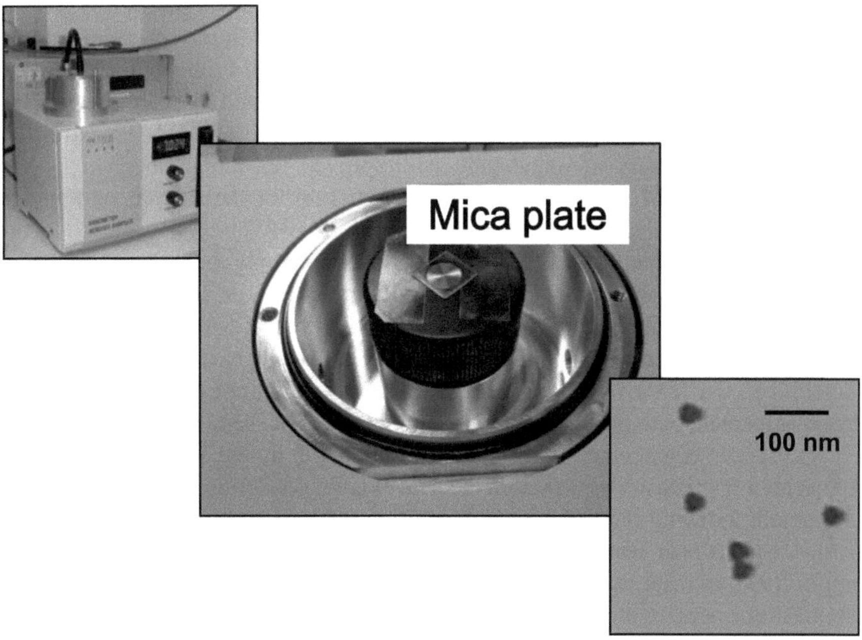

Fig. 9.10 Modular commercial electrostatic sample collector (*left top*) to the GEMMA/PDMA system exhibiting the inner collector surface with a fixed mica platelet (*center*) for subsequent AFM (tapping mode) investigation. Part of an AFM image of size-separated and collected TBEV vaccine particles (*right bottom*). Instead of the mica platelet an electron microscopy grid and a small piece of a nitrocellulose membrane (for DotBlot analysis) was fixed alternatively

verification that bio-NPs are after size classification intact and to a certain extent biological active (via the immunological DotBlot analysis – certain epitopes of the collected bio-NPs are recognized by specific antibodies). Furthermore aggregation analyses and structural characterizations by imaging MS are feasible. In case of the second generation PDMA this commercial collecting device can be permanently attached to the DMA 1 (Classifier) of the high performance system (Figs. 9.6 and 9.8.).

9.3 Results and Discussion

In the following section, we will present measurements performed with the above four described systems (two plain GEMMA systems and two PDMA systems) based on the same concept of the intact virus HRV2, a VLP decorated with numerous peptides, hepatitis B virus capsids and a tick-borne encephalitis virus (TBEV) vaccine.

The GEMMA/PDMA (latter device run in the analytical mode only) spectra of the intact virus HRV2 [12] (including not only the virus capsid, but also the RNA part and other components as spermidine in the inner part of HRV2). Figure 9.9 now shows (from the top to the bottom) clearly how the separation performance of the applied DMAs is increasing from the prototype GEMMA [6] to the second generation PDMA illustrated with HRV2 samples [10]. Figure 9.9, top spectrum was the first successful analysis of the intact virus HRV2 related to a human disease by the prototype GEMMA system [6]. The first virus at all analyzed by such a device was the analysis of a purified MS2 bacteriophage [13]. The GEMMA spectrum in Fig. 9.9, middle part shows already an improved resolution obtained by the commercial GEMMA system. Finally, the spectrum at the bottom of Fig. 9.9 exhibits the dramatically increased resolution for the HRV2 bio-NP achieved by the so-called Vienna-type DMA [2]. In conclusion it is not only the different design of the nano DMA that helps to improve the separation power, but the significantly increased (from a maximum of 20 Lpm with other three devices) gas flow rate of 50 Lpm in this second generation device makes this improvement possible. The determined size in the GEMMA spectrum (bottom) at the apex was 29.75 nm and in good agreement with data derived for example from cryo EM [14].

VLPs, i.e. engineered bio-NPs are widely used in the pharmaceutical and biotechnology industry as well as in medical diagnostics during the last decade with a dramatic growing field of applications. Their unique partly size-dependent and surface-specialized features make them ideally suited for numerous biomedical applications as drug carriers or vaccines [15, 16]. An increasing number of these applications need rigorous quality control and prior to that physico-chemical characterization of these engineered bio-NPs.

For most applications, the sizes as well as surface characteristics of bio-NPs need to be well characterized. In our example, we measured the apex of the peak derived from a VLP with attached bioactive peptides and the size distribution of the engineered bio-NP. Usually such information is obtained by SEC combined with MALS or DLS alone showing often a significant bias. The GEMMA spectrum shown in

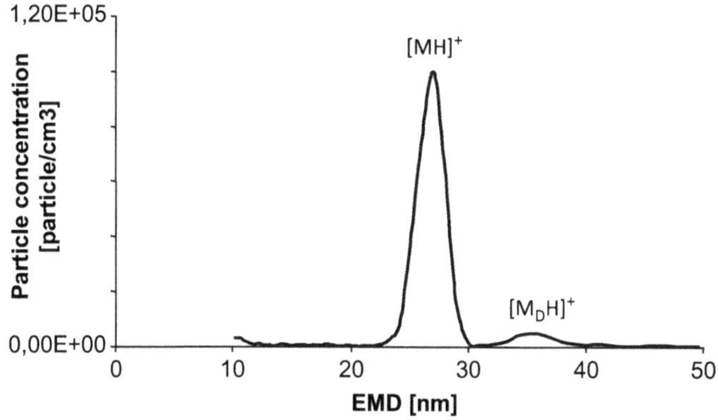

Fig. 9.11 GEMMA (size) spectrum of a VLP (covered by peptides) out of a 20 mM ammonium acetate (pH 7) solution obtained by the commercial GEMMA system. 20 ng of analyte was consumed to acquire the shown spectrum. The determined bio-NP size was 26.7 nm

Fig. 9.11 is a median of 10 single spectra where the maximum of the size distribution (protonated VLP $[MH]^+$) is at an EM diameter of 26.7 nm. The obtained standard deviation for the apex was ±0.4 % and the consumed material for the presented GEMMA spectrum was 20 ng. Furthermore a dimer (consisting of two VLP units, $[M_DH]^+$) was detected, too. This dimer turned out to be a specific dimer and not just a concentration-dependent aggregate. This could be verified by increasing and decreasing the VLP concentration. This confirmed the presence of the dimer, which of importance due to the fact that aggregation of VLPs might have the potential to induce an enhanced immunogenic response. It should be mentioned also that the number of analyzed bio-NPs was in the range 10^5 VLPs generating good statistics.

A further example is the engineered hepatitis B virus capsid (belonging also to the group of VLPs) which comes all the time as a doublet (T3 and T4, exhibiting different size and different number of protein monomer units), which was originally determined by native MS (electrospray ionization coupled to a special QqRTOF (quadrupole-RF-quadrupole-reflectron) MS [17]. By means of the commercial GEMMA system (Fig. 9.12) the size of T3 was determined to be 24.35 nm and of T4 26.90 nm [18]. The determined ratio between both nano-objects in the GEMMA spectrum corroborated the data obtained by native MS and AFM images.

Finally an example from the area of vaccine characterization will be presented – tick-borne encephalitis (transmitted by TBEV) is a disease found in large parts of central and eastern Europe and the best handling of this disease is vaccination. Figure 9.13 shows the GEMMA spectrum of SEC-purified TBEV vaccine particles. Prior SEC purification was necessary to remove a variety of additives and contaminants. The broad peak from 5 to 22 nm is derived from these still present compounds and their unspecific aggregates. Dilution of the sample and afterward GEMMA analysis would show the drop in intensity of this broad hump as well as a shift to lower size. The size of the major vaccine-related peak was determined to be 47.6 ± 0.4 nm at the apex [19]. As next step the CPC was removed from the commercial GEMMA system and replaced by the collection device.

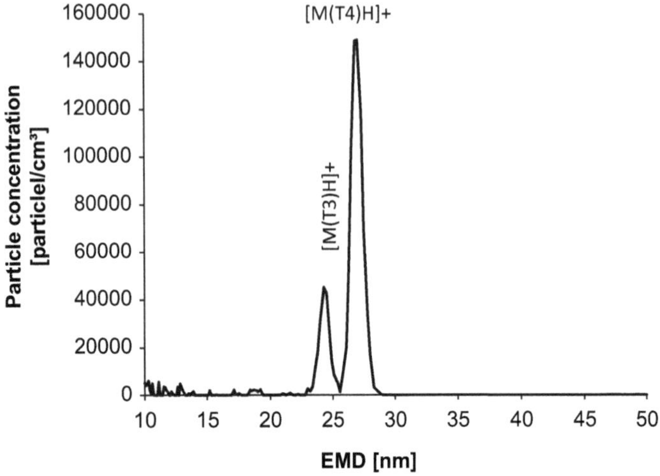

Fig. 9.12 GEMMA (size) spectrum of hepatitis B virus capsid (VLP) out of a 50 mM ammonium acetate (pH 6.8) solution obtained by the commercial GEMMA system. The determined bio-NP sizes of the two variants were 24.35 (T3) and 26.90 (T4) nm

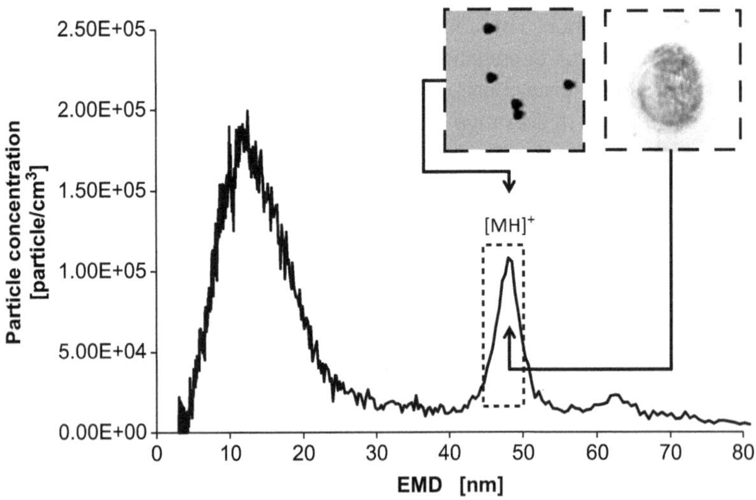

Fig. 9.13 GEMMA (size) spectrum of SEC-purified TBEV vaccine particles showing a bio-NP size of 47.6 nm (at peak apex). The broad peak below 22 nm derived from salt and additive contaminations. The analysis was performed out of a 50 mM ammonium acetate (pH 6.8) solution by means of the commercial GEMMA system. Afterwards the vaccine bio-NPs in the size window 45–50 nm (*box of dotted line*) were collected on a mica platelet (for AFM imaging) and on nitrocellulose membrane (for immunological DotBlot analysis). The *left inset* shows the AFM image of part of the collected intact vaccine particles and the *right inset* shows the DotBlot image

For TBEV vaccine particle collection, the electrostatic particle collector (Fig. 9.10, left top) was used first with a mica platelet (Fig. 9.10, center) for subsequent AFM analysis in the tapping mode and second with a nitrocellulose membrane for subsequent DotBlot experiments. The nano DMA was operated at a size range of 45–50 nm (dotted box in Fig. 9.13) for 2 h. Vaccine NPs were deposited directly on a freshly prepared mica platelet or a washed sheet of nitrocellulose. Subsequently both surfaces were analyzed via AFM and DotBlot with a TBEV-specific antibody (see insets of Fig. 9.13, left inset AFM image and right inset DotBlot image). In the image obtained by means of AFM the intact TBEV vaccine particles could be clearly determined as well as the slightly distorted (the bio-NP is adsorbed to the mica surface at many sites transforming the globular shape into a slight ellipsoid shape) size and shape could be obtained (see insets of Fig. 9.13, left inset AFM image). The DotBlot technique is used to identify specific proteins by antibodies without using any prior planar gel electrophoresis. However, due to this fact it offers no information on the size of the target protein except the sizing of the GEMMA system is used which is much more accurate than planar gel electrophoresis. In our case the DotBlot is based on a mouse-derived anti-virus (TBEV) antibody which was detected by a second anti-mouse IgG alkaline phosphatase conjugate with a dye conversion resulting in a blue-violet color on the membrane (see insets of Fig. 9.13, right inset DotBlot image). This immunological analysis confirmed the presence of a TBEV protein in the collected bio-NPs with high specificity. This can open up the possibility to detect attached contaminating proteins on the vaccine particle which is of high general interest in vaccine quality and safety.

9.4 Conclusions

The described two methods derived from aerosol science have proven to be extremely useful and versatile tools for a number of bioanalytical applications beyond the original field of research. One could say that electrostatic size analysis and classification of bio-NPs was initiated in aerosol science, but matures in bioanalytics, pharmaceutical analysis and a number of biotechnological applications. The analysis of bio-NPs as intact viruses, VLPs and vaccines by electrical means as discussed here seems to offer the possibility to complement established bioanalytical techniques as SEC, AF4, MALS or EM [20–25]. The shown technique of nES GEMMA and PDMA fills the gap in bioanalytical techniques for sizes between 5 and 150 nm where a lot of other techniques are not optimal in many aspects. Due to the fact that measurements are performed at ambient pressure, it also opens up a wide spectrum of new applications much closer to reality than techniques requiring a vacuum. The gentle generation of single charged functional bio-NPs combined with electrophoretic mobility analysis is a sizing tool with huge potential for lifesciences and biotechnology, especially with instrumental designs delivering increased sizing accuracy as well as sizing resolution. Furthermore GEMMA and more elegantly PDMA provide unique opportunities for bio-NP collection/

enrichment (maybe up to preparative dimensions), manipulation and other sophisticated characterization methods (from AFM or TEM images to imaging MS and specific immunological tests) as demonstrated for vaccine NPs.

Acknowledgments The work regarding the PDMA was supported by grant of the Austrian Science Foundation (TRP29-N20 to W.W.S and G.A.). Furthermore we appreciate for making the following samples available – the hepatitis B virus capsid (A. Heck), TBEV vaccine (C. Tauer) and HRV2 (D. Blaas). The authors express their thanks for the AFM picture of the sample to G. Friedbacher and E. Eitenberger.

References

1. De la Mora JF, de Juan L, Eichler T, Rosell J (1998) Differential mobility analysis of molecular ions and nanometer particles. Trends Anal Chem 17(6):328–339. doi:10.1016/S0165-9936(98)00039-9
2. Steiner G, Attoui M, Wimmer D, Reischl GP (2010) A medium flow, high-resolution Vienna DMA running in recirculating mode. Aerosol Sci Tech 44(4):308–315
3. Chen D-R, Pui DYH, Hummes D, Fissan H, Quant FR, Sem GJ (1998) Design and evaluation of a nanometer aerosol differential mobility analyzer (Nano-DMA). J Aerosol Sci 29(5–6):497–509. doi:10.1016/S0021-8502(97)10018-0
4. Chen D-R, Pui DYH (1997) Numerical modeling of the performance of differential mobility analyzers for nanometer aerosol measurements. J Aerosol Sci 28(6):985–1004. doi:10.1016/S0021-8502(97)00004-9
5. Allmaier G, Laschober C, Szymanski WW (2008) Nano ES GEMMA and PDMA, new tools for the analysis of nanobioparticles – protein complexes, lipoparticles, and viruses. J Am Soc Mass Spectrom 19(8):1062–1068. doi:10.1016/j.jasms.2008.05.017
6. Bacher G, Szymanski WW, Kaufman SL, Zöllner P, Blaas D, Allmaier G (2001) Charge-reduced nano electrospray ionization combined with differential mobility analysis of peptides, proteins, glycoproteins, noncovalent protein complexes and viruses. J Mass Spectrom 36(9):1038–1052. doi:10.1002/jms.208
7. Loo JA, Berhane B, Kaddis CS, Wooding KM, Xie Y, Kaufman SL, Chernushevich IV (2005) Electrospray ionization mass spectrometry and ion mobility analysis of the 20S proteasome complex. J Am Soc Mass Spectrom 16(7):998–1008. doi:10.1016/j.jasms.2005.02.017
8. Knutson EO, Whitby KT (1975) Aerosol classification by electric mobility: apparatus, theory, and applications. J Aerosol Sci 6(6):443–451. doi:10.1016/0021-8502(75)90060-9
9. Collins DR, Nenes A, Flagan RC, Seinfeld JH (2000) The scanning flow DMA. J Aerosol Sci 31(10):1129–1144. doi:10.1016/S0021-8502(99)00576-5
10. Kallinger P, Steiner G, Szymanski WW (2012) Characterization of four different bipolar charging devices for nanoparticle charge conditioning. J Nanopart Res 14(6). doi:10.1007/s11051-012-0944-z
11. Laschober C, Kaddis CS, Reischl GP, Loo JA, Allmaier G, Szymanski WW (2007) Comparison of various nano-differential mobility analysers (nDMAs) applying globular proteins. J Exp Nanosci 2(4):291–301. doi:10.1080/17458080701660550
12. Fuchs NA (1963) On the stationary charge distribution on aerosol particles in a bipolar ionic atmosphere. Geofisica Pura e Applicata 56:185–193. doi:10.1007/BF01993343
13. Wick CH, McCubbin PE (1999) Characterization of purified MS2 bacteriophage by the physical counting methodology used in the integrated virus detection system (IVDS). Toxicological Methods 9(4):245–252. doi:10.1080/105172399242591
14. Verdaguer N, Blaas D, Fita I (2000) Structure of human rhinovirus serotype 2 (HRV2). J Mol Biol 300(5):1179–1194. doi:10.1006/jmbi.2000.3943

15. Zhao L, Seth A, Wibowo N, Zhao C-Z, Mitter N, Yu C, Middelberg APJ (2014) Nanoparticle vaccines. Vaccine 32:327–337. doi:10.1016/j.vaccine.2013.11.069
16. Ashley CE, Carnes EC, Phillips GK, Durfee PN, Buley MD, Lino CA, Padilla DP, Phillips B, Carter MB, Willman CL, Brinker CJ, Caldeira JC, Chackerian B, Wharton W, Peabody DS (2011) Cell-specific delivery of diverse cargos by bacteriophage MS2 virus-like particles. ACS Nano 5(7):5729–5745. doi:10.1021/nn201397z
17. Heck AJR (2008) Native mass spectrometry: a bridge between interactomics and structural biology. Nat Methods 5:927–933. doi:10.1038/nmeth.1265
18. Bereszczak JZ, Havlik M, Weiss VU, Marchetti-Deschmann M, van Duijn E, Watts NR, Wingfield PT, Allmaier G, Steven AC, Heck AJR (2014) Sizing up large protein complexes by electrospray ionisation-based electrophoretic mobility and native mass spectrometry: morphology selective binding of Fabs to hepatitis B virus capsids. Anal Bioanal Chem 406(5):1437–1446. doi:10.1007/s00216-013-7548-z
19. Havlik M, Marchetti-Deschmann M, Friedbacher G, Messner P, Winkler W, Perez-Burgos L, Tauer C, Allmaier G (2014) Development of a bio-analytical strategy for characterization of vaccine particles combining SEC and nano ES GEMMA. Analyst 139:1412–1419. doi:10.1039/C3AN1962D
20. Pease LF (2012) Physical analysis of virus particles using electrospray differential mobility analysis. Trends Biotechnol 30(4):216–224. doi:10.1016/j.tibtech.2011.11.004
21. Seyfried BK, Siekmann J, Turecek PL, Schwarz HP, Scheiflinger F, Zappe H, Bossard ML et al (2011) PEGylated recombinant von Willebrand factor analyzed by means of MALDI-TOF-MS, CGE-on-a-chip and nES-GEMMA. Int J Mass Spectrom 305(2–3):157–163. doi:10.1016/j.ijms.2010.10.028
22. Weiss VU, Subirats X, Pickl-Herk A, Bilek G, Winkler W, Kumar M, Allmaier G et al (2012) Characterization of rhinovirus subviral A particles via capillary electrophoresis, electron microscopy and gas-phase electrophoretic mobility molecular analysis: Part I. Electrophoresis 33(12):1833–1841. doi:10.1002/elps.201100647
23. Fuchs R, Blaas D (2010) Uncoating of human rhinoviruses. Rev Med Virol 20(5):281–297. doi:10.1002/rmv.654
24. Kallinger P, Weiss VU, Lehner A, Allmaier G, Szymanski WW (2013) Analysis and handling of bio-nanoparticles and environmental nanoparticles using electrostatic aerosol mobility. Particuology 11(1):14–19. doi:10.1016/j.partic.2012.09.004
25. Laschober C, Wruss J, Blaas D, Szymanski WW, Allmaier G (2008) Gas-phase electrophoretic molecular mobility analysis of size and stoichiometry of complexes of a common cold virus with antibody and soluble receptor molecules. Anal Chem 80(6):2261–2264. doi:10.1021/ac702463z

Chapter 10
Mass Spectrometric Target Analysis and Proteomics in Environmental Toxicology

Ksenia J. Groh and Marc J.-F. Suter

Abstract Mass spectrometric techniques are widely used in environmental toxicology. One major application is the quantitative determination of chemical pollutants in environmental compartments. This is increasingly linked with biological effects assessment in an approach called effect-directed analysis, which, as the term says, allows focusing on samples that cause an effect in *in vitro* or *in vivo* test systems. Identification of the chemical(s) causing an effect is done by submitting the active sample to a classical target analysis using established methods. If the causative agent is not part of the list of target analytes, scan-dependent MS/MS analyses have to be performed and active samples compared to controls. This then allows to narrow-down the elemental composition of compounds primarily found in active samples, find functional groups and substructures, and potentially identify the unknowns.

Equally important for a refined risk assessment is the determination of actual internal concentrations in organisms, which reduces uncertainties in predicting toxicity thresholds across chemicals and species.

An entirely new level in environmental toxicology has been reached with the application of novel techniques such as proteomics and metabolomics. They allow investigating the molecular response of a model organism to environmental challenge. Ideally this leads to the identification of robust biomarkers of exposure and the identification of conserved stress response pathways which can be used to

K.J. Groh
Department of Environmental Toxicology, Eawag, Swiss Federal Institute of Aquatic Science and Technology, 8600 Dübendorf, Switzerland

Department of Chemistry and Applied Biosciences, ETH Zürich, Swiss Federal Institute of Technology, 8093 Zürich, Switzerland

M.J.-F. Suter (✉)
Department of Environmental Toxicology, Eawag, Swiss Federal Institute of Aquatic Science and Technology, 8600 Dübendorf, Switzerland

Department of Environmental Systems Science, ETH Zürich, Swiss Federal Institute of Technology, 8092 Zürich, Switzerland
e-mail: suter@eawag.ch

J. Banoub (ed.), *Detection of Chemical, Biological, Radiological and Nuclear Agents for the Prevention of Terrorism*, NATO Science for Peace and Security Series A: Chemistry and Biology, DOI 10.1007/978-94-017-9238-7_10

extrapolate to other species and predict adverse effects of novel chemical stressors or even their mixtures.

This chapter gives an introduction into effect-directed analysis and environmental proteomics.

Keywords Aquatic organisms • Ecotoxicology • Effect-directed analysis • Chemical target analysis • Proteomics

10.1 Introduction

In the early days of environmental research, analytical chemists developed techniques for the determination of known toxic chemicals in various matrices. With the advent of GC-MS in the 1970s, very low concentration levels became detectable [1], allowing to analyze the environmental fate and behavior of chemical contaminants. GC-MS also enabled the simultaneous determination of hundreds of chemicals in a high-throughput manner. This was the basis for monitoring the pollution state of environmental compartments such as soils, surface waters and human health-relevant ground waters used for drinking water production. One classical example of chronic exposure linked to a high risk of cancer is perchloroethylene (PER) contaminated drinking water [2]. In the study presented by Giger and Molnar-Kubica, PER was found at concentrations of up to 450 μg/L in tap water and traced to contaminated soils behind a dry cleaning facility that leached into the groundwater. Even higher concentrations (8 mg/L) were found in Massachusetts drinking water, originating from vinyl-lined water pipes [3]. Both cases required immediate action because of their potential high health impact on humans.

With increasing availability of monitoring data and the observation of adverse effects of extensive pesticide use, as documented 1962 by Rachel Carson in "Silent Spring" [4], growing public concern forced governments to address environmental pollution. Major incidents further increased public awareness of the detrimental effects of heavy pollution and major accidents. Famous examples are the accidental release of (i) one kilogram of TCDD (2,3,7,8-tetrachlorodibenzodioxin), the most toxic dioxin congener, in Seveso, Italy in 1976 [5], (ii) of methyl-isocyanate in Bhopal, India in 1984, considered the world's worst industrial disaster with thousands dead [6], and (iii) the fire in a Sandoz storehouse in Schweizerhalle, Switzerland 1986 that heavily polluted the river Rhein, killing most of the fish [7]. Another thing these examples demonstrate is that exposure is very dynamic, with accidents being the worst case of a pulse exposure. But also other entryways into the environment show a very dynamic behavior. WWTP effluent composition for instance shows daily and weekly cycles, while run-offs from urban or agricultural surfaces will peak with rain events [8]. Since there is a constant replenishment of these inputs the net result is a chronic exposure, even though discharge is counteracted by degradation, sorption and other elimination processes.

Presently it is recognized that organisms of an ecosystem typically are not exposed to high concentrations of one toxicant alone, as sometimes is the case with accidents. Instead, a chemical cocktail of compounds is present at low concentrations, originating from point sources like wastewater treatment plant effluent points of discharge, leachates from landfills, remobilization from polluted sediments, or diffuse sources like run-off from urban and agricultural lands, and dry deposition.

Hence, the challenge in environmental toxicology is not so much investigating accidents and acute toxic events, but rather understanding the dynamic nature and the effects of chemical mixtures on organisms. Sensitive and precise mass spectrometric techniques are available today for analyzing organic pollutants at ultra-trace levels in water, soil and tissue samples. Similarly important is enrichment, tailored for the target analyte's physico-chemical properties. Apolar compounds like the dioxins mentioned above, polycyclic aromatic hydrocarbons, polychlorinated biphenyls and other hydrophobic pollutants are routinely measured using GC-MS, because they are relatively volatile and thermally stable up to 300 °C, the hot end of the temperature gradient typically used in a GC run. Additionally, dioxin analysis requires high resolution mass spectrometers in order to unambiguously quantify all congeners, a total of 210 including the polychlorodibenzodifuranes, in an environmental sample [9]. For analytes that are polar or even ionic, such as pesticides, personal care products and pharmaceuticals, the method of choice since the 1990s is LC-MS with electrospray (ESI) or atmospheric pressure chemical ionization [10]. A multitude of publications has appeared since then. However, considering the fact that more than 72 million chemicals are commercially available, of which more than 300,000 are regulated substances [11], clearly new monitoring concepts are needed, since these compounds can potentially enter the environment and negatively affect wildlife. Furthermore, microbial transformation products of these chemicals would also have to be monitored, since very often they are biologically active as well [12]. Clearly, this makes it very difficult to find a chemical cause for an observed adverse effect, such as declining fish populations in Swiss rivers [13], or unexplained gonadal abnormalities in whitefish from an alpine oligotrophic lake in Switzerland [14]. Faced with this problem, environmental scientists have two options: they can either try to do a global high-throughput chemical analysis and hope to find biologically active compounds [15], or they can evaluate the potential to cause adverse effects in aquatic organisms using *in vitro* or *in vivo* assays [16, 17]. With the chemical analysis, chemicals present in an environmental sample are determined and, based on previously published information, statements can be made regarding their toxicity or adverse effects. With the bioassays, toxic mechanisms induced by the sample are identified, but the causative agent remains unknown.

One of the main caveats of chemical analysis is that depending on the physico-chemical properties of the target analytes, different extraction and enrichment techniques have to be used, or in other words, there is no method that simultaneously and sensitively enriches the universe of chemicals. However, there are efforts in this direction, using multimode solid-phase cartridges that combine different adsorption mechanisms, capturing ionic, polar and lipophilic compounds in one step [15]. An approach for simultaneously enriching chemicals known to interfere with the gluco- and mineralocorticoid signaling pathway has recently been developed [18]. Another problem is that the number of analytes monitored with a multi-residue method also directly affects sensitivity, in that high numbers reduce the time spent for detecting each individual component. This sensitivity loss can be partly addressed by the use of retention time windows or scan-dependent experiments. The latter performs single reaction monitoring (SRM) or MS/MS only when a target analyte appears in a survey scan, which reduces the time lost by monitoring all transitions in a classical SRM experiment. Furthermore, instrumental development has greatly improved the possibilities offered by today's mass spectrometers. High resolution and high accuracy achievable with orbitraps and time-of-flight instruments for instance results in very high selectivity and in combination with scan-dependent and scan-independent approaches allows detecting suspects or non-target compounds. Moschet et al. demonstrated that known pesticides could be identified in environmental samples based on their accurate mass and retention time only (suspect screening), using LC separation and an orbitrap for detection [19].

10.2 Effect-directed Analysis

As mentioned above, one major drawback of the chemical approach is that all samples have to be analyzed, providing data on chemical composition but no information on the effect of the individual components or the whole mixture on aquatic organisms, unless ecotoxicological data can be found in the literature. Alternatively, testing the samples for a biological effect first, using *in vitro* or *in vivo* assays, will allow reducing the number of chemical analyses significantly. The drawback here is that biological assays test for a specific molecular effect, like receptor binding, unless more integrative tests, based for instance on growth, photosynthetic yield, or reproductive fitness, are used. The yeast estrogen screen (YES) developed in John Sumpter's lab measures estrogen receptor binding only [20], thereby disregarding any non-receptor mediated effects, such as interference with steroid synthesis and degradation, which can cause an imbalance in estrogen concentrations and corresponding detrimental effects. Consequently, ecotoxicologists have to select available assays that answer their research question best. The US Environmental Protection Agency (EPA) on the other hand has launched a program called Toxicity Forecaster (ToxCast) which aims at high-throughput testing of chemicals using over 700 assays that cover a range of high-level cell responses and roughly 300 signaling pathways. Ultimately this should lead to decision support tools that help prioritize chemicals for further investigation, while at the same time reducing animal testing [21].

When testing environmental samples the combination with chemical analysis is needed in order to identify the active component(s), unlike in testing of single chemicals as done by EPA. This combination is today called effect-directed analysis (EDA) but has been used in various forms since the 1980s. This procedure, as proposed by Werner Brack [22], starts with the testing of an environmental extract in a biological assay, for instance the YES, which measures estrogen receptor binding (see Fig. 10.1).

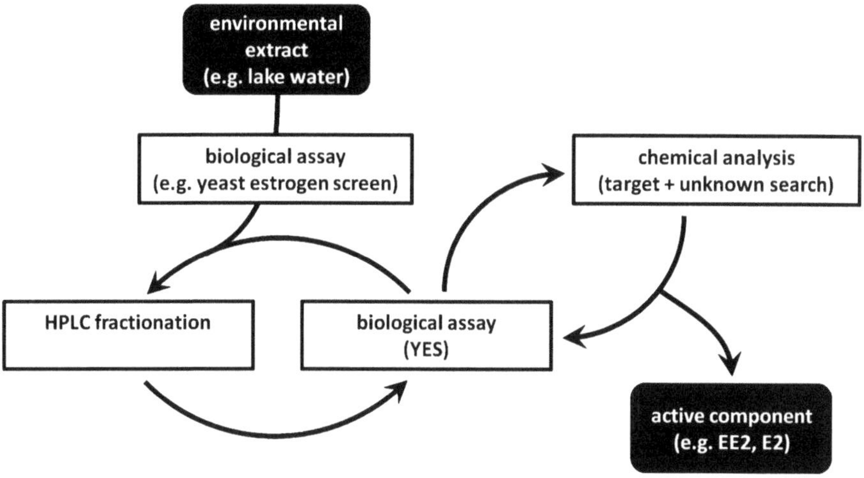

Fig. 10.1 Schematic representation of Effect-Directed Analysis (*EDA*) (Adapted from W. Brack [22])

Only active samples are then fractionated using a chromatographic separation method, most of the time LC, and the fractions tested again. Active fractions now enter the second loop in Fig. 10.1 and undergo chemical analysis, which normally means suspect screening for known environmental contaminants (target analysis, e.g. ethinlyestradiol EE2 and estradiol E2), but also can start a search for unknown active components, if none of the suspected targets is found. If target analytes have been found in the sample, the corresponding (commercially available or synthesized) reference standards will then be tested in the bioassay, for validation purposes.

While this procedure seems quite straightforward, it nevertheless very often does not succeed [23]. This is because only classical target analysis identifies compounds unambiguously, since it makes use of reference standards. When the active component is not included in the target list, i.e. isn't a suspect, assigning a structure and chemical name to the unknown becomes difficult. The only information directly available then is (i) the knowledge that it interacted with the biological assay, which can give preliminary structural information, (ii) the unknown's accurate mass and isotope pattern, and (iii) it's retention time. Scan-dependent and scan-independent methods will additionally provide MS/MS information and with that data on functional groups and substructures.

Depending on the separation technique used, the retention time already provides clues on the polarity and size of the unknown, if for instance a C18 column has been used [24]. This can later be used for validation purposes, once a candidate structure has been proposed. The accurate mass allows narrowing down the possible chemical sum formulas, better so at low molecular weight, as less elemental combinations become possible. A highly accurate mass spectrometer should easily reach a mass accuracy of 10 ppm. After using filtering rules [25], this still corresponds to 78 possible formulas at mass 400 Da and an even higher number of structural isomers. However, using a mass accuracy of 3 ppm reduces the number of candidates to 23. If the isotopic abundance matches the experimental data to within 2 %, only two candidates are left at 3 ppm mass accuracy [25].

Even with all restrictions mentioned above, biologically active chemicals have been successfully identified by EDA. Many of these positive identifications have been done in the field of endocrine disruption, and more specifically so in the case of xenoestrogens. Since many estrogenic compounds are incompletely eliminated in wastewater treatment plants they enter the aquatic environment through the point of effluent discharge.

The most likely estrogens based on catchment models are endogenous estradiol (E2), and estrone (E1), and the active component of contraceptive pills, ethinylestradiol (EE2). We have quantified these three in treated effluents, together with the xenoestrogens nonylphenol (NP) and nonylphenolmono- and –diethoxylate by GC-MS. Using chemical concentration data and relative potencies in the YES, we then calculated estrogenic activity expressed in E2-equivalents and compared the resulting activity with that actually measured in the YES. A good correlation was found, explaining most of the estrogenicity [26]. Similar results were obtained in a study in UK rivers, where E2 was identified as the most dominant estrogen [27]. As mentioned above, the YES is a cell based reporter gene assay and as such rather artificial.
In the search for a system that produces whole organism responses to chemical challenge, the zebrafish embryo toxicity assay (zFET) has been put forward as a useful alternative [28, 29]. Using the zFET in an effect-directed analysis of soil samples from a former municipal landfill, Legler et al. identified several previously unknown developmental toxicants [30].

We have used EDA in an attempt to unravel mysterious gonad malformations observed in an indigenous fish (*Coregonus* sp.) found in the oligotrophic Lake Thun in the Swiss alps [14]. Significantly high incidences of these malformations appeared around the year 2000 and one hypothesis was that they were caused by xenostrogens. Since the fish could ingest these xenoestrogens, hatchlings from Lake Thun were raised during 3–4 years, feeding on the collected zooplankton. The result of this study was that zooplankton from Lake Thun is the key factor in the development of gonad malformations [31]. For this reason zooplankton extracts were collected and analyzed in the YES [32]. The zooplankton extracts gave a positive response in the YES. Following the EDA scheme they were then separated into ten fractions on a C18 column. Only fractions 1, 3 and 6 were estrogenic, while none were in extracts from a reference lake upstream of Lake Thun [33].

Target analysis of the active fractions using a LC-MS method [34], excluded the presence of E1, E2, EE2 and NP as the cause for the positive response. Hence, we did a differential analysis of fractions 1, 3 and 6 of estrogenic and inactive samples with the goal of identifying the unknown estrogen. This was done using a scan-dependent MS/MS analysis, in which the MS is asked to collisionally activate every new component that elutes from the LC column and acquire the corresponding fragments. Components with significantly different concentrations in the estrogenic versus inactive samples were then investigated further. In this way one compound was found that was present only in the Lake Thun extract. The mass of its protonated molecular ion was 180.1018. Based on the isotope pattern, the halogens Cl and Br could clearly be excluded from the elemental composition. Furthermore, the even molecular weight indicated an odd number of N (nitrogen rule for protonated molecules). The accuracy of the mass assignment in spectra acquired with a freshly calibrated orbitrap XL (Thermo Scientific) is better than 2 ppm. Including the elements H, C, N, O, F, Si, P, S and I only allowed three candidate sum formulas for the mass 180.1018: $C_{10}H_{14}O_2N$, $C_5H_{13}O_2N_4F$ and $C_7H_{16}NF_2Si$. The corresponding ring and double-bond equivalents (RDB) are 4.5, 1.0 and 0.5 respectively. Because the unknown induced an estrogenic response in the YES, it can be expected that it contains a phenolic substructure, common to most estrogenic compounds. Hence, due to the aromatic ring, RDB has to be greater than 4, excluding the second and third elemental composition. Further confirmation of a positive match was the fitting isotopic pattern of the first candidate.

Assigning a structure to the postulated elemental composition is relatively straightforward when trying to e.g. identify transformation products, because only a limited number of transformations starting from the precursor are possible. For estrogenic compounds, a reasonable assumption would be that it contains a phenolic substructure. One possible compound, fulfilling all requirements so far would then be N-acetyltyramine (CAS 1202-66-0). Using fragmentation prediction software (Mass Frontier, Thermo Scientific) produces fragments m/z 138.0910 ($C_8H_{12}ON$) and m/z 121.0648 (C_8H_9O), both found in the MS/MS spectra of m/z 180.1018. Furthermore, the predicted $\log K_{ow}$ of 1 is lower than that of E2 ($\log K_{ow}$ 4) which means that it would elute earlier than E2 [35]. Unfortunately, N-acetyltyramine, synthesized in-house tested negative in the YES and for that reason could not be the suspected estrogen. Nevertheless, the Lake Thun investigation illustrates how EDA could help identify potential candidates for an observed adverse effect.

The above-described EDA framework relies on the existence of bioassays that measure specific biological responses relevant for the pollutant or pollutant groups in question. However, suitable bioassays are currently lacking for many toxicologically relevant endpoints, especially those characterizing the responses to chronic low-level exposures. Focusing the research on molecular modes of action carries the potential to advance our understanding of toxicological pathways activated in the organisms as well as to identify and develop novel biomarkers.

10.3 Environmental Proteomics

In every organism, exposure to a chemical stressor leads to the activation of diverse defense mechanisms aimed at adaptation to stressful conditions and eventual reestablishment of homeostasis. However, upon exposure to higher or more prolonged stress levels, cellular defenses become overwhelmed, which in the end would result in damage due to chemical toxicity. It is assumed that any cellular or physiological alterations occurring in response to exposures are preceded and/or accompanied by diverse molecular and biochemical changes in the organism. These include the effects on the levels of mRNA and protein products of diverse genes, leading to activation or inhibition of various pathways and processes in the cells. This in turn results in alterations in the levels of small metabolites involved in biochemical reactions in the functional organism. Thus, molecular and biochemical alterations occurring in the organism in response to exposure may serve as a "footprint" of the stressor in question. Therefore, they can provide the linkages to diverse toxicity phenotypes as well as the means to monitor them.

Traditionally, molecular responses to chemicals have been studied only for a few targets at a time. However, it has been recognized that, in order to be able to build up an understanding of complex interactions between cellular constituents and members of regulatory networks, it is necessary to acquire the information on gene expression and metabolic changes for multiple targets simultaneously. It is anticipated that detailed knowledge on the molecular changes induced by chemical exposure may provide valuable insights into the mechanisms of toxicity, thus supporting the search for biomarkers of exposure and effects [36–38]. With the advent of "omics" technologies aimed at high-throughput measurement of gene products and cellular metabolites, science has moved one step closer to this ultimate goal.

The abundance of multiple mRNA transcripts in an organism has traditionally been assessed by microarray technology and, more recently, by application of diverse next-generation sequencing techniques [39]. With transcriptomics, the expression of multiple genes has been characterized during normogenesis as well as following the exposure to diverse stressors [39–42]. However, in most of the cases the cellular functions of a particular gene are carried out by its protein, not mRNA, product. Thus, compared to mRNA, changes in protein expression would provide a more direct link between chemical effects on gene expression and resulting phenotypes. Although mRNA levels are often used as a proxy for the corresponding protein expression, this approach may lead to unreliable conclusions, since it has been repeatedly demonstrated that the correlation between transcript and protein levels is rather weak [43–45]. This is due to the fact that, apart from transcriptional control, protein expression can also be regulated on post-transcriptional, translational and post-translational levels [46, 47]. Therefore, in order to reliably assess and interpret the occurrence of pollutant-induced gene expression changes, as well as their involvement in the observed toxicity phenotypes, transcriptomics data should ideally be complemented with information on the accompanying proteome changes, which are addressed by proteomics [48, 49]. Finally, changes in the abundance of low-molecular-weight (<1,000 Da) primary and secondary metabolites can be assessed using metabolomics approaches. In fact, the cellular metabolome is considered to represent the most "functional" among all "omics" targets, because metabolites are the end products of all cellular regulatory processes [50, 51]. Proteomics analyses heavily rely on MS-based techniques, while small metabolites can be identified by both nuclear magnetic resonance (NMR) and MS-based approaches.

Overall, integration of "omics" data obtained on different levels of the gene expression cascade may be expected to provide a more complete picture of the status of the cellular machinery and its perturbation in response to a particular stress, compared to the approaches that rely on a single data source. For example, complementation of proteomics data with the information on metabolites has proven instrumental in the elucidation of yet unknown gene functions [52, 53]. However, as the performance of "omics" experiments is still associated with significant resource investment, examples of ecotoxicological studies that successfully combine the information obtained by different techniques are still rather rare. Most of the "omics" studies in ecotoxicology so far have focused on transcriptomics [39] and proteomics [54] analyses, while the examples of metabolomics applications in this field are more limited. Nonetheless, metabolome changes have already proven to be valuable indicators of cellular perturbations induced by pollutants [55–57] and thus are expected to be increasingly applied to elucidation of molecular stress response mechanisms in various species in the future. Below we will discuss some MS applications for proteomics studies in environmental toxicology, as this particular "omic" technique is the main focus of this chapter. It is noted that this review does not intend to provide an exhaustive list of all proteomics studies performed in the field, but rather aims to highlight some common approaches and challenges, as well as outline future research needs.

In proteomics, analytes are usually separated by one- or two-dimensional gel electrophoresis (2D-GE; gel-based approaches) or LC (gel-free approaches). In gel-based techniques, protein abundances are quantified by densitometry either on different gels (classical 2D-GE) or within the same gel with use of differential fluorescent dye labeling (two-dimensional difference gel electrophoresis, 2D-DiGE). The latter technique delivers more reliable results because it corrects for inter-gel variability [58]. Following quantification, proteins identified as differentially expressed can be excised out of the gel and subjected to MS-based identification [59]. In gel-free proteomics analyses, top-down and bottom-up approaches can be distinguished [60]. In the former, intact proteins are being directly analyzed by MS [61], while in the latter, also called shotgun proteomics, digestion of proteins is followed by LC separation of the resulting peptide mixture and MS analysis [62]. Top-down proteomics is rather challenging due to difficulties associated with separation, ionization and detection of large molecules, making this method less suitable for high-throughput analysis [49]. On the other hand, more robust bottom-up shotgun proteomics techniques, along with still frequently used gel-based approaches, have found diverse applications in the analysis of proteome responses in diverse model and even non-model species [63, 64]. The quantification approaches used in shotgun proteomics include label-free quantification based on spectral counting as well as diverse labeling methods, such as isotope-coded affinity tags (ICAT) and isobaric tagging for relative and absolute quantification (iTRAQ) [65–67].

To obtain sufficient coverage in shotgun proteomics, extensive separation of the peptide mixture before introduction into the mass spectrometer is required in order to decrease the mixture complexity and thus increase the detection of less abundant species during MS analysis. Such pre-fractionation of peptide mixtures can be performed either offline or online. Often, two-dimensional LC (2D-LC) separation is used for this purposes by combining strong cation exchange and reverse-phase chromatography [68]. An online 2D-LC setup, Multidimensional Protein

Identification Technology (MudPIT), initially applied in yeast [69], has proven useful in proteomics studies with aquatic model species, including the green algae *Chlamydomonas reinhardtii* [70] or zebrafish *Danio rerio* [45]. In the latter study, MudPIT-based characterization of zebrafish ovary and testis delivered a significantly higher number of protein identifications than those obtained previously by gel-based techniques. Several novel protein groups differentially expressed in zebrafish gonads were identified, demonstrating that a global proteome characterization is capable of delivering information useful for identifying the genes carrying out particular functions in a specific tissue [45]. However, an important drawback of gel-free global proteomics studies is that, similar to gel-based techniques, the analysis is biased towards higher abundance proteins, especially in complex mixtures with high dynamic range. Because of the stochastic nature of a scan-dependent acquisition, more abundant peptides generated from higher abundance proteins are more likely to be detected during global MS analysis, while less abundant peptides would be selected and analyzed only rarely. This results in an under sampling of less abundant species and limits proteome coverage. Thus, the relatively low sensitivity of global MS analysis makes it less suitable for studying low abundance proteins. Taking this into account is especially important when planning proteomics studies aimed at identification of detailed molecular mechanisms of toxicity in complex organisms such as fish. It is likely that in the complex samples obtained, for example, from fish embryos or larvae, many proteins possibly responding to a particular stress, especially those involved in the regulation, would be present in amounts too low to be reliably monitored by global proteomics.

On the other hand, in less complex species, such as bacteria or unicellular algae, as well as animal cell lines, a global proteomics analysis appears to be more suitable for studying molecular stress response mechanisms [71]. Compared to multicellular species, typically higher overall proteome coverage can be expected in the unicellular ones [49]. A MudPIT-based analysis of *C. reinhardtii* exposed to the herbicides paraquat, diuron and norflurazon resulted in the identification of several hundred differentially expressed proteins involved in a wide variety of cellular metabolic pathways. This allowed to establish the links between the observed proteome responses and previously characterized physiological and biochemical changes induced by herbicides [70, 72]. Moreover, proteome responses were found to be more sensitive than the physiological and biochemical endpoints, suggesting the potential usefulness of some of the identified proteins as early and sensitive biomarkers of herbicide exposure [70]. Recently, combined transcriptomics and proteomics analyses have allowed detailed insights into the mechanisms of silver toxicity in *C. reinhardtii* [73]. This work demonstrated that silver is promoting oxidative stress through its effects on ATP and photosynthesis. The recovery from this initial damage can occur due to the activation of antioxidant defenses as well as possible elimination of silver via efflux transporters. MudPIT can also be employed to obtain precise information on protein synthesis and degradation dynamics by using pulse-labeling with isotopically labeled substrate [74]. For example, using a ^{15}N-enriched nutrient solution to label the proteins during algal growth, kinetics of diuron-induced degradation of an important photosystem II component, the D1 protein, have been recently characterized by MudPIT [75]. As a valuable addition to label-free quantification approaches, this method expands the array of proteome characterization tools available for studying mechanisms of toxicity in green algae and similar organisms.

Overall, global proteomics appears to be particularly suitable for quantitative studies aimed at monitoring the responses of well characterized biomarkers of general and oxidative stress, such as for example heat shock proteins, superoxide dismutase, catalase, glutathione peroxidases and glutathione-S-transferases, since these relatively abundant proteins are readily detectable by this method [67]. In addition, some insights into fundamental cellular processes such as energy metabolism, translation or cytoskeleton organization, can typically be obtained [76]. Thus, proteome changes commonly monitored by global proteomics may in fact represent only general stress responses [77], while the usefulness of this approach for the elucidation of toxicant-specific responses may be rather limited due to the insufficient proteome coverage presently delivered by state-of-the-art methods. This implies that careful consideration of the significance and specificity of the observed changes is necessary when interpreting the results as showing a link between a specific stressor and a particular cellular function that it perturbs.

Compared to global proteomics, the so-called targeted proteomics offers the possibility to increase the sensitivity of MS detection. Here, data acquisition is aimed at detecting specific proteins of interest instead of attempting to monitor all potentially present analytes [78]. For example, SRM, a method originally developed for ultratrace chemical analysis, can be readily applied in proteomics analyses as well [79]. For this, specific fragmentation transitions are monitored for selected peptides representative of proteins to be measured [80]. As a first step in SRM assay development, the proteins to be monitored are selected and proteotypic peptides are chosen for each protein based on global profiling data (if available) or in silico prediction. SRM detection conditions are then optimized for each peptide either directly for the endogenous peptides or by using synthesized analogs. The respective endogenous peptides are then studied across different samples using the developed SRM protocols. In scheduled acquisition mode, up to several hundred targets can be monitored within one run [81]. Moreover, since the completion of one technical replicate in MudPIT typically requires up to 22 h instrument time, while an SRM run only lasts one hour, the latter technique allows for much higher sample throughput. Thus, SRM analysis can be very helpful not only for fundamental studies looking at the lower abundance proteins undetectable by global proteomics, but also for rapid monitoring of proteins established as biomarkers of a particular exposure or effect, within a wide dynamic range.

For quantification by SRM, isotopically labeled counterparts of each peptide target are often used [82]. Although this method is known to provide most accurate quantitative information, the associated costs may limit its broader application in environmental toxicology, especially during the initial screening phases in biomarker discovery projects. It has been demonstrated recently that SRM can also be used to obtain semi-quantitative data for target proteins by normalizing their abundance to the levels of "housekeeping" proteins measured in the same sample, similar to the relative quantification approach used in RT-PCR [83] In this way, rapid and relatively inexpensive initial screening can be performed for proteins of interest across multiple samples obtained from specimens subjected to different exposure regimes. The ability to identify the minimum concentration at which protein changes would result in an adverse outcome, as opposed to purely adaptive or beneficial changes, is recognized as one of the challenges of ecotoxicogenomics [84]. Thus, the possibility to generate detailed time- and concentration-dependent profiles with a relatively low investment would greatly benefit the establishment of dose-response relationships between exposure levels and resulting molecular, as well as phenotypic changes. For those proteins that exhibit promising expression patterns, e.g. sensitive and specific responses to a particular stressor, more precise label-based quantification can then be performed if needed. Such a setup may prove to be instrumental for the identification and validation of protein biomarkers for various pollutants in different species, as well as the elucidation of their relative significance for the manifestation of toxicity. However, it has to be noted that biomarker discovery by SRM largely depends on the existence of additional information obtained by other means, i.e. transcriptomics, or the formulation of viable hypotheses that could guide the initial selection of target proteins to be monitored.

Another proteomics method called SWATH combines the screening capabilities of global proteomics with the sensitivity of SRM by activating all analytes eluting at a given moment and acquiring all fragment ions afterwards [85]. For this, the precursor ion selection window (swath) is increased to 25 Da instead of typically one for SRM. All compounds found in this window are activated and all fragments from 400 to 1,200 m/z are then acquired. The process is then repeated with the 25 Da window each time moving to a higher mass, until the full mass range has been covered. The fragments of a precursor ion are then identified by matching their chromatographic profile to the precursor. An important aspect of this novel approach is that there is no bias caused by intensity thresholds or the stochastic nature of the scan-dependent selection of precursors. Thus, fragment ions from all precursors are theoretically available. This allows to mine the SWATH data at later time points for compounds that were not the focus of the initial study. The usefulness of SWATH for biomarker discovery in clinical research has been demonstrated [86]. Certainly, the application of this technique to both pollutant screening and molecular effects characterization holds a great promise for environmental research as well.

It has to be noted that, although applying targeted proteomics (SRM) to monitor sex-related genes in zebrafish during gonadal differentiation indeed significantly increased MS detection sensitivity compared to MudPIT-based global profiling, several low abundance proteins of particular interest for sexual differentiation in zebrafish still could not be detected [45, 83]. Notably, these proteins included various transcription factors, typically expressed at a low level and only in a few specific cell types. Since the overall sensitivity of the SWATH technique is also not expected to substantially surpass that of SRM, it can be said that monitoring such proteins in complex samples still is a significant challenge.

Regardless of the type of downstream MS analysis applied, the detection of low abundance proteins can be facilitated by decreasing the sample complexity, along with the enrichment of particular analytes of interest. Separate analyses carried out for different organs typically improve overall proteome coverage compared to whole-body extracts [87], but performing dissection in small specimens such as fish larvae or insects is rather tedious. Moreover, even within one organ, multiple cell types with widely differing proteomes are usually present. Thus the overall level of sample complexity often remains rather high. Laser capture microdissection allows targeted isolation of material from single organs or even specific cell types [88]. However, this costly technique requires extensive resources and still needs further development to improve its compatibility with downstream MS analysis. Diverse fractionation, enrichment and depletion procedures can also be performed [89, 90], but such extensive sample processing leads to an increase in the technical variability as well as analysis costs. Moreover, the need for higher amounts of starting material may also represent a significant challenge for most types of ecotoxicological studies. Thus, the advantages of increased protein detection sensitivity need to be balanced on a case by case basis against the costs and feasibility of performing extensive sample processing, particularly critical for studies requiring assessment of multiple samples.

The functions and activity of a protein are often regulated by post-translational modifications (PTMs). Gel-based methods are well suited for resolving protein isoforms carrying different PTMs [59] and diverse gel-free approaches exist to study PTMs by MS as well, including both global and targeted techniques [91–93]. However, the efficiency and robustness of PTM characterization methods may still need further development before their application for high-throughput analyses in model and non-model species, relevant for environmental toxicology, becomes more widely used.

Compared to human toxicology, the unique challenge of environmental toxicology is that it has to deal with multiple ecologically relevant species. The peptide-protein identification based on the acquired mass spectra relies on the use of sufficiently populated and well-annotated sequence databases. Therefore, the quality and completeness of sequence information available for a given species can be viewed as a critical factor determining the success of proteomics analyses, in particular the depth of proteome coverage that can be obtained. Presently, databases from closely related species are often used when performing proteomics analyses in non-sequenced species. In this case, peptide-protein identification relies on the existence of homologous peptides conserved between different species. However, since such peptides tend to originate from the regions that are also well conserved among different members of the same protein families, unambiguous assignment to one particular protein often becomes impossible. Alternatively, *de novo* sequencing algorithms can be used to extract the peptide sequence from MS/MS spectra by searching all possible sequence combinations that could explain experimentally observed ions [94]. This approach could also allow identifying peptides resulting from unannotated alternative splicing transcripts or post-translational modifications often not included in databases typically used for matching MS/MS data. However, due to the required very high spectral quality and increased analysis time, *de novo* sequencing is still applied rather rarely. Thus, the greatest improvement in the proteomics analyses of non-model species is expected to be brought about by the increasing availability of annotated genomic sequences, since the DNA sequencing instrumentation and associated bioinformatics solutions for genome annotation have undergone major developments in the recent years [38, 63].

Apart from its application for detecting changes in protein abundance, MS can also be used to characterize the direct interactions between chemicals and their protein targets. For example, matrix-assisted laser desorption/ionization time-of-flight (MALDI-TOF) MS has been successfully applied to study the binding of Cd to the rainbow trout estrogen receptor [95] and of silver nanoparticles (AgNP) to *Escherichia coli* proteins [96]. In the latter study, several fragments of the enzyme tryptophanase (TNase) were found to preferentially and strongly bind to AgNP, as opposed to a simple association with ionic silver. Moreover, the loss of TNase enzymatic activity upon associating with AgNPs could be explained by the fact that the active site of the enzyme included a high-binding protein fragment. Thus, MALDI-TOF MS allowed to suggest a probable mechanism for AgNP-protein interactions as well as AgNP effects on enzyme activities [96]. In another study, the formation of Pb-phytochelatin (PC) and Zn-PC complexes has been studied by nano-ESI-MS [97]. PCs are oligopeptides presumably involved in metal detoxification due to their ability to bind metals and thus prevent their contact with other biomolecules. Although the metals are known to induce PC synthesis, the direct evidence for binding of the inducing metal to PC has been scarce. Using nano-ESI-MS, the coordination of lead through the thiol and possibly carboxylic groups could be demonstrated, confirming the existence of previously postulated Pb-PC complexes [97, 98].

10.4 Conclusion

As discussed above, MS provides the basis for versatile and dynamic tools that can be successfully applied to the characterization of chemical exposure and effects in environmental toxicology. Because EDA delivers a comprehensive view of the real life exposure situation, its application will undoubtedly increase in the future. An improved understanding of molecular mechanisms leading to chemical exposure-caused adverse outcomes is expected to result from the development of systems biology models integrating diverse omics data with the physiological phenotypes.

These novel developments will contribute to the refinement of current risk assessment strategies for hazard evaluation and prediction of risks posed by chemicals in the environment.

References

1. Hites RA, Biemann K (1968) Mass spectrometer-computer system particularly suited for gas chromatography of complex mixtures. Anal Chem 40:1217–1221
2. Giger W, Molnar-Kubica E (1978) Tetrachloroethylene in contaminated ground and drinking waters. Bull Environ Contam Toxicol 19:475–480

3. Aschengrau A, Ozonoff D, Paulu C, Coogan P, Vezina R, Heeren T, Zhang Y (1993) Cancer risk and tetrachloroethylene-contaminated drinking water in Massachusetts. Arch Environ Health 48:284–292
4. Carson R (1962) Silent spring. Houghton Mifflin, Boston
5. Hay A (1979) Seveso: the crucial question of reactor safety. Nature 281:521
6. Broughton E (2005) The Bhopal disaster and its aftermath: a review. Environ Health 4:6
7. Giger W (2009) The Rhine red, the fish dead – the 1986 Schweizerhalle disaster, a retrospect and longterm impact assessment. Environ Sci Pollut Res 16:S98–S111
8. Vermeirssen ELM, Suter MJ-F, Burkhardt-Holm P (2006) Estrogenicity patterns in the Swiss midland river Lützelmurg in relation to treated domestic sewage effluent discharges and hydrology. Environ Toxicol Chem 25:2413–2422
9. Schecter A, Tiernan T (1985) Occupational exposure to polychlorinated dioxins, polychlorinated furans, polychlorinated biphenyls, and biphenylenes after an electrical panel and transformer accident in an office building in Binghamton, NY. Environ Health Perspect 60:305–313
10. Kolpin DW, Furlong ET, Meyer MT, Thurman EM, Zaugg SD, Barber LB, Buxton HT (2002) Pharmaceuticals, hormones, and other organic wastewater contaminants in U.S. streams, 1999–2000: a national reconnaissance. Environ Sci Technol 36:1202–1211
11. http://www.cas.org/content/counter
12. Helbling DE, Hollender J, Kohler H-PE, Singer H, Fenner K (2010) High-throughput identification of microbial transformation products of organic micropollutants. Environ Sci Technol 44:6621–6627
13. Burkhardt-Holm P, Giger W, Guettinger H, Ochsenbein U, Peter A, Scheurer K, Segner H, Staub E, Suter MJ-F (2005) Where have all the fish gone? The reasons why fish catches in Swiss rivers are declining. Environ Sci Technol 39:441A–447A
14. Bernet D, Wahli T, Küng C, Segner H (2004) Frequent and unexplained gonadal abnormalities in whitefish (central alpine *Coregonus* sp.) from an alpine oligotrophic lake in Switzerland. Dis Aquat Org 61:137–148
15. Huntscha S, Singer HP, McArdell CS, Frank CE, Hollender J (2012) Multiresidue analysis of 88 polar organic micropollutants in ground, surface and wastewater using online mixed-bed multilayer solid-phase extraction coupled to high performance liquid chromatography-tandem mass spectrometry. J Chromatogr A 1268:74–83
16. Cheshenko K, Brion F, Le Page Y, Hinfray N, Pakdel F, Kah O, Segner H, Eggen RI (2007) Expression of the zebra fish aromatase cyp19a and cyp19b genes in response to the ligands of estrogen receptor and aryl hydrocarbon receptor. Toxicol Sci 96:255–267
17. Steinberg P (2013) In: Steinberg P (ed) High-throughput screening methods in toxicity testing. Wiley, Hoboken
18. Ammann AA, Macikova P, Groh KJ, Schirmer K, Suter MJ-F LC-MS/MS determination of potential endocrine disruptors of gluco- and mineralocorticoid signalling in river and waste water. Anal Bioanal Chem (under revision)
19. Moschet C, Piazzoli A, Singer HP, Hollender J (2013) Alleviating the reference standard dilemma using a systematic exact mass suspect screening approach with liquid chromatography-high resolution mass spectrometry. Anal Chem 85:10312–10320
20. Routledge EJ, Sumpter JP (1996) Estrogenic activity of surfactants and some of their degradation products assessed using a recombinant yeast screen. Environ Toxicol Chem 15:241–248
21. Kavlock R, Chandler K, Houck K, Hunter S, Judson R, Kleinstreuer N, Knudsen T, Martin M, Padilla S, Reif R, Richard A, Rotroff D, Sipes N, Dix D (2012) Update on EPA's ToxCast program: providing high throughput decision support tools for chemical risk management. Chem Res Toxicol 25:1287–1302
22. Brack W (2003) Effect-directed analysis: a promising tool for the identification of organic toxicants in complex mixtures? Anal Bioanal Chem 337:397–407
23. Reemtsma T (2001) Prospects of toxicity-directed wastewater analysis. Anal Chim Acta 426:279–287
24. Boswell PG, Schellenberg JR, Carr PW, Cohen JD, Hegeman AD (2011) A study on retention "projection" as a supplementary means for compound identification by liquid chromatography–mass

spectrometry capable of predicting retention with different gradients, flow rates, and instruments. J Chromatogr A 1218:6732–6741
25. Kind T, Fiehn O (2006) Metabolomic database annotations via query of elemental compositions: mass accuracy is insufficient even at less than 1 ppm. BMC Bioinforma 7:234
26. Aerni H-R, Kobler B, Rutishauser BV, Wettstein FE, Fischer R, Giger W, Hungerbühler A, Marazuela MD, Peter A, Schönenberger R, Vögeli AC, Suter MJ-F, Eggen RIL (2004) Combined biological and chemical assessment of estrogenic activities in wastewater treatment plant effluents. Anal Bioanal Chem 378:688–696
27. Thomas KV, Hurst MR, Matthiessen P, Waldock MJ (2001) Characterization of estrogenic compounds in water samples collected from United Kingdom estuaries. Environ Toxicol Chem 20:2165–2170
28. Volz DC, Belanger S, Embry M, Padilla S, Sanderson H, Schirmer K, Scholz S, Villeneuve D (2011) Adverse outcome pathways during early fish development: a conceptual framework for identification of chemical screening and prioritization strategies. Toxicol Sci 123:349–358
29. Knoebel M, Busser FJM, Rico-Rico A, Kramer NI, Hermens JLM, Hafner C, Tanneberger K, Schirmer K, Scholz S (2012) Predicting adult fish acute lethality with the zebrafish embryo: relevance of test duration, endpoints, compound properties, and exposure concentration analysis. Environ Sci Technol 46:9690–9700
30. Legler J, van Vetzen M, Cenijn PH, Houtman CJ, Lamoree MH, Wegener JW (2011) Effect-directed analysis of municipal landfill soil reveals novel developmental toxicants in the zebrafish *Danio rerio*. Environ Sci Technol 45:8552–8558
31. Bernet D, Liedtke A, Bittner D, Eggen RIL, Kipfer S, Küng C, Largiader CR, Suter MJ-F, Wahli T, Segner H (2008) Gonadal malformations in whitefish from Lake Thun: defining the case and evaluating the role of EDCs. CHIMIA 62:383–388
32. Liedtke A, Schoenenberger R, Eggen RIL, Suter MJ-F (2009) Internal exposure of whitefish (*Coregonus lavaretus*) to estrogens. Aquat Toxicol 93:158–165
33. Vögeli AC (2008) Endocrine disrupting chemicals – linking internal exposure to effects in wild fish. PhD Thesis No. 17756, Federal Institute of Technology, Zürich
34. Vermeirssen ELM, Korner O, Schonenberger R, Suter MJ-F, Burkhardt-Holm P (2005) Characterization of environmental estrogens in river water using a three pronged approach: active and passive water sampling and the analysis of accumulated estrogens in the bile of caged fish. Environ Sci Technol 39:8191–8198
35. http://www.epa.gov/opptintr/exposure/pubs/episuitedl.htm
36. Eggen RIL, Behra R, Burkhardt-Holm P, Escher BI, Schweigert N (2004) Challenges in ecotoxicology. Environ Sci Technol 38:58A–64A
37. Eggen RIL, Suter MJ-F (2007) Analytical chemistry and ecotoxicology – tasks, needs and trends. J Toxicol Environ Health A 70:1–3
38. Garcia-Reyero N, Perkins EJ (2011) Systems biology: leading the revolution in ecotoxicology. Environ Toxicol Chem 30:265–273
39. Schirmer K, Fischer BB, Madureira DJ, Pillai S (2010) Transcriptomics in ecotoxicology. Anal Bioanal Chem 397:917–923
40. Voelker D, Vess C, Tillmann M, Nagel R, Otto GW, Geisler R, Schirmer K, Scholz S (2007) Differential gene expression as a toxicant-sensitive endpoint in zebrafish embryos and larvae. Aquat Toxicol 81:355–364
41. Froehlicher M, Liedtke A, Groh K, Lopez-Schier H, Neuhauss SC, Segner H, Eggen RI (2009) Estrogen receptor subtype beta2 is involved in neuromast development in zebrafish (*Danio rerio*) larvae. Dev Biol 330:32–43
42. Michaelson JJ, Trump S, Rudzok S, Graebsch C, Madureira DJ, Dautel F, Mai J, Attinger S, Schirmer K, von Bergen M, Lehmann I, Beyer A (2011) Transcriptional signatures of regulatory and toxic responses to benzo-[a]-pyrene exposure. BMC Genomics 12:502
43. Washburn MP, Koller A, Oshiro G, Ulaszek RR, Plouffe D, Deciu C, Winzeler E, Yates JR 3rd (2003) Protein pathway and complex clustering of correlated mRNA and protein expression analyses in *Saccharomyces cerevisiae*. Proc Natl Acad Sci USA 100:3107–3112

44. Link V, Carvalho L, Castanon I, Stockinger P, Shevchenko A, Heisenberg C-P (2006) Identification of regulators of germ layer morphogenesis using proteomics in zebrafish. J Cell Sci 119:2073–2083
45. Groh KJ, Nesatyy VJ, Segner H, Eggen RIL, Suter MJ-F (2011) Global proteomics analysis of testis and ovary in adult zebrafish (*Danio rerio*). Fish Physiol Biochem 37:619–647
46. Pradet-Balade B, Boulme F, Beug H, Muellner EW, Garcia-Sanz JA (2001) Translation control: bridging the gap between genomics and proteomics. Trends Biochem Sci 26:225–229
47. Schwanhaeusser B, Busse D, Li N, Dittmar G, Schuchhardt J, Wolf J, Chen W, Selbach M (2011) Global quantification of mammalian gene expression control. Nature 473: 337–342
48. Nesatyy VJ, Suter MJ-F (2007) Proteomics for the analysis of environmental stress responses in organisms. Environ Sci Technol 41:6891–6900
49. Groh KJ, Nesatyy VJ, Suter MJ-F (2011) In: de Bruijn FJ (ed) Handbook of molecular microbial ecology, vol. 1, 1st edn. John Wiley & Sons, Inc., Hoboken NJ, USA, pp 605–624
50. Wolfender JL, Rudaz S, Choi YH, Kim HK (2013) Plant metabolomics: from holistic data to relevant biomarkers. Curr Med Chem 20:1056–1090
51. Milne SB, Mathews TP, Myers DS, Ivanova PT, Brown HA (2013) Sum of the parts: mass spectrometry-based metabolomics. Biochemistry 52:3829–3840
52. De Wit M, Keil D, van der Ven K, Vandamme S, Witters E, De Coen W (2010) An integrated transcriptomic and proteomic approach characterizing estrogenic and metabolic effects of 17alpha-ethinylestradiol in zebrafish (*Danio rerio*). Gen Comp Endocrinol 167:190–201
53. Goulitquer S, Potin P, Tonon T (2012) Mass spectrometry-based metabolomics to elucidate functions in marine organisms and ecosystems. Mar Drugs 10:849–880
54. Nesatyy VJ, Suter MJ-F (2008) Analysis of environmental stress response on the proteome level. Mass Spectrom Rev 27:556–574
55. Viant MR (2008) Recent developments in environmental metabolomics. Mol Biosyst 4:980–986
56. Taylor NS, Weber RJ, White TA, Viant MR (2010) Discriminating between different acute chemical toxicities via changes in the daphnid metabolome. Toxicol Sci 118:307–317
57. Mirbahai L, Southam AD, Sommer U, Williams TD, Bignell JP, Lyons BP, Viant MR, Chipman JK (2013) Disruption of DNA methylation via S-adenosylhomocysteine is a key process in high incidence liver carcinogenesis in fish. J Proteome Res 12:2895–2904
58. Tonge R, Shaw J, Middleton B, Rowlinson R, Rayner S, Young J (2001) Validation and development of fluorescence two-dimensional differential gel electrophoresis proteomics technology. Proteomics 1:377–396
59. Gorg A, Weiss W, Dunn MJ (2004) Current two-dimensional electrophoresis technology for proteomics. Proteomics 4:3665–3685
60. Chait BT (2006) Mass spectrometry: bottom-up or top-down? Science 314:65–66
61. Syka JE, Coon JJ, Schroeder MJ, Shabanowitz J, Hunt DF (2004) Peptide and protein sequence analysis by electron transfer dissociation mass spectrometry. Proc Natl Acad Sci USA 101:9528–9533
62. Washburn MP (2008) Sample preparation and in-solution protease digestion of proteins for chromatography-based proteomic analysis. Curr Protoc Protein Sci 53:1–11
63. Sanchez BC, Ralston-Hooper KJ, Sepulveda MS (2011) Review of recent proteomic applications in aquatic toxicology. Environ Toxicol Chem 30:274–282
64. Martyniuk CJ, Popesku JT, Chown B, Denslow ND, Trudeau VL (2012) Quantitative proteomics in teleost fish: insights and challenges for neuroendocrine and neurotoxicology research. Gen Comp Endocrinol 176:314–320
65. Leithner A, Lindner W (2006) Chemistry meets proteomics: the use of chemical tagging reactions for MS-based proteomics. Proteomics 6:5418–5434
66. Martyniuk CJ, Denslow ND (2009) Towards functional genomics in fish using quantitative proteomics. Gen Comp Endocrinol 164:135–141
67. Martyniuk CJ, Denslow ND (2012) DIGE and iTRAQ as biomarker discovery tools in aquatic toxicology. Ecotoxicol Environ Saf 76:3–10

68. Fournier ML, Gilmore JM, Martin-Brown SA, Washburn MP (2007) Multidimensional separations-based shotgun proteomics. Chem Rev 107:3654–3686
69. Washburn MP, Wolters D, Yates JR 3rd (2001) Large-scale analysis of the yeast proteome by multidimensional protein identification technology. Nat Biotechnol 19:242–247
70. Nestler H, Groh KJ, Schoenenberger R, Eggen RIL, Suter MJ-F (2012) Linking proteome responses with physiological and biochemical effects in herbicide-exposed *Chlamydomonas reinhardtii*. J Proteomics 75:5370–5385
71. Lacerda CMR, Reardon KF (2009) Environmental proteomics: applications of proteome profiling in environmental microbiology and biotechnology. Brief Funct Genomics Proteomics 8:75–87
72. Nestler H, Groh KJ, Schoenenberger R, Behra R, Schirmer K, Eggen RIL, Suter MJ-F (2012) Multiple endpoint assay provides a detailed mechanistic view of responses to herbicide exposure in *Chlamydomonas reinhardtii*. Aquat Toxicol 110–111:214–224
73. Pillai S, Behra R, Nestler H, Suter MJ-F, Sigg L, Schirmer K (2014) Linking of toxicity and stress response pathways across the transcriptome, proteome and phenotype of *Chlamydomonas reinhardtii* exposed to silver. Proc Natl Acad Sci USA 111:3490–3495
74. Muehlhaus T, Weiss J, Hemme D, Sommer F, Schroda M (2011) Quantitative shotgun proteomics using a uniform ^{15}N-labeled standard to monitor proteome dynamics in time course experiments reveals new insights into the heat stress response of *Chlamydomonas reinhardtii*. Mol Cell Proteomics 10:M110.004739
75. Stamatelatos D (2013) Analysis of the half-lives of PSII proteins in *Chlamydomonas reinhardtii* using ^{15}N labeled nutrient solution. Bachelor Thesis, ETH Zurich, Zurich
76. Van Aggelen G, Ankley GT, Baldwin WS, Bearden DW, Benson WH, Chipman JK, Collette TW, Craft JA, Denslow ND, Embry MR, Falciani F, George SG, Helbing CC, Hoekstra PF, Iguchi T, Kagami Y, Katsiadaki I, Kille P, Liu L, Lord PG, McIntyre T, O'Neill A, Osachoff H, Perkins EJ, Santos EM, Skirrow RC, Snape JR, Tyler CR, Versteeg D, Viant MR, Volz DC, Williams TD, Yu L (2010) Integrating omic technologies into aquatic ecological risk assessment and environmental monitoring: hurdles, achievements, and future outlook. Environ Health Perspect 118:1–5
77. Petrak J, Ivanek R, Toman O, Cmejla R, Cmejlova J, Vyoral D, Zivny J, Vulpe CD (2008) *Déjà vu* in proteomics. A hit parade of repeatedly identified differentially expressed proteins. Proteomics 8:1744–1749
78. Domon B, Aebersold R (2006) Mass spectrometry and protein analysis. Science 312:212–217
79. Picotti P, Bodenmiller B, Mueller LN, Domon B, Aebersold R (2009) Full dynamic range proteome analysis of *S. cerevisiae* by targeted proteomics. Cell 138:795–806
80. Maiolica A, Juenger MA, Ezkurdia I, Aebersold R (2012) Targeted proteome investigation via selected reaction monitoring mass spectrometry. J Proteomics 75:3495–3513
81. Lange V, Picotti P, Domon B, Aebersold R (2008) Selected reaction monitoring for quantitative proteomics: a tutorial. Mol Syst Biol 4:222
82. Picotti P, Aebersold R (2012) Selected reaction monitoring-based proteomics: workflows, pitfalls and future directions. Nat Methods 9:555–566
83. Groh KJ, Schoenenberger R, Eggen RIL, Segner H, Suter MJ-F (2013) Analysis of protein expression in zebrafish during gonad differentiation by targeted proteomics. Gen Comp Endocrinol 193:210–220
84. Lemos MFL, Soares AMVM, Correia AC, Esteves AC (2010) Proteins in ecotoxicology – how, why and why not? Proteomics 10:873–887
85. Gillet LC, Navarro P, Tate S, Röst H, Selevsek N, Reiter L, Bonner R, Aebersold R (2012) Targeted data extraction of the MS/MS spectra generated by data-independent acquisition: a new concept for consistent and accurate proteome analysis. Mol Cell Proteomics 11, PMID 22261725
86. Liu Y, Hüttenhain R, Collins B, Aebersold R (2013) Mass spectrometric protein maps for biomarker discovery and clinical research. Expert Rev Mol Diagn 13:811–825
87. Abramsson A, Westman-Brinkmalm A, Pannee J, Gustavsson M, von Otter M, Blennow K, Brinkmalm G, Kettunen P, Zetterberg H (2010) Proteomics profiling of single organs from individual adult zebrafish. Zebrafish 7:161–168

88. Jorgensen A, Nielsen JE, Morthorst JE, Bjerregaard P, Leffers H (2009) Laser capture microdissection of gonads from juvenile zebrafish. Reprod Biol Endocrinol 7:97
89. Wang N, MacKenzie L, De Souza AG, Zhong H, Goss G, Li L (2007) Proteome profile of cytosolic component of zebrafish liver generated by LC-ESI MS/MS combined with trypsin digestion and microwave-assisted acid hydrolysis. J Proteome Res 6:263–272
90. Jiang D, Jarrett HW, Haskins WE (2009) Methods for proteomic analysis of transcription factors. J Chromatogr A 1216:6881–6889
91. Pan S, Chen R, Aebersold R, Brentnall TA (2011) Mass spectrometry based glycoproteomics – from a proteomics perspective. Mol Cell Proteomics 10:R110.003251
92. Young NL, Plazas-Mayorca MD, Garcia BA (2010) Systems-wide proteomics characterization of combinatorial post-translational modification patterns. Expert Rev Proteomics 7:79–92
93. Liebler DC, Zimmermann LJ (2013) Targeted quantitation of proteins by mass spectrometry. Biochemistry 52:3797–3806
94. Grossmann J, Roos FF, Cieliebak M, Liptak Z, Mathis LK, Muller M, Gruissem W, Baginsky S (2005) AUDENS: a tool for automated peptide *de novo* sequencing. J Proteome Res 4:1768–1774
95. Nesatyy VJ, Rutishauser BV, Eggen RIL, Suter MJ-F (2005) Identification of the estrogen receptor Cd-binding sites by chemical modification. Analyst 130:1087–1097
96. Wigginton NS, De Titta A, Piccapietra F, Dobias J, Nesatyy VJ, Suter MJ-F, Bernier-Latmani R (2010) Binding of silver nanoparticles to bacterial proteins depends on surface modifications and inhibits enzymatic activity. Environ Sci Technol 44:2163–2168
97. Scheidegger C, Suter MJ-F, Behra R, Sigg L (2012) Characterization of lead-phytochelatin complexes by nano-electrospray ionization mass spectrometry. Front Microbiol 3:1–7
98. Scheidegger C, Sigg L, Behra R (2011) Characterization of lead induced metal-phytochelatin complexes in *Chlamydomonas reinhardtii*. Environ Toxicol Chem 30:2546–2552

Chapter 11
Proteogenomics for the Enhanced Discovery of Bacterial Biomarkers

Erica M. Hartmann and Jean Armengaud

Abstract As nucleic acid sequencing technologies become faster and less expensive, whole genome sequences are increasingly available. Advances in sequencing have enhanced our ability to detect potential bacterial bioterrorism agents and to predict biomarkers for yet uncharacterized or man-made pathogens. However, speed and ease of access are no substitute for accuracy. Indeed, errors in both structural and functional annotations of genomes are still frequent. Proteogenomics, in which proteomic data complements genomic sequencing to correct errors and improve gene annotation, has already been performed on several bacteria on the Centers for Disease Control and Prevention list of bioterrorism agents, revealing non-negligible errors that could hinder the detection of known strains and be propagated in emerging strains whose sequencing and annotation depend on similarities with known sequences. In this chapter, we review the fundamentals of proteogenomics, its application in the prevention of bioterrorism and protection of public health, specific techniques for future development, and the confidence that should be afforded to resulting sequences.

Keywords Proteogenomics • Genome sequence correction • Bacterial biomarkers

11.1 Introduction

The mass spectrometry-based detection of biological agents often uses specific protein sequences or combinations thereof (*i.e.*, fingerprints) to identify unknown samples. Due to the tremendous reservoir of biodiversity on Earth, it is impossible to record fingerprints for all biological agents. Therefore, genomic information is increasingly used for the definition of specific biomarkers or the computation of theoretical fingerprints. While there are many approaches to accomplish this task,

E.M. Hartmann (✉)
Biology and the Built Environment Center, University of Oregon, Eugene, Oregon, USA
e-mail: ericamh@uoregon.edu

J. Armengaud
CEA, DSV, IBEB, Lab Biochim System Perturb, Bagnols-sur-Cèze F-30207, France

J. Banoub (ed.), *Detection of Chemical, Biological, Radiological and Nuclear Agents for the Prevention of Terrorism*, NATO Science for Peace and Security Series A: Chemistry and Biology, DOI 10.1007/978-94-017-9238-7_11

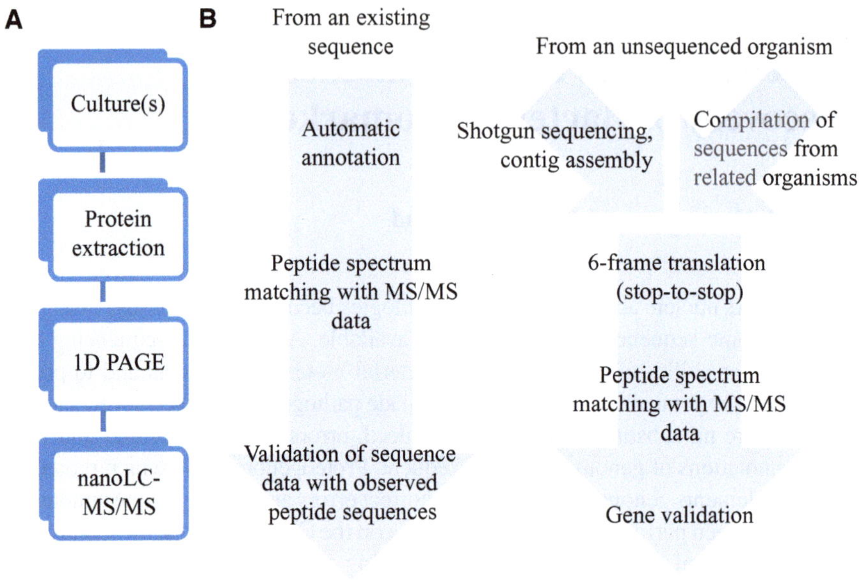

Fig. 11.1 Schematic of a generic proteogenomic protocol. (**a**) The wet lab workflow. (**b**) The bioinformatics workflow

they all share a common weakness: genome database errors. These errors have diverse sources, but they can all be corrected through proteogenomic analysis. In this chapter, we give a brief introduction to proteogenomics, explain its importance to counter bioterrorism, give an example of how it can be applied, and discuss reasonable precautions to be taken in interpreting proteogenomic data (Fig. 11.1).

11.2 Proteogenomics

Current molecular microbiology methods allow the increasingly rapid, specific detection of bacterial pathogens against the complex background noise of clinical or environmental samples [1]. Techniques like the detection of protein biomarkers rely on extensive sequence databases to avoid potential false positive signals from closely related organisms and to correctly interpret the results. Genomics has therefore traditionally been viewed as an enabler of shotgun proteomic studies and the identification of protein biomarkers [2].

Genomic databases are composed of sequences, obtained using multiple techniques with varying degrees of accuracy, wherein the location, direction, and length of genes is automatically predicted based on known start and stop codons and similarities with previously observed genes. It is becoming increasingly evident that, while manual inspection and annotation of whole genome sequences would be a herculean task, current automated algorithms insufficiently reflect the empirical products of transcription and translation, especially in novel organisms [3].

Sequencing errors, incorrectly predicted open reading frames, and post-translational modifications create a gulf between the reality of proteins and their *in silico* versions.

The increased accessibility of peptide sequences, thanks to high accuracy, high throughput next-generation tandem mass spectrometry, has made it possible to correct many of the aforementioned discrepancies [4–7]. This process, known as proteogenomics, can reveal non-silent mis-reads that occurred during genome sequencing as well as inaccurate start and stop codons, N-terminal excisions, signal peptide cleavages, and other post-translational modifications. Experimental procedures generally involve protein extraction followed by polyacrylamide gel separation and purification, enzymatic digestion, and peptide detection using tandem mass spectrometry (MS/MS), such as would normally be applied for proteomic analysis of bacterial samples [8]. However, instead of identifying peptides from gene sequence databases, sequences and annotations are vetted by comparing observed MS/MS spectra to a six-frame translation of the selected genome(s).

For an organism whose genome has already been sequenced, proteogenomic analysis can be as simple as aligning data from a proteomic experiment with the six-frame stop-to-stop *in silico* translation of the previously generated nucleic acid sequence. Any given instance of an organism's proteome is only a fraction of its whole genome, so the evaluation of multiple culture conditions and growth stages can increase the final genome coverage. One such example is the proteogenomic re-annotation of *Ruegeria pomeroyi*, a marine *alpha-proteobacteria* belonging to the *Roseobacter* clade [9]. Extensive proteomic analysis of this organism under 30 different culture conditions and growth stages, including changes in media, salt concentration, temperature, oxygen concentration, and UV illumination, resulted in the identification of 2,006 polypeptides, 36 of which were previously unannotated. The N-terminal peptides of 486 polypeptides were also identified, allowing the correction of 64 mis-annotations. By extension, 1,082 homologous genes in other members of the *Roseobacter* clade were also corrected.

Of course, not all organisms have already been sequenced. In this case, proteogenomics can aid in the construction of a shotgun nucleic acid sequence assembly, *e.g.*, when sequencing is inhibited by highly repetitive sequencing, or sequencing can be obviated if sequences from closely related species are available. The former technique was performed on *Spiroplasma melliferum*, the causative agent of spiroplasmosis in *Apis mellifera* [10]. Sequencing of members of the *Spiroplasma* genus has been hindered by the presence of long repetitive sequences, which can lead to inaccurate reads and make assembly of contigs much more difficult, if not impossible. However, using a combination of gene sequencing and peptide detection, 44 % of the genome could be confidently sequenced and annotated. For the kinetoplastid protozoan *Leishmania donovani*, 1 over 20 species that cause leishmaniasis, a custom database compiled from other members of the *Leishmania* genus allowed the identification of 9,431 proteins whose sequences, it can then be inferred, are conserved in the *L. donovani* genome [11].

To more fully survey N-terminal peptides, additional preparation and enrichment steps are necessary. Chemical tags, such as (N-succinimidyloxycarbonylmethyl) tris(2,4,6-trimethoxyphenyl)phosphonium bromide (TMPP), are used to selectively label the N-terminus of proteins prior to digestion [12]. These tags can then be used to enrich the fraction of N-terminal peptides, increasing their detection by removing other signals. In addition, because tags like TMPP can be used to definitively label N-termini, phenomena such as methionine excision and signal peptide removal, as well as the true location of start codons, can be observed with confidence. Experiments using TMPP labeling have revealed the use of non-canonical start codons in the highly radio-resistant bacteria *Deinococcus deserti*, including for the DnaA protein, which is responsible for initiating DNA replication and therefore may be implicated in *D. deserti*'s extraordinary robustness in the face of DNA-damaging radiation [12].

Proteogenomics is a way to detect novel genes, correct sequencing errors, gene reversals, inaccurate start sites and premature stop sites, and to identify N-terminal methionine excisions and signal peptide cleavages. These phenomena occur with alarming frequency: the pan-proteogenomic analysis of 46 organisms revealed 682 novel ORFs, 1,336 incorrect start sites, and 1,175 signal peptide cleavages [13]. Moreover, the observed error frequency is likely below the true frequency because proteogenomic studies only detect proteins that are present at relatively high abundance under the observed conditions and modifications, *e.g.*, acetylation, deformylation, etc., need to be specified a priori to be detected. Therefore, sequencing errors, incorrect annotations, and post-translational modifications are non-trivial and need to be addressed using proteogenomics.

11.3 The Importance of Proteogenomics in Counter Bioterrorism

The Centers for Disease Control and Prevention maintain a list of bioterrorism agents, many of which have been sequenced (Table 11.1). Of the many species and strains on the list, three have been the subject of proteogenomic evaluation. An examination of *Salmonella* Typhimurium (STM) 14028 confirmed >40 % of the predicted protein-coding genes, corrected 47 start sites, and detected 12 novel ORFs, some of which play a role in pathogenesis [14]; a study of *Shigella flexneri* 2a str.301 validated 823 protein products, including hundreds of hypothetical proteins, corrected 2 stop codons, and detected 7 novel ORFs [15]; and an investigation of *Yersinia pestis* KIM detected 1,302 proteins, corrected 6 start sites, identified 4 novel ORFs, reported 82 proteins with signal peptide cleavages [16]. Therefore, sequencing errors, incorrect annotations, and post-translational modifications are non-trivial in potential bioterrorism agents.

Furthermore, the European Commission says of public health preparedness and response, "EU preparedness focuses on all types of CBRN hazard—man-made, natural, accidental or deliberate, *e.g.* deliberate contamination of drinking water, accidental radio-nuclear contamination or the emergence of a new infectious disease including those that take the form of a pandemic." Sequencing can be performed and errors corrected for known biological agents, as has been done for *Salmonella*, *Shigella*, and *Yersinia*. However, anthropogenic alterations—intentional or otherwise—may have unpredictable results, such as antibiotic resistance or

Table 11.1 Number of sequence entries in the National Center for Biotechnology Information genome database for select bacteria on the Centers for Disease Control and Prevention list of bioterrorism agents (as of 26 April 2013)

Species	Disease	Chromosomes	Scaffolds or contigs	SRA[a] or traces	No data
Bacillus anthracis	Anthrax	7	22	0	6
Brucella abortus	Brucellosis	4	97	18	1
Brucella melitensis	Undulant fever	5	46	33	1
Burkholderia mallei	Glanders	4	6	0	1
Burkholderia pseudomallei	Melioidosis	9	25	0	53
Chlamydia psittaci	Psittacosis	18	1	28	38
Clostridium botulinum	Botulism	13	12	0	6
Clostridium perfringens	Epsilon toxin	4	7	0	6
Coxiella burnetii	Q fever	5	5	0	2
Cryptosporidium parvum	Parasite	2	0	0	0
Francisella tularensis	Tularemia	13	16	13	16
Rickettsia prowazekii	Typhus fever	10	0	0	2
Salmonella Typhi	Typhoid fever	33	280	184	582
Shigella flexneri	Shigellosis	5	20	0	10
Vibrio cholerae	Cholera	10	146	5	23
Yersinia pestis	Plague	12	199	0	18

[a]Sequence read archive

increased virulence, and emerging pathogens may not share homology with known agents. Proteogenomics is therefore necessary to ensure the accurate identification of biological agents and prediction of biomarkers.

Fortunately, it is possible to leverage the proteogenomic corrections performed on these known strains in the further correction of related strains and the future characterization of as yet unsequenced strains. In so doing, the accurate sequences of protein biomarkers can be determined, which allows the *in silico* analysis of biomarker specificity by searching for homologous known sequences and the prediction of biomarker sequences in related organisms. Furthermore, in the case of whole cell fingerprinting, proteins responsible for prominent peaks can be identified by their mass and, conversely, accurate masses for prominent peaks can be predicted. Thus, proteogenomics enhances biomarker discovery and prediction and gathers information to help interpret reference spectra.

11.4 Applications of Proteogenomics in MS-Based Bacterial Detection

Many proteogenomic studies are aimed solely at exposing and correcting errors in sequencing and annotation, which is surely a noble goal in and of itself. However, once corrections have been made, these proteogenomic discoveries can be applied to better the detection of environmental bacteria in general and potential bioterrorism agents in particular. One example where proteogenomic analysis has enabled detection of bacteria from environmental samples is the reannotation and biomarker prediction for *Ruegeria lacuscaerulensis* [17].

R. lacuscaerulensis is another marine *alpha-proteobacteria* in the *Roseobacter* clade and a relative of *R. pomeroyi*, which has also been the subject of proteogenomic investigation [9]. As of 16 July 2013, there are five genome entries for *Ruegeria* species in the National Center for Biotechnology Information database (http://www.ncbi.nlm.nih.gov/) of varying degrees of completion. The goal of this particular study was to use the available sequence for *R. lacuscaerulensis* as a model organism to predict biomarkers for related species that cannot be isolated or cultured in the laboratory [17].

Potential biomarkers were screened using whole cell matrix-assisted laser desorption/ionization time-of-flight mass spectrometry (MALDI-TOF-MS). Proteins responsible for reproducibly detected, prominent peaks were identified by running protein extracts on sodium dodecyl sulfate polyacrylamide gels followed by in-gel digestion with trypsin and high performance nano-liquid chromatography tandem mass spectrometry. These high resolution proteomic data were then used to evaluate the sequences and annotations of the corresponding genes, revealing that several of the mature proteins underwent N-terminal methionine excision and that three homologous *Ruegeria* sequences were missing between 4 and 17 amino acids at the N-terminus. By correcting these sequences, it was possible to predict a set of protein biomarkers that are conserved for the *Ruegeria* genus and easily observed by MALDI-TOF-MS. To demonstrate the utility of this process, the predicted biomarkers were used to screen unknown isolates from an environmental sample. Of the unknown strains, one contained the biomarkers; its classification as a *Ruegeria* strain was confirmed by 16S rDNA sequencing.

This technique could just as easily be applied to the screening of potential bioterrorism agents for the rapid screening of environmental or clinical samples.

11.5 Confidence

Just as errors can occur during gene sequencing, a certain amount of inaccuracy can also be produced during the mass spectrometric analysis of peptides and the interpretation thereof. Sample quality may not be ideal, instruments are not 100 % accurate or precise, and the computer algorithms that assign spectra are not infallible. These challenges notwithstanding, proteomic-corrected genome sequences and biomarkers can still be used, but reasonable precautions need to be taken.

Weaknesses in DNA sequencing are fairly well known. It is accepted that poor quality starting material will yield poor sequencing results; that highly repetitive sequences will produce errors is also an oft repeated adage; and the pitfalls of insufficient sequence coverage in shotgun experiments and subsequent automatic annotation are similarly evident. Criteria such as those outlined in the Minimum Information for Publication of Quantitative Real-Time PCR Experiments guidelines

[18] should also be considered when evaluating the quality of a genome sequence and its potential to contain errors, although specific criteria related to annotation, such as the target species' divergence from the model organisms on which the algorithms are based, should be developed.

The results of a mass spectrometric analysis are no less susceptible to the state of the starting material, especially bias and degradation during protein extraction and preparation. Furthermore, the mass error of the instrument can vary widely depending on the particular model and its maintenance and calibration. It is especially important to use a good estimate of the mass error in the assignment of peptide spectrum matches, as even small changes will affect the peptides matched and the strength of the assignment. Search engines, *e.g.*, Mascot, are notoriously fickle and extremely sensitive to changes in input parameters [19, 20]. Even given the optimal search parameters, results are based on probability and therefore have an intrinsic false discovery rate. For example, for every million MS/MS spectra assigned, 10,000 false positive peptide spectrum matches may be assigned for a *p* value below 0.01. Each of these false assignments could lead to a false discovery, i.e., the erroneous modification of a genome sequence. The actual false discovery rate can be estimated by searching decoy databases, but this process overestimates the number of false positives because of the unusual large search space [21]. Dubious assignments still need to be inspected manually, and very stringent criteria should be used before modifying a gene sequence based on an observed peptide.

The use of contrived databases composed of six-frame translations of one or a few closely related organisms makes peptide spectrum matching in proteogenomic experiments especially precarious. The dangers of using a limited database were highlighted in a study looking for pathogens of *Apis mellifera* [22]. In this example, two different peptide spectral matches are determined for the same proteomic data using two different databases, one arguably better than the other, but because that particular sequence was not included in the first database, it could not be identified. It is essential to remember that the exclusion of the correct peptide from the set of possible matches does not preclude a match with a significant score; it just precludes the correct match. The interpretation of proteogenomic data must therefore proceed with the utmost attention.

The rapid, high throughput mass spectrometry-based identification of potential bioterrorism agents hinges on accurate, specific protein sequences that can be used to detect the specified target from a complex environmental background containing untold numbers of other bacteria. While genome sequencing methods have vastly improved, an undeniable advance towards attaining this goal, errors and misannotations pervade both new and archived genome sequences, leading to inaccurate protein sequence and mass prediction. Furthermore, as an ever increasing number of environmental bacteria are characterized and sequenced, non-canonical start codons and other unprecedented modifications are seen more frequently, compounding the aforementioned errors. Proteogenomics helps compensate for imperfections in DNA sequencing and gene annotation by casting light on non-silent errors in genome sequences, confirming or correcting the location, direction, and length of coding regions, and identifying post-translational modifications that affect the mass of observed peptides. Just as DNA sequencing has its limitations, mass spectrometry is also prone to certain errors; however, the use of appropriate standards and adequate statistics can compensate for imperfections in mass spectrometry and data processing. Thus proteogenomics is an invaluable tool in the mass spectrometry-based detection of potential bioterrorism agents.

Acknowledgements The authors thank the Fulbright program for supporting EMH.

References

1. Hartmann EM, Halden RU (2010) Challenges of detecting bioterrorism agents in complex matrices. In: Banoub J (ed) NATO-science for peace and security-chemistry and biology series: detection of biological agents for the prevention of bioterrorism. Springer, pp 149–162
2. Fenselau CC (2013) Rapid characterization of microorganisms by mass spectrometry-what can be learned and how? J Am Soc Mass Spectrom 24(8):1161–1166
3. Devos D, Valencia A (2001) Intrinsic errors in genome annotation. Trends Genet 17(8): 429–431
4. Armengaud J (2009) A perfect genome annotation is within reach with the proteomics and genomics alliance. Curr Opin Microbiol 12(3):292–300
5. Armengaud J (2010) Proteogenomics and systems biology: quest for the ultimate missing parts. Expert Rev Proteomics 7(1):65–77
6. Armengaud J (2013) Microbiology and proteomics, getting the best of both worlds! Environ Microbiol 15(1):12–23
7. Armengaud J, Hartmann EM, Bland C (2013) Proteogenomics for environmental microbiology. Proteomics 13:2731–2742
8. Hartmann EM, Allain F, Gaillard J, Pible O, Armengaud J (2013) Taking the shortcut for high-throughput shotgun proteomic analysis of bacteria. Method Mol Biol 1197 (in press)
9. Christie-Oleza JA, Miotello G, Armengaud J (2012) High-throughput proteogenomics of Ruegeria pomeroyi: seeding a better genomic annotation for the whole marine Roseobacter clade. BMC Genomics 13:73
10. Alexeev D, Kostrjukova E, Aliper A, Popenko A, Bazaleev N, Tyakht A, Selezneva O, Akopian T, Prichodko E, Kondratov I, Chukin M, Demina I, Galyamina M, Kamashev D, Vanyushkina A, Ladygina V, Levitskii S, Lazarev V, Govorun V (2012) Application of Spiroplasma melliferum proteogenomic profiling for the discovery of virulence factors and pathogenicity mechanisms in host-associated Spiroplasmas. J Proteome Res 11(1):224–236
11. Pawar H, Sahasrabuddhe NA, Renuse S, Keerthikumar S, Sharma J, Kumar GSS, Venugopal A, Sekhar NR, Kelkar DS, Nemade H, Khobragade SN, Muthusamy B, Kandasamy K, Harsha HC, Chaerkady R, Patole MS, Pandey A (2012) A proteogenomic approach to map the proteome of an unsequenced pathogen – Leishmania donovani. Proteomics 12(6):832–844
12. Baudet M, Ortet P, Gaillard JC, Fernandez B, Guerin P, Enjalbal C, Subra G, de Groot A, Barakat M, Dedieu A, Armengaud J (2010) Proteomics-based refinement of Deinococcus deserti genome annotation reveals an unwonted use of non-canonical translation initiation codons. Mol Cell Proteomics 9(2):415–426
13. Venter E, Smith RD, Payne SH (2011) Proteogenomic analysis of bacteria and archaea: a 46 organism case study. Plos One 6(11):e27587
14. Ansong C, Tolic N, Purvine SO, Porwollik S, Jones M, Yoon H, Payne SH, Martin JL, Burnet MC, Monroe ME, Venepally P, Smith RD, Peterson SN, Heffron F, McClelland M, Adkins JN (2011) Experimental annotation of post-translational features and translated coding regions in the pathogen Salmonella Typhimurium. BMC Genomics 12:433
15. Zhao LN, Liu LG, Leng WC, Wei CD, Jin Q (2011) A proteogenomic analysis of Shigella flexneri using 2D LC-MALDI TOF/TOF. BMC Genomics 12:528

16. Payne SH, Huang ST, Pieper R (2010) A proteogenomic update to Yersinia: enhancing genome annotation. BMC Genomics 11:460
17. Christie-Oleza JA, Pina-Villalonga JM, Guerin P, Miotello G, Bosch R, Nogales B, Armengaud J (2013) Shotgun nanoLC-MS/MS proteogenomics to document MALDI-TOF biomarkers for screening new members of the Ruegeria genus. Environ Microbiol 15(1):133–147
18. Bustin SA, Benes V, Garson JA, Hellemans J, Huggett J, Kubista M, Mueller R, Nolan T, Pfaffl MW, Shipley GL, Vandesompele J, Wittwer CT (2009) The MIQE guidelines: minimum information for publication of quantitative real-time PCR experiments. Clin Chem 55(4):611–622
19. Hartmann EM, Colquhoun DR, Halden RU (2010) Identification of putative biomarkers for toluene-degrading Burkholderia and pseudomonads by matrix-assisted laser desorption/ionization time-of-flight mass spectrometry and peptide mass fingerprinting. Biosci Biotechnol Biochem 74(7):1470–1472
20. Fox K, Fox A, Rose J, Walla M (2011) Speciation of coagulase negative staphylococci, isolated from indoor air, using SDS page gel bands of expressed proteins followed by MALDI TOF MS and MALDI TOF-TOF MS-MS analysis of tryptic peptides. J Microbiol Methods 84(2):243–250
21. Blakeley P, Overton IM, Hubbard SJ (2012) Addressing statistical biases in nucleotide-derived protein databases for proteogenomic search strategies. J Proteome Res 11(11):5221–5234
22. Knudsen GM, Chalkley RJ (2011) The effect of using an inappropriate protein database for proteomic data analysis. Plos One 6(6):e20873

Chapter 12
Laser-Based Detection of Explosives and Related Compounds

Itamar Malka, Salman Rosenwaks, and Ilana Bar

Abstract Laser-based spectroscopic methods offer the possibility of detecting compounds of interest with high sensitivity, in real time, without or with minimal sample preparation and therefore have a potential to be readily transferred for use in the field. A variety of methods, including laser photolysis/laser-induced fluorescence, laser-induced breakdown spectroscopy, as well as vibrational spectroscopies: Raman and coherent anti-Stokes Raman scattering, have been studied and tested over the years for explosives detection. The fundamentals and preferred applications of these methods for standoff and point detection of traces or bulk compounds in the gas and condensed phases are discussed, pointing to their advantages and drawbacks with the hope that recent progress will bring them closer to find their niche.

Keywords Explosives detection • Laser photolysis/laser-induced fluorescence • Laser-induced breakdown spectroscopy • Raman scattering • Coherent anti-Stokes Raman scattering

12.1 Introduction

There appears to be almost no day that the electronic or print media does not inform us on the latest suicide bombing or improvised explosive device (IED) attack somewhere in the world. Therefore, it is not surprising that this rising use of explosives and other hazardous materials by terrorists place the protection of civilians as well as soldiers at the forefront of efforts of improving existing and developing new security measures. This, in turn, leads to a rapidly expanding area of research directed toward development of various methods for detection and identification of threats by explosives.

Indeed, different means are already available for military, counterterrorism and homeland security applications. For example, metal detectors in combination with

---

I. Malka • S. Rosenwaks • I. Bar (✉)
Department of Physics, Ben-Gurion University of the Negev, Beer-Sheva 84105, Israel
e-mail: ibar@bgu.ac.il

J. Banoub (ed.), *Detection of Chemical, Biological, Radiological and Nuclear Agents for the Prevention of Terrorism*, NATO Science for Peace and Security Series A: Chemistry and Biology, DOI 10.1007/978-94-017-9238-7_12

X-ray machines are used at airports to identify concealed weapons and for viewing baggage contents, respectively [1]. Other approaches for screening items of hand baggage include ion mobility spectrometry combined with swabbing. Nevertheless, a major difficulty is that often explosive compounds are not easily detectable using these conventional approaches and under some scenarios, detection of explosives in gas phase at low concentrations, or of contaminants or residues left on surfaces is required. Furthermore, IEDs can contain a variety of explosives, i.e., military, commercial, or homemade, with different compositions, demanding detection and identification of various compounds and consequently posing great challenges for the methods to be used for their detection.

Indeed, as previously reviewed [1–9], various laser-based spectroscopic methods were considered to satisfy these requirements and were tested for this purpose. These methods raise considerable interest, since they are expected to provide several advantages, including high sensitivity and selectivity, as well as the possibility to perform in-situ measurements, without any, or with minimal sample preparation and in real time. In principle, these methods use laser beams of certain wavelengths that are shone on the sample and the resulting signal, related to different light-matter interactions can lead to distinctive spectral signatures, which are collected and acquired by specific detectors. Here we discuss some of these methods and particularly those that our group was involved in assessing their feasibility, while presenting some aspects of the current state of research.

12.2 Spectroscopic Methods

12.2.1 Laser Photolysis/Laser-Induced Fluorescence

Polyatomic analytes are usually characterized by broad and structureless absorption features in the ultraviolet (UV). Therefore, it is usually difficult to directly detect them by absorption, or to electronically excite them and detect their fluorescence by laser-induced fluorescence (LIF). However, since these molecules are considered to be very fragile, their bonds can be cleaved by laser photolysis (LP) and hence be converted to small species, which can be easily detected by LIF. This is due to the release of photofragments with characteristic strong and well-defined transitions, which lead to unique and well-resolved spectra that may improve significantly their identification.

Many explosives, being referred as nitrocompounds (NCs), contain the common functional nitro group. These compounds can be photodissociated in a relatively broad UV range allowing the release of typical photofragments, i.e., NO and NO_2, where the latter is rapidly predissociated to NO molecules and O atoms. Indeed, experiments based on LP, followed by resonance-enhanced multiphoton ionization (REMPI), where ions were detected by miniature electrodes or by mass spectrometry (MS), as well as by LIF, were performed [9, 10]. Particularly, for REMPI/LIF detection of the NO photofragment, a one-color UV laser was utilized, which photodissociated the analyte effectively and subsequently excited its $A^2\Sigma^+(v' = 0) \cdot \leftarrow$

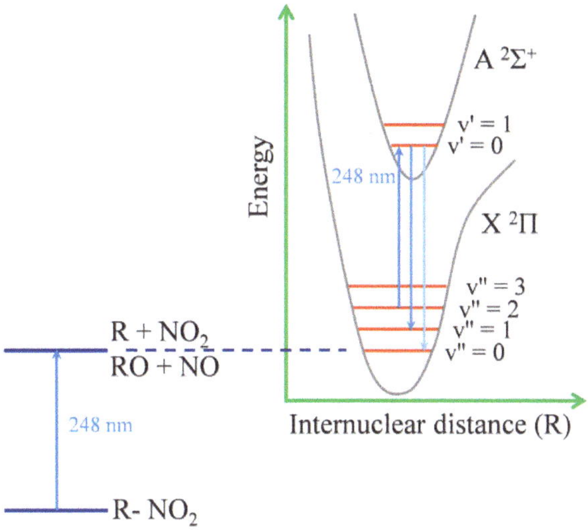

Fig. 12.1 Schematic of the concept used in detection of nitrocompounds by the laser photolysis/laser induced fluorescence method, where vibrationally excited NO is probed to avoid background fluorescence

$X^2\Pi(v''=0)$ rovibrational transitions at wavelengths <226 nm, while capturing the corresponding signal.

Considering that NCs are non-volatile and with very low vapor pressures [2, 4], it is obvious that their concentrations in ambient air under normal conditions are very low, implying the necessity for low limits of detection (LODs) for NO photofragments. In case of LIF detection, one of the factors that hinder the possibility of obtaining low LODs is the background fluorescence from other gases and aerosols, occurring at equal or longer wavelengths than the exciting radiation. Therefore, we proposed and tested the use of rovibrationally excited NO photofragments as indicators for an NC analyte presence [11–14]. As can be seen from Fig. 12.1 that displays the concept of the experiment, excitation of higher vibrational states enables capturing the NO fluorescence at shorter wavelengths than the exciting one, thus extensively reducing the background fluorescence and the associated scattering. An additional advantage offered by this concept is that it prevents the interference from ambient NO, which can be at high concentration.

Our experiments on the simplest nitroaromatics, i.e., nitrobenzene (NB), 1,4 dinitrobenzene (DNB) and 2,4 dinitrotoluene (DNT) have shown that not only vibrationless ground state NO is produced, but also states with initial excess of higher vibrational energy $(X, v''=1,2)$ are substantially populated. This allowed detection of NCs by a one-color laser, scanned around 248 nm, coinciding with the LIF excitation wavelength of the $A^2\Sigma^+(v'=0) \leftarrow X^2\Pi_{1/2,3/2}(v''=2)$ rovibrational transitions (see Fig. 12.1), while monitoring signal photons at shorter wavelengths. This scheme enabled the detection of NO with high sensitivities, with LODs of the order of several parts per billion (ppb) by weight of NC diluted in air and at short integration times.

However, as mentioned above, explosives are characterized by low vapor pressures and since they might be hidden in IEDs, or their vapor might be dispersed by normal airflow, the concentration of specific NCs in air can be very low, turning detection in the gas phase extremely difficult. Yet, NC residues are likely to be present on surfaces since considerable amount of material could be transferred to any touched object during preparation, transportation or handling of IEDs [15]. Therefore, it was conjectured that by heating the sample existing on surfaces, detectable amount of vapor could be obtained.

Indeed, coupling of thermal [16, 17], or laser desorption [18–20] of the sample with LP/LIF, enabled standoff detection of explosives in the condensed phase from short distances. This was achieved by measuring the resulting "on" and "off" NO LIF signal at wavelengths corresponding to specific NO transitions, or at the background (where no transitions occur), respectively. For example, Heflinger et al. placed 2,4,6-trinitrotoluene (TNT) flakes on a temperature-controlled sample holder in ambient air and adapted the experiment to standoff detection from 2.5 m [17]. They reported detection limits in the low ppb region with a response time of 15 s. In a more recent experiment, a single 7 ns pulse of a tunable laser near 236.2 nm was used to perform the multiple processes of vaporization of the compound, of inducing its photodissociation and of exciting the vibrationally "hot" NO indicator [18]. In this experiment the blue-shifted fluorescence was collected from a distance of 6 cm for various nitro-bearing compounds, including 2,6-dinitrotoluene (DNT), TNT, pentaerythritol tetranitrate (PETN), and hexahydro-1,3,5-trinitro-1,3,5-triazine (RDX) with a signal-to-noise of 25 dB.

12.2.2 Laser-Induced Breakdown Spectroscopy

In view of the fact that NCs are likely to be present on surfaces, it seemed to us that laser-induced breakdown spectroscopy (LIBS) [2, 3, 8, 21], could be a very promising detection technique. In this technique a pulse of high intensity is focused on the surface of a material, ablating a small quantity of it. The ablated material is vaporized and then partially broken to atomic and small molecular species, which are excited and ionized within the laser-generated plasma. Actually, the generation of the plasma on the surface sample is followed by several processes, including plasma expansion, building up of a shock wave, continuum emission (Bremsstrahlung), and ions, atoms, and molecules de-excitation through optical emission, see Fig. 12.2.

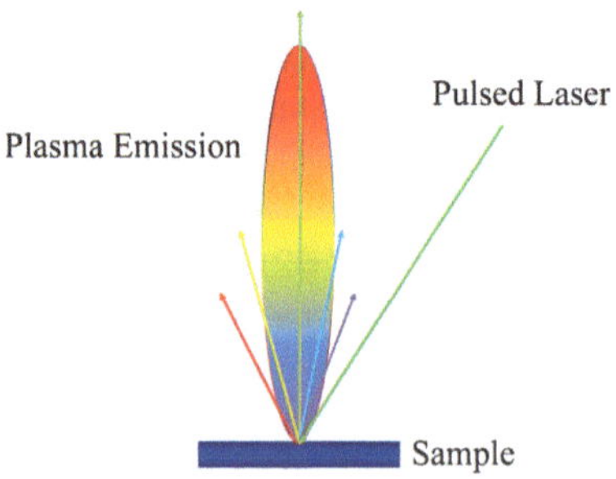

Fig. 12.2 Schematic representation of the idea behind laser induced breakdown spectroscopy, showing the plume that is generated as a result of the interaction of the beam laser with the target. The plume contains different excited species, which emit light at different wavelengths

Although most early LIBS applications involved metal targets, we conjectured and assessed the feasibility of using this method for detection of organic compounds, including NCs [22, 23]. In particular, we have shown that the emission resulting from different compounds exhibits sequences of the $CN(B^2\Sigma^+ \rightarrow X^2\Sigma^+)$ violet system and of the $C_2(d^3\Pi_g \rightarrow a^3\Pi_u)$ Swan system. These species originated not only from the decomposition of the target compounds, but also from the interaction of the C_2-containing plume with N_2 from the surrounding air and from C_2H_x fragments directly released from the aromatic rings in the target compounds or from carbon-carbon recombination in the hot plasma, respectively. Furthermore, emission of H, N, and O atomic lines could also be observed.

Considering that the composition of the different compounds is quite similar and that the LIB spectra exhibit the same emission lines, we suggested to base the identification of the target compounds on the integrated intensity ratios of the molecular (C_2, CN) and atomic (O, N) fragments [22, 23]. Since a correlation was found between the ratio of the characteristic emission features in the plasma and the target structure, we proposed that these ratios could indicate the presence of the compound of interest and therefore identify it. This should be particularly so, if the spectral features and the obtained ratios could be compared to a previously acquired database, while using pattern recognition algorithms.

Several groups have investigated whether LIBS can be used as an analytical tool for standoff analysis of energetic materials. For example, Miziolek and co-workers [21] investigated the LIB spectra obtained from explosive target materials, including highly purified RDX, cyclotetramethylene tetranitramine (HMX), TNT, and PETN, and from operational explosives and propellants C-4, A-5, M-43, LX-14, and JA2, using a 10 ns pulse with energy of 30 mJ from an actively Q-switched Nd:YAG laser. The observed LIBS spectra contained the expected C_2 and CN molecular emission peaks as well as C, H, N, and O emission lines, and atomic and ionic emission resulting from impurities (Ca, Na, K, and Mg). Furthermore, Lopez-Moreno et al. [24] used a Q-switched Nd:YAG laser operating at 1,064 nm and producing 350 mJ single pulses at 20 Hz and an open-path system working under a coaxial configuration to demonstrate detection of energetic materials in the field. They have shown promising results for the discrimination of explosive residues placed on a vehicle surface located at 45 m from the instrument. Miziolek and co-workers [21] described the development of algorithms for classification of compounds by LIBS, while comparing many different algorithms. Initially, they used an algorithm similar to that employed in Ref. [24], but they compared peak ratios with peak ratios, instead of fixed numbers, to discriminate explosives from non-explosives. The algorithm showed good performance for pure samples, however, it was suggested that for real-world applications with sample mixtures, a more sophisticated algorithm should be used.

As can be seen from the above, quite high energies are required for obtaining the analyte LIB signal. Later on, even higher laser energies were used to enhance the LIB signal, however, this approach has some complications, particularly if LIBS is employed for residue detection. In this case, the use of higher laser energy leads to increased substrate penetration with an associated increase in substrate signal and damage to the target surface [25]. This in turn, might be problematic when LIBS is used for detection of explosive residues deposited on high-value substrate materials. Therefore, different methods were tested for this purpose [2, 3, 8], including, preferential ablation of the residue using specific laser wavelengths [26–28], the use of ultrashort laser pulses [29, 30], maintaining the LIBS plasma through addition of non-ablating energy [31], or resonance signal enhancement of the ablated material in the laser-induced plasma [32, 33].

An additional means that should be mentioned is the use of double-pulses, where the first laser pulse produces a "laser-generated vacuum" [34] and the second one, a few μs later, allows obtaining a LIB spectrum without the influence of ambient air. This method was employed for detection of explosives at a standoff distance of 20 m [35]. It was shown that the double-pulse LIBS increases the sensitivity, while increasing the ablation rate, plasma volume, temperature and ion density, and therefore allows a larger standoff distance and enhanced selectivity.

12.2.3 Raman Spectroscopy

One of the vibrational spectroscopic methods, namely, spontaneous Raman spectroscopy, has recently seen a dramatic increase for detection of explosives. In contrast to LP/LIF, which is based on the detection of NO photofragments, resulting from the functional groups in compounds and to LIBS, which shows the same features, but with somewhat different intensities, Raman spectroscopy provides unique signatures reflecting specific vibrations of a particular compound. This is due to the fact that the observed spectral features are related to the vibrational frequencies of molecules, which depend on the masses of the atoms involved in the motion and the strength of the involved bond(s) and therefore provide unique spectral signatures, implying that they are well-suited for molecular identification and for studying molecular structures and interactions. The vibrational spectrum provided by Raman scattering is a molecular signature that can be used to differentiate the sample from a complex media or even differentiate very similarly structured molecules from one another.

As can be seen from Fig. 12.3a, Raman spectroscopy is related to inelastic scattering of photons, where some energy is lost to (Stokes), or gained (anti-Stokes) from the compound, leading to scattered light with a different wavelength depending on the energy of the molecular vibrational modes. In most cases the Stokes scattering is utilized for detection, but even the intensity of this signal is very low. This is so since the scattering cross section per unit volume for Raman Stokes scattering in condensed matter is only approximately 10^{-6} cm^{-1} and therefore, in propagating through 1 cm of the scattering medium, only approximately one in 10^6 photons incident upon the sample is scattered into the Stokes frequency [36].

Even though, standoff Raman explosives detection and spatial offset Raman spectroscopy (SORS) for detection of concealed content in distant objects has been tested and shown to be possible by several groups [37–41]. For example, by using a

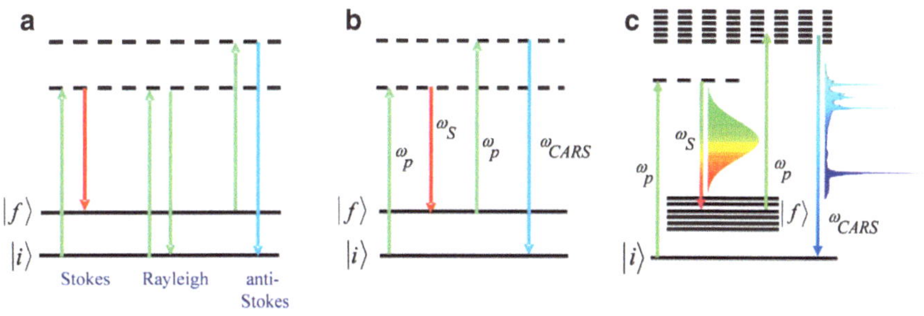

Fig. 12.3 Energy diagram, including the initial, |*i*>, final, |*f*>, and virtual states (*dashed lines*) involved in the (**a**) elastic (Rayleigh) and inelastic (Stokes and anti-Stokes) spontaneous scattering, (**b**) coherent anti-Stokes Raman scattering (*CARS*), where a beam, which is used as both pump and probe at frequency ω_p and a narrowband Stokes beam of frequency ω_S interact with the sample to generate a coherent signal at the anti-Stokes frequency (ω_{CARS}), and (**c**) CARS with a broadband Stokes beam, which allows generation of the coherent signal for all the vibrations at the corresponding anti-Stokes frequencies

frequency doubled pulsed Nd:YAG laser, a Schmidt-Cassegrain telescope, a spectrometer for wavelength dispersion, and an intensified charge-coupled device (ICCD) for gated detection it was shown that it is possible to detect explosives through different container materials (transparent or semitransparent) and outdoor environment at distances of up to 55 m [39].

Moreover, it was shown that Raman scattering could be obtained by excitation with UV wavelengths [42, 43], taking advantage of several characteristics of the signal. In particular, the intensity of the Raman signal scales as $1/\lambda^4$ (λ is the excitation wavelength), therefore this results in improvement of the inherently weak signal. Besides, if a UV wavelength is close to an electronic transition of the sample, resonance enhancement can occur, leading to significant enhancement of Raman lines related to the chromophore. For example, a Raman system for field detection and identification of minimal amounts of explosives on relevant surfaces at a distance of up to 30 m was demonstarated [43]. The use of energetic photons in the UV Raman system resulted in Raman lines in a spectral range close to the incident excitation frequency, where the luminescence was still very weak. This allowed interference free detection of Semtex by gated UV Raman spectroscopy, while exciting at 248 nm. It should be noted that the identification of this explosive with 532 nm excitation was not possible due to the accompanying high fluorescence [43].

The approach of resonance-enhanced Raman spectroscopy was also recently used for detection of explosives vapor at standoff distances [44]. It was shown that the Raman cross sections for DNT and TNT in vapor phase are enhanced up to 250,000 times and up to 60,000, respectively, compared to the nonresonant signal at 532 nm. Furthermore, the outdoor measurements performed on nitromethane (NM) in vapor phase at a 13 m distance, indicated the potential of resonance Raman spectroscopy as a standoff technique for detection of vapor phase explosives.

Since Raman and LIBS standoff detection share similar pieces of equipment, systems that are able to detect both types of signals were also tested [45, 46].

For instance, it was demonstrated that by using a standoff fused Raman-LIBS sensor [46] with well-defined timing conditions, simultaneous acquisition of molecular and multielemental spectral data, from a sample located at 20 m, was possible while monitoring the same laser event. In this approach, the energy gradient created along the target surface by the laser beam was exploited to extract vibrational data from the outer part of the interacting surface and atomic information from the ablated mass. It was suggested that the complementary information of Raman spectroscopy and LIBS holds promise for situations requiring simultaneous spectral and compositional information from the same spot of a sample.

Despite the substantial success of standoff detection by Raman or Raman/LIBS systems in the lab and in field trials, it is to be recalled that the above-mentioned methods have to overcome intrinsic difficulties while detecting explosives at large distances, including eye safety of the excitation beam, free beam and signal paths, absorption and scattering losses in air (wavelength-dependent), decrease in signal intensity with inverse squared distance as well as complication and high cost of systems that could not always be considered to be appropriate for field measurements.

At any rate, at least under some scenarios, point and proximal detection or remote sensing of explosives or other substances is required. For this purpose, we constructed and evaluated the performance of a compact Raman spectrometer, based on 180° backscattering configuration and a low-power frequency doubled green laser pointer for sample excitation [47]. This allowed measurement of the spectra of several liquids and solid particles of explosives and related compounds at short acquisition times. The various samples presented different spectral signatures, consisting of several peaks, corresponding to the multiple Raman resonances in a specific compound. The Raman spectra were found to differ even when the structural differences between the compounds were minimal. Hence, even the spectra obtained for two isomers of the same compound (2,4 and 3,4 DNT) were somewhat different, allowing discrimination between the samples. This implies that identification of a compound in a sample would be possible by comparing the measured Raman spectra to those appearing in a database, while finding the highest correlation.

Moreover, since this system included a x,y motorized translational stage that sustained the sample (dispersed powder on glass microscopic slides), it had the capability to first scan the sample optically, while sweeping from side to side, till a specific particle was found to strongly "glow" and then to measure the Raman spectrum at this location. But even more so it enabled to chemically image a sample of residues from latent fingerprint of potassium nitrate (KNO_3). By raster scanning, while recording successive Raman spectra at different measurement points over an array, it was possible to obtain the spectral signatures at particular positions through the fingerprint. The spectral signatures were found to correspond to the Raman spectra of KNO_3 and therefore by measuring the height of the dominant peak at 1,054 cm^{-1} with respect to the background enabled obtaining an intensity matrix, which could be transformed into an x, y Raman intensity map. This map clearly showed the existence of KNO_3 particles, smaller than 10 μm diameter, in the measured region of the fingerprint, which corresponded very well with the positions of the particles in an optical image.

The performance of the system was further improved by integrating it with a smartphone camera, which enabled photo-guided sampling and proximal detection [48]. This was achieved by first photographing the target that contained the particles of interest and then by employing a Matlab code for particle recognition to find the exterior boundaries of the individual particles. This approach led to coordinates and required area for scanning in the region of each particle and finally to measurement of the Raman spectra in these limited regions, allowing to acquire Raman maps. This selective monitoring of the spatially resolved particles allowed mapping at reduced sampling times, compared to raster-scanning.

The ability to chemically image traces of a specific compound by these compact systems is of much importance, since such a task is usually performed by large and cumbersome Raman microscopes. It should be noted that although different portable Raman spectrometers were developed by various companies, they typically demonstrate lower sensitivities and are used to identify only unknown bulk samples [5, 49]. Moreover, considering that the spontaneous Raman scattering cross sections are very low, we expected that it will be possible to increase the detection sensitivity by replacing spontaneous Raman with coherent anti-Stokes Raman scattering (CARS).

12.2.4 Coherent Anti-stokes Raman Scattering

CARS, as shown in Fig. 12.3b, is a four-wave parametric process in which three waves, one at a pump frequency, ω_p, the second at a Stokes frequency, ω_s, and the third at a probe frequency, which in most cases (including here) is the same as ω_p (the same laser beam is employed), are mixed in a sample to produce a new coherent beam at the anti-Stokes frequency, i.e., $\omega_{CARS}=2\omega_p-\omega_s$ when the wavevectors fulfill the phase matching condition, $2k_p–k_S–k_{CARS}=0$ [36, 50, 51]. The mixing occurs for all frequencies, but is greatly enhanced when the $\omega_p-\omega_s$ frequency difference corresponds to Raman-active vibrational resonances of the detected species. This implies that by using a broadband laser or varying ω_s it is possible to detect the vibrational frequencies of a compound and monitor a spectrum, reminding that obtained by spontaneous Raman.

Considering its characteristics, it was suggested that this non-linear Raman spectroscopy can enhance the signal relative to that obtained with spontaneous Raman. CARS can be obtained by applying pulses ranging from ultrashort (fs) to ns and indeed in the last years different schemes of fs CARS, were applied for various applications, including standoff detection of different species [52–56]. Nevertheless, when fs lasers are used, usually a nonresonant background signal, which is orders of magnitude larger than the resonant CARS signal, is accompanying it. Therefore, for measurements of the resonant CARS, suppression of the nonresonant component is required and different schemes, relying on the properties of this nonlinear four-wave mixing interaction, have been developed for this aim. These approaches include tailoring of the temporal width and delay of the probe-pulse [57] as well as polarization pulse shaping techniques [52, 54] for performing background-free measurements. In another scheme [55, 56], the non-resonant background was used as a local oscillator for homodyne amplification of the CARS signal, allowing standoff detection of sub-milligram quantities of explosives.

One more way to reduce extensively the nonresonant CARS contribution is by performing the excitation with ns pulses. Thus, using a relatively simple system, we were interested in finding the capabilities and testing this approach. A system that consisted of a ns optical parametric oscillator (OPO) pumped by the third harmonic of a Nd:YAG laser operating at 10 Hz was built [58, 59]. In this case, the tunable OPO signal beam provided the ω_s beam and the second harmonic residual of the Nd:YAG laser the ω_p beam and both beams were spatiotemporally overlapped, to fulfill the requirement for CARS generation. The system enabled measurement of the CARS signal, resulting from diffuse reflections of the forward-generated CARS, as well as the spontaneous Raman scattering. This was achieved by blocking the ω_s beam and replacing the short wave pass filters ahead of the monochromator by long wave pass ones. Measurement of CARS and Raman spectra of different explosives allowed comparing their sensitivities, showing that the CARS signal for the dominant peaks is about 25 to 1.6×10^4-fold larger than for the corresponding Raman signal, depending on the compound. Furthermore, by measuring the dependence of the CARS signal on the distance, the standoff detection capabilities of the method were roughly estimated.

It seemed that using this approach would be beneficial for detection since only the resonant CARS is observed under this conditions and in addition it was found to be favorable compared to spontaneous Raman spectroscopy. Nevertheless, one of its main drawbacks is that the CARS spectra can be monitored only by the time consuming process of tuning ω_s beam over the excitation frequencies of the sampled compound. Therefore, we suggested to use a system that would still provide ~ ns pulses, but with a broadband ω_s beam, Fig. 12.3c, instead of the tunable one. Such a system could provide a multiplex CARS signal in short acquisition times.

Since this system was built for proof of concept studies and was not described in our previous work, a short description will be given here [60]. In the broadband ns CARS experiments, a *Q*-switched Nd:YAG microchip laser at 1,064 nm was used. The output of this laser was split into two, so that a part of it was amplified and doubled to provide the second harmonic (532 nm) beam and the other portion was coupled into a photonic crystal fiber (PCF) to generate a supercontinuum (SC) beam. This laser operated at a repetition rate of 10.6 kHz and delivered pulses of ~900 ps duration. The 532 nm beam with a power of about 200 mW served as the pump beam, ω_p, while the SC, in the 450–1,850 nm range with about 40 mW power, was used for obtaining the Stokes beam, ω_S. As can be seen from Fig. 12.4a, which shows a portion, up to 1,000 nm, of the SC beam spectrum [due to the limited response of the spectrometer used to measure the spectrum of the SC output] most intensity is in the range of 450–660 nm. However, since only the 532–660 nm is required for generating the CARS signal, the SC beam was filtered by a long wave-pass edge filter and a color glass filter to obtain the spectrum shown in Fig. 12.4b. This filtered beam was used as the ω_S beam. Spatiotemporal overlap of the ω_p and ω_S beams was required for generating the CARS signal with high efficiency.

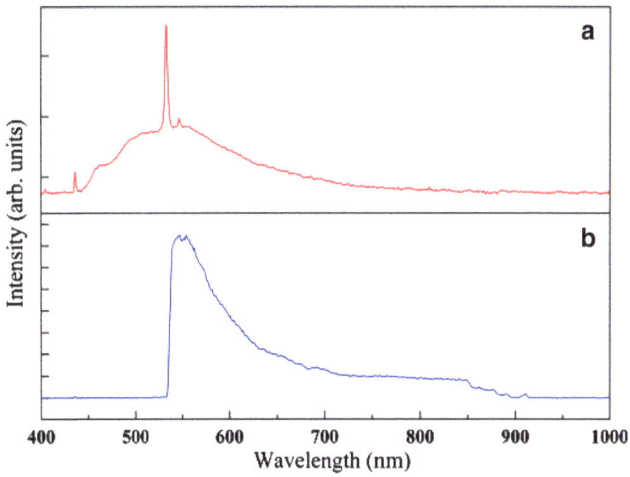

Fig. 12.4 The spectrum of the supercontinuum beam in the 400–1,000 nm spectral range, as measured (**a**) immediately at the output of the photonic crystal fiber and (**b**) following filtration

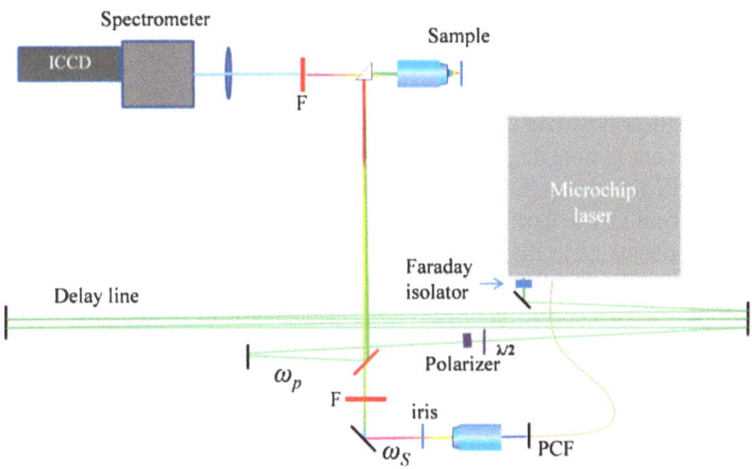

Fig. 12.5 Schematic of the experimental setup for the sub-ns ultra-broadband CARS spectrometer, including a microchip laser, a photonic crystal fiber (*PCF*), mirrors and dichroic mirrors, prism, objectives, lens and a detector [spectrometer/intensified charge-coupled device (*ICCD*)]

Therefore, as seen from Fig. 12.5, a delay line for the ω_p beam and a dichroic mirror was used to recombine the beams and to copropagate them toward the sample. The laser beams were then turned by a 90° tiny prism and tightly focused through ×40/0.65 microscope objectives to provide powers of about 15 mW for the vertically polarized ω_p and ω_S beams on the sample. Attenuation of the ω_p beam was achieved by transferring the beam through a λ/2 waveplate mounted on a rotational

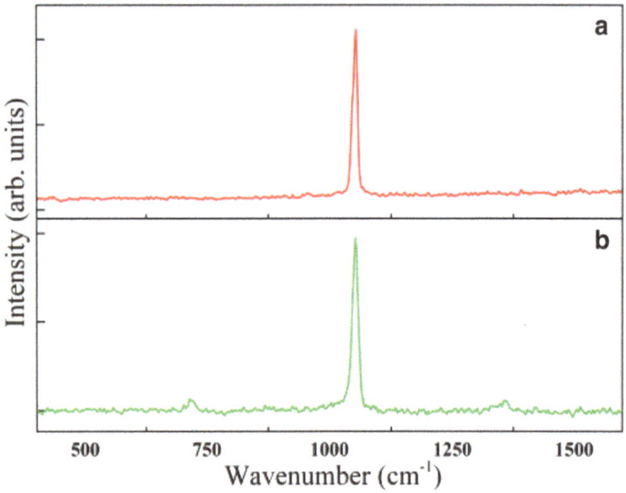

Fig. 12.6 (**a**) Multiplex CARS spectrum measured with the newly built CARS spectrometer with exposure time of 1 s and (**b**) Raman spectrum monitored with the compact Raman spectrometer (exposure 15 s)

stage, which allowed continuous rotation of the polarization direction of the beam, combined with a polarizer. The power for the ω_p beam was adjusted to this low value, to prevent sample degradation during the measurement.

The CARS signal was collimated by the same objective and following blocking of the laser beam by a short wavepass edge filter it was directed toward the ICCD (same components as in the Raman system) [59]. The CARS signal was acquired while operating the ICCD in the continuous mode. Figure 12.6 shows in panel (a) the multiplex CARS spectrum of KNO_3, as measured with an exposure time of 1 s. It is clearly seen that a very high signal with a high SNR was obtained with the CARS spectrometer, even for short integration times of 1 s.

It is important to note that the comparison of the CARS spectrum, Fig. 12.6a, to the Raman spectrum, Fig. 12.6b, shows that in the former only the dominant peak, attributed to symmetric stretching of the nitrate (NO_3^-) ions, is observed. The other bands, at lower and higher frequencies, do not show up. This might be attributed to the Raman and CARS signal dependencies on the Raman cross section, where for the former it shows a linear dependence, while for the latter a quadratic one [58, 59]. These dependencies imply that the ratio between the dominant and weak bands should be higher in CARS, leading to difficulties in obtaining weak signals, while using similar sensitivities for the detector. It is anticipated that after optimization of the system, we will be able to obtain even higher performance. This feasibility study is a first step in showing that the concept of broadband CARS works, however, additional experiments that will compare the sensitivities obtained by Raman and CARS are required for testing the capabilities of this newly developed approach.

12.3 Summary

Despite substantial success in laboratory and field trials, it is obvious that the laser-based spectroscopic methods that were mentioned above, namely LP/LIF, LIBS, Raman and CARS, as well as additional ones, still need further refinement before they can be considered for use in real environments. Since the detection and identification of explosives is a field that poses great challenges with a variety of very complex requirements, further research and development is required. In particular, since various threats require different means for detection, it would be advantageous to find their capabilities, regarding unique scenarios and technical potential. It seems that by using different methods and combining them into single setups the possibility to detect explosive residues as well as a greater number of explosives or their mixtures will be improved. Although many efforts were devoted to standoff detection, it seems that the use of point detection or detection from short distances with compact systems that provide high selectivity and sensitivity might find their slot for use in security applications, as well for identifying unknown compounds [1].

Acknowledgments Financial support from NATO Science for Peace (SfP) Project 983789, the Technion – The Institute for Future Security Research, the Israel Science Foundation founded by The Israel Academy of Science and Humanities and the James Franck Binational German-Israeli Program in Laser-Matter Interaction is gratefully acknowledged.

References

1. Caygill JS, Davis F, Higson SPJ (2012) Current trends in explosive detection techniques. Talanta 88:14
2. Willer U, Schade W (2009) Photonic sensor devices for explosive detection. Anal Bioanal Chem 395:275
3. Wallin S, Pettersson A, Östmark H, Hobro A (2009) Laser-based standoff detection of explosives: a critical review. Anal Bioanal Chem 395:259
4. Moore DS (2004) Instrumentation for trace detection of high explosives. Rev Sci Instrum 75:2499
5. Moore DS, Scharff RJ (2009) Portable Raman explosives detection. Anal Bioanal Chem 393:1571
6. Nambayah M, Quickenden TI (2004) A quantitative assessment of chemical techniques for detecting traces of explosives at counter-terrorist portals. Talanta 63:461
7. Steinfeld JI, Wormhoudt J (1998) Explosives detection: a challenge for physical chemistry. Annu Rev Phys Chem 49:203
8. Gottfried JL, De Lucia FC, Jr MCA, Miziolek AW (2009) Laser-induced breakdown spectroscopy for detection of explosives residues: a review of recent advances, challenges, and future prospects. Anal Bioanal Chem 395:283
9. Simeonsson JB, Sausa RC (1998) Laser photofragmentation/fragment detection techniques for chemical analysis of the gas phase trends. Anal Chem 17:542
10. Simeonsson JB, Sausa RC (1996) A critical review of laser photofragmentation fragment detection techniques for gas phase chemical analysis. Appl Spectrosc Rev 31:1

11. Daugey N, Shu J, Bar I, Rosenwaks S (1999) Nitrobenzene detection by one-color laser-photolysis/laser-induced fluorescence of NO ($\nu'' = 0$–3). Appl Spectrosc 53:57
12. Shu J, Bar I, Rosenwaks S (1999) Dinitrobenzene detection by use of one-color laser photolysis and laser-induced fluorescence of vibrationally excited NO. Appl Opt 38:4705
13. Shu J, Bar I, Rosenwaks S (2000) The use of rovibrationally excited NO photofragments as trace nitrocompounds indicators. Appl Phys B 70:621
14. Shu J, Bar I, Rosenwaks S (2000) NO and PO photofragments as trace analyte indicators of nitrocompounds and organophosphonates. Appl Phys B 71:665
15. Gresham GL, Davies JP, Goodrich LD, Blackwood LG, Liu BYH, Thimsem D, Yoo SH, Hallowell SF (1994) Development of particle standards for testing detection systems – mass of rdx and particle-size distribution of composition-4 residues. Proc SPIE Int Soc Opt Eng 2276:34
16. Arusi-Parpar T, Heflinger D, Lavi R (2001) Photodissociation followed by laser-induced fluorescence at atmospheric pressure and 24 degrees C: a unique scheme for remote detection of explosives. Appl Opt 40:6677
17. Heflinger D, Arusi-Parpar T, Ron Y, Lavi R (2002) Application of a unique scheme for remote detection of explosives. Opt Commun 204:327
18. Wynn CM, Palmaccí S, Kunz RR, Clow K, Rothschild M (2008) Detection of condensed-phase explosives via laser-induced vaporization, photodissociation, and resonant excitation. Appl Opt 47:5767
19. Smith GP, Krancevic B, Huestis DL, Oser H (2009) Laser desorption studies using laser-induced fluorescence of large aromatic molecules. Appl Phys B 94:127
20. Wen B, Eilers H (2012) Potential interference mechanism for the detection of explosives via laser-based standoff techniques. Appl Phys B 106:473
21. Gottfried JL, De Lucia FC, Munson CA, Miziolek AW (2008) Strategies for residue explosives detection using laser-induced breakdown spectroscopy. J Anal At Spectrom 23:205
22. Portnov A, Rosenwaks S, Bar I (2003) Identification of organic compounds in ambient air via characteristic emission following laser ablation. J Luminesc 102:408
23. Portnov A, Rosenwaks S, Bar I (2003) Emission following laser-induced breakdown spectroscopy of organic compounds in ambient air. Appl Opt 42:2835
24. Lopez-Moreno C, Palanco S, Laserna JJ, DeLucia F, Miziolek AW, Rose J, Walters RA, Whitehouse AI (2006) Test of a stand-off laser-induced breakdown spectroscopy sensor for the detection of explosive residues on solid surfaces. J Anal At Spectrom 21:55
25. Gottfried JL (2013) Influence of metal substrates on the detection of explosive residues with laser-induced breakdown spectroscopy. Appl Opt 52:B10
26. Wang Q, Jander P, Fricke-Begemann C, Noll R (2008) Comparison of 1064 nm and 266 nm excitation of laser-induced plasmas for several types of plastics and one explosive. Spectrochim Acta B 63:1011
27. Bauer C, Geiser P, Burgmeier J, Holl G, Schade W (2006) Pulsed laser surface fragmentation and mid-infrared laser spectroscopy for remote detection of explosives. Appl Phys B 85:251
28. Wong DM, Dagdigian PJ (2008) Comparison of laser-induced breakdown spectra of organic compounds with irradiation at 1.5 and 1.064 μm. Appl Opt 47:G149
29. Dikmelik Y, McEnnis C, Spicer JB (2008) Femtosecond and nanosecond laser-induced breakdown spectroscopy of trinitrotoluene. Opt Express 16:5332
30. De Lucia FC, Gottfried JL, Miziolek AW (2009) Evaluation of femtosecond laser-induced breakdown spectroscopy for explosive residue detection. Opt Express 17:419
31. Kearton B, Mattley Y (2008) Laser-induced breakdown spectroscopy – sparking new applications. Nat Photon 2:537
32. Lui SL, Cheung NH (2005) Minimally destructive analysis of aluminum alloys by resonance-enhanced laser-induced plasma spectroscopy. Anal Chem 77:2617
33. Lui SL, Cheung NH (2003) Resonance-enhanced laser-induced plasma spectroscopy: ambient gas effects. Spectrochim Acta B 58:1613
34. Babushok VI, Delucia FC, Gottfried JL, Munson CA, Miziolek AW (2006) Double pulse laser ablation and plasma: laser induced breakdown spectroscopy signal enhancement. Spectrochim Acta B 61:999

35. Gottfried JL, De Lucia FC, Munson CA, Miziolek AW (2007) Double-pulse standoff laser-induced breakdown spectroscopy for versatile hazardous materials detection. Spectrochim Acta B 62:1405
36. Boyd RW (2008) Nonlinear optics, 3rd edn. Academic, San Diego, p 19
37. Carter JC, Angel SM, Lawrence-Snyder M, Scaffidi J, Whipple RE, Reynolds JG (2005) Standoff detection of high explosive materials at 50 meters in ambient light conditions using a small Raman instrument. Appl Spectrosc 59:769
38. Sharma SK, Misra AK, Sharma B (2005) Portable remote Raman system for monitoring hydrocarbon, gas hydrates and explosives in the environment. Spectrochim Acta A 61:2404
39. Pettersson A, Johansson I, Wallin S, Nordberg M, Ostmark H (2009) Near real-time standoff detection of explosives in a realistic outdoor environment at 55 m distance. Propell Explos Pyrotech 34:297
40. Zachhuber B, Gasser C, Chrysostom E, Lendl B (2011) Stand-off spatial offset Raman spectroscopy for the detection of concealed content in distant objects. Anal Chem 83:9438
41. Izake EL, Cletus B, Olds W, Sundarajoo S, Fredericks PM, Jaatinen E (2012) Deep Raman spectroscopy for the non-invasive standoff detection of concealed chemical threat agents. Talanta 94:342
42. Ray MD, Sedlacek AJ, Wu M (2000) Ultraviolet mini-Raman lidar for stand-off, in situ identification of chemical surface contaminants. Rev Sci Instrum 71:3485
43. Gaft M, Nagli L (2008) UV gated Raman spectroscopy for standoff detection of explosives. Opt Mater 30:1739
44. Ehlerding A, Johansson I, Wallin S, Östmark H (2012) Resonance-enhanced Raman spectroscopy on explosives vapor at standoff distances. Int J Spectrosc 2012:158715
45. Sharma SK, Misra AK, Lucey PG, Lentz RCF (2009) A combined remote Raman and LIBS instrument for characterizing minerals with 532 nm laser excitation. Spectrochim Acta A 73:468
46. Moros J, Lorenzo JA, Lucena P, Tobaria LM, Laserna JJ (2010) Simultaneous Raman spectroscopy-laser-induced breakdown spectroscopy for instant standoff analysis of explosives using a mobile integrated sensor. Platform Anal Chem 82:1389
47. Malka I, Petrushansky A, Rosenwaks S, Bar I (2013) Detection of explosives and latent fingerprint residues utilizing laser pointer–based Raman spectroscopy. Appl Phys B 113:511
48. Malka I, Rosenwaks S, Bar I (2014) Photo-guided sampling for rapid detection and imaging of traces of explosives by a compact Raman spectrometer. Appl Phys Lett 104:221103
49. Mogilevsky G, Borland L, Brickhouse M, Fountain AW III (2012) Raman spectroscopy for homeland security applications. Int J Spectrosc 2012:808079
50. Nibler JW, Knighten GV (1979) In: Weber A (ed) Raman spectroscopy of gases and liquids. Springer, Berlin, pp 253–99
51. Valentini JJ (1987) In: Radziemski LJ, Solarz RW, Paisner JA (eds) Laser spectroscopy and its applications, vol 11, Optical Science and Engineering Series. Marcel Dekker, New York, pp 507–64
52. Li H, Harris DA, Xu B, Wrzesinski PJ, Lozovoy VV, Dantus M (2008) Coherent mode-selective Raman excitation towards standoff detection. Opt Express 16:5499
53. Li H, Harris DA, Xu B, Wrzesinski PJ, Lozovoy VV, Dantus M (2009) Standoff and arms-length detection of chemicals with single-beam coherent anti-Stokes Raman scattering. Appl Opt 48:B17
54. Bremer MT, Wrzesinski PJ, Butcher N, Lozovoy VV, Dantus M (2011) Highly selective standoff detection and imaging of trace chemicals in a complex background using single-beam coherent anti-Stokes Raman scattering. Appl Phys Lett 99:101109
55. Katz O, Natan A, Silberberg Y, Rosenwaks S (2008) Standoff detection of trace amounts of solids by nonlinear Raman spectroscopy using shaped femtosecond pulses. Appl Phys Lett 92:171116
56. Natan A, Levitt JM, Graham L, Katz O, Silberberg Y (2012) Standoff detection via single-beam spectral notch filtered pulses. Appl Phys Lett 100:051111

57. Pestov D, Murawski RK, Ariunbold GO, Wang X, Zhi MC, Sokolov AV, Sautenkov VA, Rostovtsev YV, Dogariu A, Huang Y, Scully MO (2007) Optimizing the laser-pulse configuration for coherent Raman spectroscopy. Science 316:265
58. Portnov A, Rosenwaks S, Bar I (2008) Detection of particles of explosives via backward coherent anti-Stokes Raman spectroscopy. Appl Phys Lett 93:041115
59. Portnov A, Rosenwaks S, Bar I (2010) Highly sensitive standoff detection of explosives via backward coherent anti-Stokes Raman scattering. Appl Phys B 98:529
60. Malka I, Rosenwaks S, Bar I (unpublished)

Chapter 13
Detection of Metals and Radionuclides Using Rapid, On-site, Antibody-Based Assays

Diane A. Blake and Bhupal Ban

Abstract This chapter introduces immunosensors as an alternative analytical tool for rapid and portable analysis of both biological and radiological threats and describes most widely used formats for these analyses. The strengths and weaknesses inherent in antibody-based procedures are discussed and common QA/QC practices for antibody-based assays are described. Finally examples of antibody-based assays for radionuclides are provided from the authors' laboratory.

Keywords Immunosensors • Real-tine radionuclide detection • Uranium immunoassays

13.1 Introduction

Society faces a wide range of complex challenges and threats to its security, including risks presented by both state and non-state entities. Current threats include the proliferation of weapons of mass destruction (WMD) and the increasingly sophisticated methods for shielding the radiological signatures of special nuclear materials from current detectors [1, 2]. In addition, rapid advances in biological science and technology continue to increase the threat of bioterrorism [3, 4]. While instrument-intensive methods, including spectral analysis of gamma and neutron radiation for WMD and mass spectrometry for biological agents, currently dominate the techniques used for the detection of these threats [5, 6], development of alternative detection methods for both radionuclides and bioterrorism agents will provide a wider array of choices for personnel that must detect these weapons and respond to their deployment. Employing a variety of detection modalities based on different analytical principles will also serve to thwart those who may wish to circumvent current monitoring and detection processes.

D.A. Blake (✉) • B. Ban
Department of Biochemistry and Molecular Biology, Tulane University School of Medicine, 1430 Tulane Ave., New Orleans, LA 70112, USA
e-mail: blake@tulane.edu

J. Banoub (ed.), *Detection of Chemical, Biological, Radiological and Nuclear Agents for the Prevention of Terrorism*, NATO Science for Peace and Security Series A: Chemistry and Biology, DOI 10.1007/978-94-017-9238-7_13

This chapter introduces immunosensors as an alternative analytical tool for rapid and portable analysis of both biological and radiological threats and describes most widely used formats for these analyses. The strengths and weaknesses inherent in antibody-based procedures are discussed and common QA/QC practices for antibody-based assays are described. Finally examples of antibody-based assays for radionuclides are provided from the authors' laboratory.

13.2 Biosensors and Immunosensors

A **biosensor** is an analytical tool consisting of a biologically active material (enzyme, nucleic acid, antibody, receptor) that is used in close conjunction with a device that will convert a biological recognition event into a quantifiable electrical signal (light or an electrical pulse). An **immunosensor** is a biosensor that uses an **antibody** as the biological recognition element; signal is transduced upon binding of the antibody to the analyte being measured. Antibodies (or immunoglobulins) are soluble proteins found in the blood plasma. They are generated by the mammalian immune system, usually in response to an infection by bacteria or viruses. All immunoglobulins have the same general structure: Y-shaped molecules composed of four polypeptide chains held together by disulfide bonds, as shown in Fig. 13.1. Within the general category of immunoglobulins, five different classes can be distinguished by both their structures and their functions in the body: IgG, IgA,

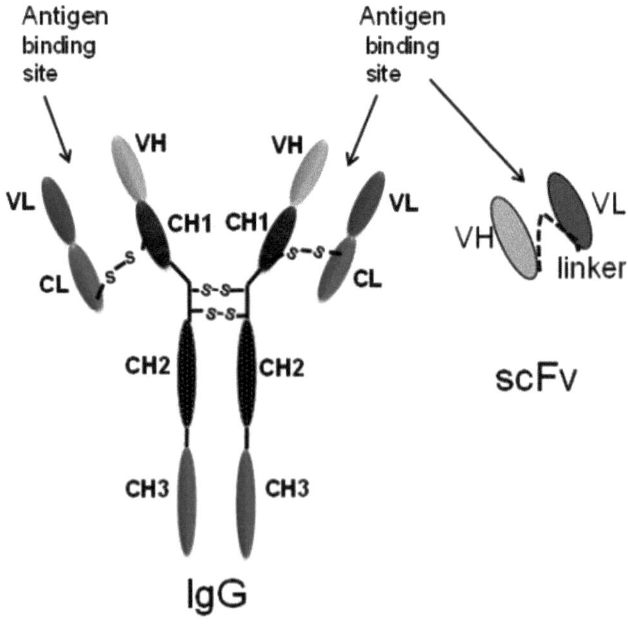

Fig. 13.1 Structure of IgG and single chain recombinant antibody (scFv)

IgM, IgD and IgE. IgG is by far the most abundant immunoglobulin in plasma, and is also the immunoglobulin class most widely used as the biological recognition element in biosensors. The popularity of antibodies for sensor applications is due to their unique properties: antibodies, especially purified IgGs, are relatively stable biomolecules [7]; the structure of the antibody molecule is understood in great detail, which makes molecular engineering of these molecules relatively straightforward [8, 9]; many commercially available reagents are available that permit facile coupling of an antibody binding event to quantifiable signal [10–12]; and, finally, antibodies can be generated for almost any kind of antigen.

Antibodies used in immunosensors can be of polyclonal, monoclonal or recombinant origin. Polyclonal antibodies are generated in response to any molecule recognized by the immune system as "non-self". They are produced by immunizing an animal with an antigen (usually a high molecular weight protein, carbohydrate or nucleic acid) and then collecting serum from the animal weeks to months after immunization. A polyclonal antibody preparation consists of a large number of different antibodies that bind to the antigen with differing affinities. These antibodies may also bind at more than one location (called an epitope) on a large antigen. Polyclonal antibodies are easier to generate than monoclonal or recombinant antibodies and may be more stable than other forms. However, polyclonal antibody preparations are usually less specific for the antigen than monoclonal or recombinant antibodies. In addition, because of the nature of the immune response, antibody affinity and specificity may change with each sample of serum collected from the animal, which means that the sensor must be re-validated every time a new preparation of polyclonal antibody is required. Monoclonal antibodies are synthesized by a population of identical immortalized B lymphocytes [13]. A monoclonal antibody preparation consists of a single, homogenous group of proteins. Because all of the antibodies in a monoclonal preparation are identical, they all recognize the same epitope with an identical affinity. In general, monoclonal antibodies have higher specificity for the antigen than do polyclonal preparations. These antibodies arise from an immortalized cell line, and with proper storage, these lines will continue to produce antibody indefinitely, thus providing a stable source of antibody for sensor applications. However, monoclonal antibodies are usually more expensive and time-consuming to generate than are polyclonal preparations, and, because all the antibody molecules in the population are identical, they may show more sensitivity to changes to pH or ionic strength than a mixed, polyclonal population. Recombinant antibodies are those produced by molecular cloning; preparation of recombinant antibodies usually involves the expression of the antibody in viruses, yeast and/or bacterial cells. For ease of cloning, recombinant antibodies are often expressed as single chain fragments called scFv's (also shown in Fig. 13.1), in which the antigen-binding binding domains of the antibody are linked via a flexible polypeptide chain.

A very large number of antibodies are available commercially and many search engines exist to assist in selecting antibodies for specific applications [14]. If the desired antibody is not available, many commercial facilities also exist for the generation of both polyclonal and monoclonal antibodies to protein antigens. In some cases, these facilities, if provided with a protein sequence, can use protein folding algorithms to predict the peptide sequence likely to generate the highest affinity or highest specificity antibodies, synthesize the immunogen, and prepare the antibody. However, any antibody preparation used for an analytical method (immunolocalization, immunoprecipitation, CHIP, ELISA, immunosensor, etc.) must be characterized and validated for individual applications. Facilities specializing in the generation of new antibodies usually do not have the capabilities for such validations, and antibodies should therefore be validated by the laboratory developing the analytical method.

13.3 Antibody Validation

Antibodies are a very important tool in both basic science research and clinical assays and a key component of immunosensors. Unfortunately, no universally accepted guidelines or standardized methods exist for determining the validity of these reagents. An antibody must be specific, selective, and reproducible in the context for which it is to be used. Incomplete antibody validation can lead to erroneous results and lab-to-lab irreproducibility [15–18]. Often investigators use vendor data with no validation or perform control experiments (substitution of pre-immune serum or buffer for primary antibody, blocking the primary antibody with molar excess of immunogen/antigen) that are not rigorous enough to detect cross-reactivity. A number of recent reviews [19–22] have addressed the problems of antibody specificity and outlined comprehensive validation procedures for use in immunohistochemical protocols. For antibodies to human proteins, a site is also available for sharing data on antibody and antigen validation [23].

Antibodies used in immunosensors often bind to low molecular weight analytes (metals, pesticides, herbicides, polycyclic aromatic hydrocarbons, etc.) rather than to proteins, and the protocols outlined above are not useful for validation of such antibodies. Our laboratory has developed a series of procedures designed for the validation of such antibodies, as shown in Fig. 13.2 and described in [24–26]. Competitive ELISAs are useful in preliminary selection of candidate antibodies for use in immunosensors; however, a more detailed validation process is necessary before choosing the antibodies to be incorporated into sensors. This process includes (1) determining the binding affinity of the antibody for the antigen of choice plus the antibody's affinity for analytes with related structures (these experiments define sensitivity and specificity); (2) comparing the binding affinity of the antibody for the soluble analyte to its binding to the analyte-protein conjugate used to generate the antibody (these experiments will help to identify the sensor format that provides maximum sensitivity) [27]; (3) measuring on- and off-rates of the antibody for the analyte of choice (this will define incubation times required during sensor operation); (4) insuring that any modification of the antibody required for the signal transduction process does not change its binding properties [28]; and (5) identifying interferences from substances likely to be found in the sample matrix [29].

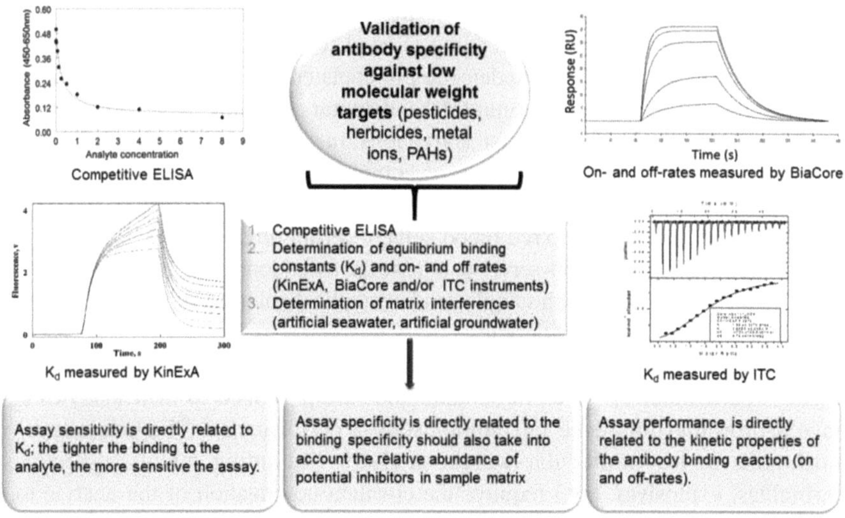

Fig. 13.2 Flow chart for validation of antibodies to be used in sensors

In some cases, even a well-characterized antibody can provide surprising results when subjected to such validation procedures. For example, the binding of prostate specific antigen (PSA) to a commercially available anti-PSA monoclonal antibody (clone M612166) had been previously studied by 22 participants representing 13 different institutions/companies using three different surface plasmon resonance platforms and the resulting 22 separate data sets were in remarkable agreement [30]. However, when this antibody was validated on our laboratories, we discovered that the M612166 antibody had an anomalously low on-rate when soluble PSA was added to the antibody. Subsequent experiments using a combination of kinetic exclusion analysis and electrospray ionization-ion mobility spectrometry revealed a heretofore unrecognized specificity for this antibody, that it bound preferentially to a PSA dimer [30].

13.4 Immunosensors that Detect Radionuclides and Heavy Metals

The key components of immunosensors that bind metal ions and radionuclides are antibodies that recognize chelated forms of the metal ions. The general procedures involved in the generation and characterization of these antibodies have been previously reviewed [31]. Briefly, metals complexed to bifunctional chelators are covalently attached to proteins and the resulting metal-chelate-protein conjugates are used as immunogens. After the immunized animal displays a measurable immune response, it is sacrificed, and the immune tissue is harvested. Hybridoma or recombinant technologies are then used to generate monoclonal or recombinant antibodies [32–34] and high throughput screening is used to isolate antibodies with useful properties in sensor formats. The detection of chelated rather than free metals has a number of advantages in radionuclide and metal ion detection. Metal ions in environmental samples do not exist as free ions; in soil, metals are tightly bound to anionic centers and in water they exist as complexes with chloride, nitrate, sulfate, phosphate, humic acids or other complexants [35]. Metals in blood and serum samples are usually bound to red blood cells or serum proteins [36]. The treatment of such samples with chelators serves to release the metal ions from the wide variety of complexants that exist in such samples and bring the divalent metals into complexes that are recognized by the antibodies.

Once these antibodies have been validated, they must then be incorporated into a sensor. The manner in which these antibodies are presented to their analytes can have a profound influence on immunosensor performance. The generation of antibodies to low molecular weight analytes (including metals, pesticides, herbicides, explosives, etc.) requires the covalent conjugation of the analyte to a carrier protein, because the analyte itself it too small to elicit an immune response. The antibodies thus generated almost always bind with higher affinity to the analyte-protein conjugate than to the actual contaminate that you wish to measure. In studies where these binding affinities have been quantified and compared, the differences in affinities have been between 150 and 1,000-fold in favor of the conjugated antigen [24, 32, 34]. When such differences in affinity exist, an assay system that requires a direct competition between the conjugated and unconjugated antigen will inevitably yield an assay with a relatively low sensitivity because a high concentration of soluble analyte is necessary to compete with the protein-analyte conjugate for antibody binding sites [37]. Because of these differences in affinity, our laboratory has preferred an analytic approach called kinetic exclusion analysis, and our most successful sensors have employed this method.

In this format, the antibody is pre-mixed with the soluble analyte and the binding reaction is allowed to come to equilibrium (2–5 min for antibodies that bind to their analyte with nanomolar affinity). The equilibrated binding reaction is then passed rapidly through a capillary tube that contains beads with immobilized protein-analyte conjugate. The design of the kinetic exclusion system minimizes the time that the antibody is in contact with the immobilized protein-analyte conjugate (usually 240–500 ms) and thus limits the time in which antibody can dissociate from the soluble antigen and rebind to the more attractive (from an affinity point of view) protein-analyte conjugate. Other factors that increase efficiency of this kinetic exclusion format include the rapid flow of assay mixtures through the beads and the high surface area of the beads in the capillary column, which minimize the diffusion limitations at the bead surface and maximize capture of free antibody on the bead surfaces [25, 38, 39]. Detailed explanations of the basis for the kinetic exclusion method and its theoretical limits of detection have been reported [24, 25, 40, 41]. In environmental analyses, the kinetic exclusion format can enhance the sensitivity of immunosensor for low molecular weight antigens from 20 to 1,000-fold, depending upon the antibody.

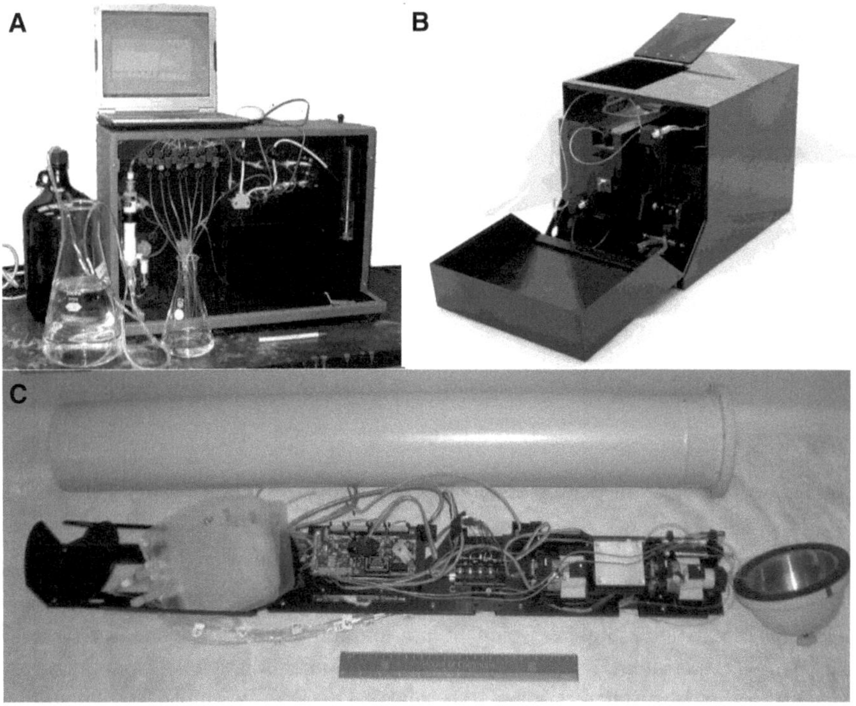

Fig. 13.3 In-line (**a**), field-portable (**b**) and submersible sensors for metal and radionuclide analysis

Our laboratory has developed three different field-ready immunosensors using this analytical method, in collaboration with our industrial partner, Sapidyne Instruments Inc [42–44]. These sensors are shown in Fig. 13.3, and all three sensors have been proven for the measurement of uranium. The an in-line sensor (Fig. 13.3a), collects a sample from a process line, amends the sample with assay reagents, and determines the amount of a specific antigen based on comparisons to an instrument-generated standard curve. The sensor's multiple sample lines facilitate assay development and the associated software is easily modified to accommodate the requirements of specific antibodies and/or analytes. The field-portable sensor (Fig. 13.3b) is battery operated. The device interfaces with a laptop computer through a wireless connection and weighs less than 6 kg. The submersible sensor (Fig. 13.3c) is designed for underwater use and has several unique features adapted for this environment. The sensor has a aluminum case with a waterproof gasket to keep out moisture (shown above the interior components). The sensor draws samples based on displacement of the plunger in the draw syringe, and the optical unit is contained behind the flow cell bracket. All reagents and wastes must be self-contained inside the case; this is accomplished via the reagent bags located at the back end of the sensor. The bags are connected via tubing to the appropriate valve to make a watertight flow path. The nose is equipped with a filter to remove particulates, preventing clogging of the interior tubing. Power is supplied to the sensor with a rechargeable battery pack containing NiMH C-type batteries.

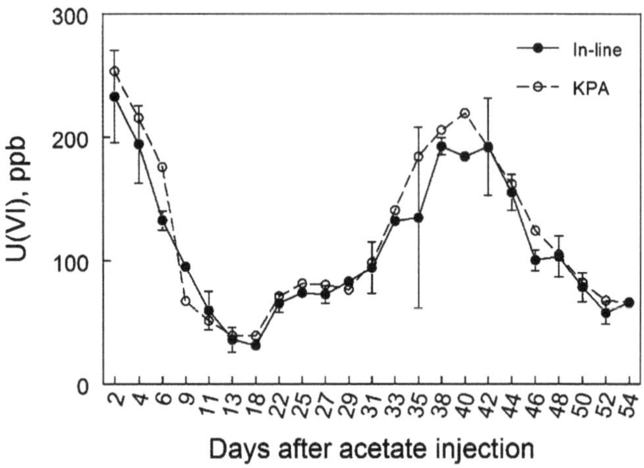

Fig. 13.4 A comparison of U(VI) concentrations measured in groundwater during a bioremediation experiment

In addition to lab validations, the in-line and field portable sensors were also tested during an actual field experiment at a uranium mine tailing site in Colorado. During this experiment, acetate was pumped into injections wells to stimulate the bacteria in the subsurface to reduce the soluble U(VI) to insoluble U(IV) and thus remove it from the groundwater. Figure 13.4 compares the U(VI) measurement in groundwater from downfield monitoring wells, measured using two methods: the in-line immunosensor and Kinetic Phosphorescence Analysis (KPA), considered the 'gold standard' for measurement of U(VI) in environmental samples [45]. The initial 14 day period of acetate injection was followed by an 8 day acetate-free groundwater flush, after which time the level of uranium began to rise in the first tier of monitoring wells approximately 22 days after the start of the injection. Uranium levels peaked at 35–40 days and then began to fall again, once acetate injection resumed (44–54 days). During this field study, the immunosensor level of detection (MDL) for individual assays ranged from 0.014 to 0.260 nM (3–62 ppt). The average assay MDL for experiments performed during this summer experiment was 0.12 nM (29 ppt). Because the samples were diluted to 1:200 or 1:400 before analysis, the MDL for the environmental samples was 24–48 nM (5.7–11.4 ppb), which is below the drinking water limit for uranium of 30 ppb. During these experiments, a global comparison of U(VI) values determined using the immunosensors with U(VI) measurements using the KPA was excellent, with a correlation coefficient of 0.94 [43].

13.5 Future Directions

Future directions in our laboratory are focused on two areas: (1) development and validation of new antibodies for novel markers of environmental contamination; (2) development of smaller and simpler-to-use sensors for these contaminants.

Our efforts to generate new antibodies have focused upon molecular techniques to engineer antibodies with superior binding properties [33]. New sensor technologies currently under development in our laboratory include lateral flow immunoassays (like home pregnancy test kits) that can be read with the naked eye or quantified using the camera of a cell phone [46]. In addition to their value in monitoring the environment, such new antibodies and immunosensors could also be easily adapted for use in bioterrorism, including rapid screening for uranium in hair, saliva and blood. Such new screening tools could be used identify individuals who have been machining or enriching uranium, or to monitor the health of uranium mine workers in countries lacking the infrastructure for complicated and expensive analytical methods.

Acknowledgements This work was supported in part by the National Science Foundation (OISE-1253272).

References

1. Runkle RC, Smith LE, Peurrung AJ (2009) The photon haystack and emerging radiation detection technology. J Appl Phys 106 041101-041101-21
2. Cester D, Nebbia G, Stevanato L, Viesti G, Neri F, Petrucci S, Selmi S, Tintori C et al (2012) Special nuclear material detection with a mobile multi-detector system. Nucl Instrum Methods Phys Res 663:55–63, Sect. A
3. Arun KR, Nishanth T, Ravi TY, Sathish KD (2011) Biothreats – bacterial warfare agents. J Bioterrorism Biodef 2:112
4. Montecucco C (2012) Bioterrorism and biological toxins. Toxicon 60:99, 5
5. Sandrin TR, Goldstein JE, Schumaker S (2013) MALDI TOF MS profiling of bacteria at the strain level: a review. Mass Spectrom Rev 32:188–217
6. Woodring ML, Rodriguez DC, Runkle RC, Hansen RR, Milbrath BD, Ely JH (2013) A large detector laboratory for the development and testing of radiation detection systems. IEEE Trans Nucl Sci 60:1151–1155
7. Ewert S, Honegger A, Pluckthun A (2004) Stability improvement of antibodies for extracellular and intracellular applications: CDR grafting to stable frameworks and structure-based framework engineering. Methods 34:184–199
8. Buss NAPS, Henderson SJ, McFarlane M, Shenton JM, de Haan L (2012) Monoclonal antibody therapeutics: history and future. Curr Opin Pharmacol 12:615–622
9. Vincent KJ, Zurini M (2012) Current strategies in antibody engineering: Fc engineering and pH-dependent antigen binding, bispecific antibodies and antibody drug conjugates. Biotechnol J 7:1444–1450
10. Chakravarty S, Zou Y, Lai W-C, Chen RT (2012) Slow light engineering for high Q high sensitivity photonic crystal microcavity biosensors in silicon. Biosens Bioelectron 38:170–176
11. Thakur MS, Ragavan KV (2013) Biosensors in food processing. J Food Sci Technol 50:625–641
12. Han KN, Li CA, Seong GH (2013) Microfluidic chips for immunoassays. Annu Rev Anal Chem 6:119–141
13. Zhang C (2012) Hybridoma technology for the generation of monoclonal antibodies. Methods Mol Biol 901:117–135
14. Antibody Resource page. http://www.antibodyresource.com/
15. Hansson SF, Korsgren S, Ponten F, Korsgren O (2013) Enteroviruses and the pathogenesis of type 1 diabetes revisited: cross-reactivity of enterovirus capsid protein (VP1) antibodies with human mitochondrial proteins. J Pathol 229:719–728

16. Anagnostou VK, Welsh AW, Giltnane JM, Siddiqui S, Liceaga C, Gustavson M, Syrigos KN, Reiter JL, Rimm DL (2010) Analytic variability in immunohistochemistry biomarker studies. Cancer Epidemiol Biomarkers Prev 19:982–991
17. Bucur O, Pennarun B, Stancu AL, Nadler M, Muraru MS, Bertomeu T, Khosravi-Far R (2013) Poor antibody validation is a challenge in biomedical research: a case study for detection of c-FLIP. Apoptosis 18:1154–1162
18. Milner R, Wombwell H, Eckersley S, Barnes D, Warwicker J, Dorp E, Dearden S, Hughes G et al (2013) Validation of the BRCA1 antibody MS110 and the utility of BRCA1 as a patient selection biomarker in immunohistochemical analysis of breast and ovarian tumours. Virchows Arch 462:269–279
19. Bordeaux J, Welsh AW, Agarwal S, Killiam E, Baquero MT, Hanna JA, Anagnostou VK, Rimm DL (2010) Antibody validation. BioTechniques 48:197–198, 200, 202, 204, 206, 208–209
20. Signore M, Reeder KA (2012) Antibody validation by western blotting. Methods Mol Biol 823:139–155
21. Ambroz K (2011) Impact of blocking and detection chemistries on antibody performance for reverse phase protein arrays. Methods Mol Biol 785:13–21
22. Mandell JW (2008) Immunohistochemical assessment of protein phosphorylation state: the dream and the reality. Histochem Cell Biol 130:465–471
23. Björling E, Uhlén M (2008) Antibodypedia, a portal for sharing antibody and antigen validation data. Mol Cell Proteomics 7:2028–2037
24. Blake DA, Chakrabarti P, Khosraviani M, Hatcher FM, Westhoff CM, Goebel P, Wylie DE, Blake RC 2nd (1996) Metal binding properties of a monoclonal antibody directed toward metal-chelate complexes. J Biol Chem 271:27677–27685
25. Blake RC 2nd, Pavlov AR, Blake DA (1999) Automated kinetic exclusion assays to quantify protein binding interactions in homogeneous solution. Anal Biochem 272:123–134
26. Blake RC 2nd, Blake DA (2003) Kinetic exclusion assay to study high-affinity binding interactions in homogeneous solutions. In: Lo BCK (ed) Antibody engineering: methods and protocols. Humana Press, Totowa, pp 417–430
27. Kusterbeck AW, Blake DA (2008) Flow immunosensors. In: Ligler FL, Taitt CR (eds) Optical biosensors, 2nd edn. Elsevier, Amsterdam, pp 243–285
28. Blake RC 2nd, Li X, Yu H, Blake DA (2007) Covalent and noncovalent modifications induce allosteric binding behavior in a monoclonal antibody. Biochemistry 46:1573–1586
29. Lopez MAM, Pons J, Blake DA, Merkoci A (2013) High sensitive gold-nanoparticle based lateral flow Immunodevice for Cd2+ detection in drinking waters. Biosens Bioelectron 47:190–198
30. Blake RC 2nd, Blake DA (2012) Electrospray ionization-ion mobility spectrometry identified monoclonal antibodies that bind exclusively to either the monomeric or a dimeric form of prostate specific antigen. Anal Chem 84:6899–6906
31. Blake DA, Blake RC 2nd, Abboud ER, Li X, Yu H, Kriegel AM, Khosraviani M, Darwish IA (2007) Antibodies to heavy metals: isolation, characterization and incorporation into microplate-based assays and immunosensors. In: Van Emon JM (ed) Immunoassay and other bioanalytical techniques. Taylor & Francis, Boca Ratan, pp 93–111
32. Blake RC 2nd, Pavlov AR, Khosraviani M, Ensley HE, Kiefer GE, Yu H, Li X, Blake DA (2004) Novel monoclonal antibodies with specificity for chelated uranium(VI): isolation and binding properties. Bioconjug Chem 15:1125–1136
33. Zhu X, Kriegel AM, Boustany CA, Blake DA (2011) Single-chain variable fragment (scFv) antibodies optimized for environmental analysis of uranium. Anal Chem 83:3717–3724
34. Khosraviani M, Blake RC 2nd, Pavlov AR, Lorbach SC, Yu H, Delehanty JB, Brechbiel MW, Blake DA (2000) Binding properties of a monoclonal antibody directed toward lead-chelate complexes. Bioconjug Chem 11:267–277
35. Roundhill DM (2001) Extraction of metals from soils and water. Kluwer Academic/Plenum Publishers, New York
36. Kriegel AM, Soliman AS, Zhang Q, El-Ghawalby N, Ezzat F, Soultan A, Abdel-Wahab M, Fathy O et al (2006) Serum cadmium levels in pancreatic cancer patients from the East Nile Delta region of Egypt. Environ Health Perspect 114:113–119

37. Velanki S, Kelly S, Thundat T, Blake DA, Ji HF (2007) Detection of Cd(II) using antibody-modified microcantilever sensors. Ultramicroscopy 107:1123–1128
38. Blake RC 2nd, Blake DA (2005) Quantitative analysis of antibody-antigen interactions using immobilized ligands: Kinetic exclusion assays are more accurate than surface plasmon resonance. In: Simons MA (ed) Progress in antibody research. Nova Science, Hauppauge, pp 1–36
39. Glass TR, Saiki H, Blake DA, Blake RC 2nd, Lackie SJ, Ohmura N (2004) Use of excess solid-phase capacity in immunoassays: advantages for semicontinuous, near-real-time measurements and for analysis of matrix effects. Anal Chem 76:767–772
40. Sasaki K, Oguma S, Glass T, Namiki Y, Sugiyama H, Ohmura N, Blake DA (2008) Simple method to reduce interference from excess magnesium in cadmium immunoassays. J Agric Food Chem 56:7613–7616
41. Glass TR, Ohmura N, Saiki H (2007) Least detectable concentration and dynamic range of three immunoassay systems using the same antibody. Anal Chem 79:1954–1960
42. Fisher RA, Melton SJ, Blake DA (2011) A submersible immunosensor. Int J Environ Anal Chem 91:123–137
43. Melton SJ, Yu H, Williams KH, Morris SA, Long PE, Blake DA (2009) Field-based detection and monitoring of uranium in contaminated groundwater using two immunosensors. Environ Sci Technol 43:6703–6709
44. Yu H, Jones RM, Blake DA (2005) An immunosensor for autonomous in-line detection of heavy metals: validation for hexavalent uranium. Int J Environ Anal Chem 85:817–830
45. Brina R, Miller AG (1993) Determination of uranium and lanthanides in real-world samples by kinetic phosphorescence analysis. Spectroscopy Duluth MN United States 8:25–28, 30–21
46. Lopez MAM, Pons J, Blake DA, Merkoci A (2013) All-integrated and highly sensitive paper based device with sample treatment platform for Cd2+ immunodetection in drinking/tap waters. Anal Chem 85:3532–3538

Chapter 14
Identification of Fraudulently Modified Foods

Giovanni Sindona

Abstract The official methods for the certification of food content are often based on very old and aged procedures. These can be easily manipulated.-A case study will be represented for the assessment of protein content in foods. Unfortunately, the official method of analysis used worldwide is still represented by the determination of the nitrogen content of a given aliment, following the brilliant, but nearly 200 years old, Kjeldahl approach. In this new century, it is now time to introduce worldwide protocols for food certification based on high tech analytical methods. In this respect Mass Spectrometry plays a fundamental role.

Keywords Tandem mass spectrometry (MS/MS) • Multiple reaction monitoring (MRM) • Isotope dilution method • Ambient ionization methods • Food quality • Food safety • Food origin

14.1 Introduction

Adulteration of foods due to deliberately added contaminants can be easily established by using high tech analytical techniques such as mass spectrometry. A great variety of instruments are available in the armory of this peculiar methodology which can be applied for the assay of any complex molecule independently from its physical state.

The official methods for the certification of food content are, unfortunately, very often based on very old and aged procedures which can be easily mistaken when a food sample is manipulated by "informed" technician.-A case study could be represented by the assessment of protein content in foods. The official method of analysis used worldwide is still represented by the determination of the nitrogen content of a given aliment, following the brilliant, but nearly 200 years old, Kjeldahl approach (Fig. 14.1)

The 2008 Chinese milk scandal was a food fraud incident involving milk and other related edible derivatives that caused the death of many infants. The foods have been deliberately treated with melamine **1** (Chart 14.1), a poisoning molecule

G. Sindona (✉)
Department of Chemistry and Chemical Technologies, University of Calabria, Italy
e-mail: sindona@unical.it

J. Banoub (ed.), *Detection of Chemical, Biological, Radiological and Nuclear Agents for the Prevention of Terrorism*, NATO Science for Peace and Security Series A: Chemistry and Biology, DOI 10.1007/978-94-017-9238-7_14

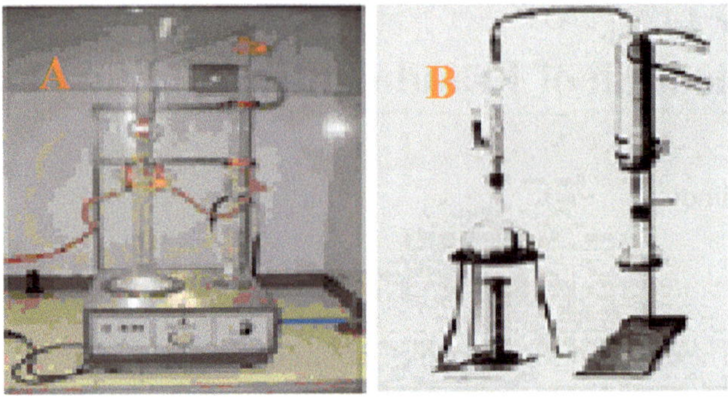

Fig. 14.1 Kjeldahl approach. (**a**) Modern and (**b**) original instrumental setup

Chart 14.1 (**1**) Melamine; MW 126.1199

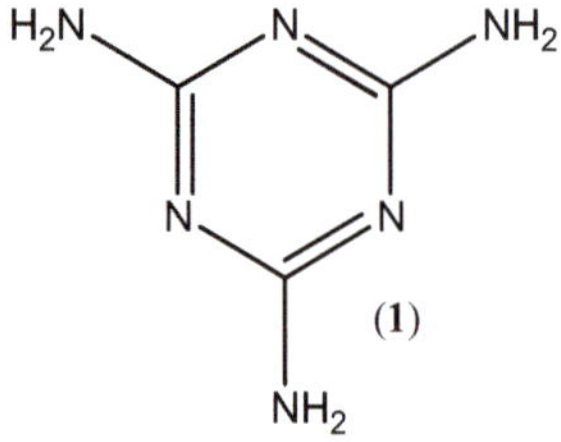

(1)

whose nitrogen content was attributed, by the Kjeldahl method, to milk natural ingredients [1].-Few months later scientists from the University of Shanghai (China) published a comprehensive scientific article whereby melamine content of foods can be determined without any ambiguity by means of Reverse Phase liquid Chromatography linked to Electrospray Ionization (ESI), by applying the isotope dilution method in MRM mode, using a $^{13}C_3$ $^{15}N_3$ melamine isomer [1].

It is now time to introduced worldwide protocols for food certification based on high tech analytical methods. In this respect Mass Spectrometry plays a fundamental role.

14.2 Quality Safety and Origin of Foods

An European project (QUASIORA) launched in southern Italy has exploited research protocols based on a number of integrated working packages aiming at introducing new high tech methodologies for the control of quality, safety and origin of agrifoods. The mission was represented by the exploitation of scientific methods to be offered to farms involved at any level in the production and trading of foodstuff for a high tech certification of their products. Hence, the real quality of foods can be now definitely assessed by methodologies based on Tandem Mass Spectrometry (MS/MS) and Magnetic Resonance Spectroscopy (NMR). Accordingly, it is now time that they replace the younger, with respect to Kjeldhal, but still old EEC Regulation regulations of the eighties and nineties of the last century. The results achieved in different food chains will be described in the following sections [2].

14.2.1 Food Quality Olive Oil

The QUASIORA laboratory performs all the routine analysis requested by EU regulations in relation to the characterization of extra virgin olive oil. The nutritional and healthy claims associated to olive oil are usually related to the fatty acid (FA) composition of the aliment regardless the real structure of the triglycerides (TAG) "sharing" the single FA. Many MS and MS/MS methods, associated to soft ionization procedures, are now available to assess the real distribution of TGA in a given oil, regardless the identification, after chemical hydrolysis and derivatization, of the FA as methyl and ethyl esters derivatives which do not provide any information on fraudulent manipulation of the food. Freshly prepared olive oil contains enzymes, such as lipase [2] which could catalyze the formation of TAG isomers and oligomers under mild thermal treatment.

A case study is represented by the incubation of tripalmitin (**2**) and triolein (**3**) with lipase. The formation of the new TAGs 1(2)-oleoyl-2(1), 3-dipalmitoylglycerol and 1(2),3-dioleoyl-2(1)-palmitoylglycerol (Fig. 14.2) is a proof that the oil TAGs can isomerizes in mild conditions [3] as a consequence of the presence [2] of lipase as a normal constituent of the aliment.

An obvious consequence is that lipase plays a notable role in the many chemical processes that occur during the storage and/or heating of the olive oil. TAG oligomers

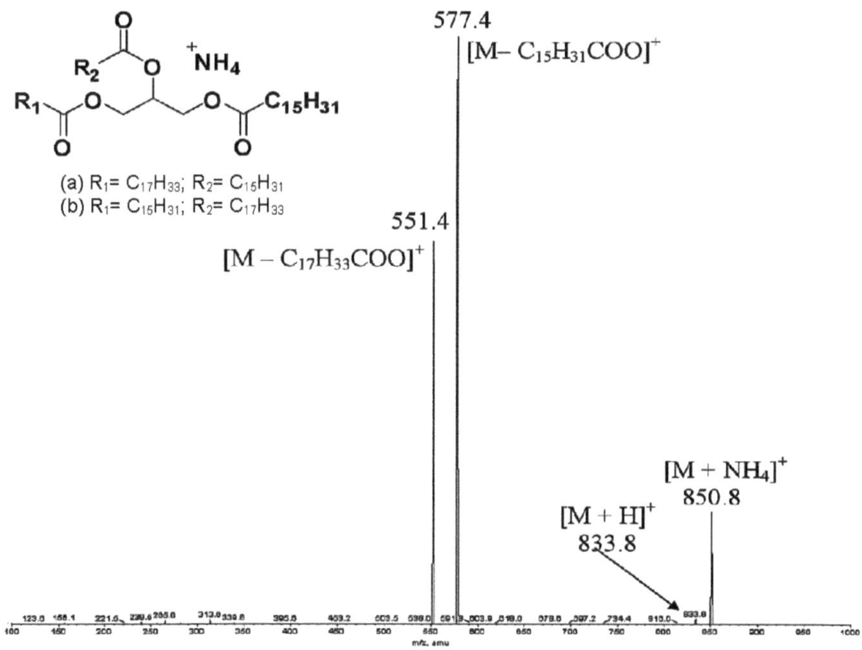

Fig. 14.2 MS/MS spectra of the $[M+NH_4]^+$ ammonium adduct of isomeric (**a**) and (**b**) di-palmitoyl, oleyl TAG formed when mixtures of tripalmitin (**2**) and triolein (**3**) are incubated with lipase

have been, in fact, identified [3]. The MS method here discussed can be, therefore, used to unveil improper treatment of this peculiar aliment which frequently corresponds to a fraudulent manipulation in a such peculiar food chain of strategic importance both for producers and consumers. It is compulsory, therefore, that this and other MS methods will be used to better identify the quality of oil through the identification of the real composition in triglycerides.

The healthy effect of the *Mediterranean diet*, recently inserted in list of the Intangible Cultural Heritage of Humanity, has been often associated to the presence in extra virgin olive oil of active principles with antioxidant and anti-inflammatory properties. The new EU directives require that each single micro component likely present in oil should be identified and assay following the accepted analytical rules, i.e., consumers should know the structure and absolute amount of these species without any ambiguity. This proposal guarantees, among others, that adulterated oils should not fraudulently introduced into the market: once more mass spectrometry is the method of choice. Hydroxytyrosol (**4**) and Tyrosol (**5**) are markers of quality of olive oil their determination have been classically performed by LC–MS/MS under MRM condition and isotope dilution method, using d_2-labelled internal standards obtained by simple synthetic procedures [4] (Chart 14.2).

The assay has been performed by monitoring two transitions for each analyte to improve the specificity. The content of this active phenolic compounds in virgin olive oil have been detected down to a limit of a few hundreds of parts per billion. Tyrosol and Hydroxytyrosol ranged from 10 to 47 ppm and from 5 to 25 ppm in commercial olive oil, respectively (Table 14.1). The accuracy (98–107 %) and analytical parameters values confirm the reliability of the proposed approach. The method can be extended to any natural matrices, including mill wastes, after a simple step of sample preparation.

The use of labeled internal standard in the MS analytical procedure to assess the real quantity of given analyte in a complex mixture can not be replaced by any other

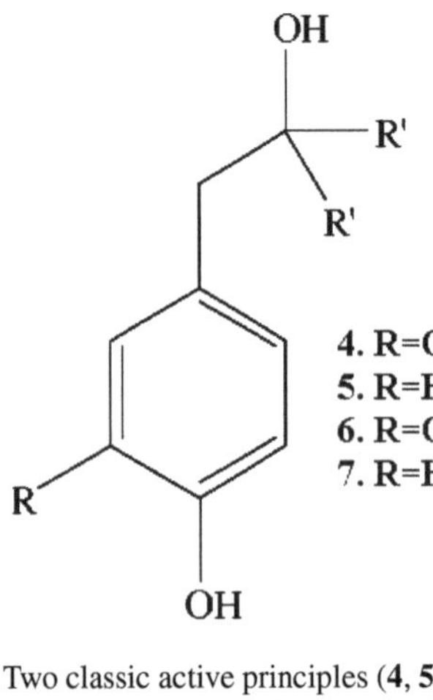

4. R=OH; R'=H Hydroxytyrosol
5. R=H; R'=H Tyrosol
6. R=OH; R'=D deuterated standard
7. R=H; R?=D deuterated standard

Chart 14.2 Two classic active principles (**4**, **5**) determined in olive oil by tandem mass spectrometry and isotope dilution method [4]

Table 14.1 Amount of HTyr and Tyr in extra virgin olive oil samples 1–6

Extra virgin olive oil samples	*Tyrosol* (ppm)	RSD %	*Hydroxytyrosol* (ppm)	RSD %
1	47.10±2.91	6.2	25.08±1.05	4.1
2	31.66±1.29	4.21	24.93±0.97	3.9
3	29.51±1.94	6.6	13.74±0.94	6.8
4	9.75±0.44	4.5	6.51±0.41	6.4
5	11.75±0.59	5.1	4.02±0.34	7.0
6	12.20±0.87	7.1	9.30±0.66	2.5

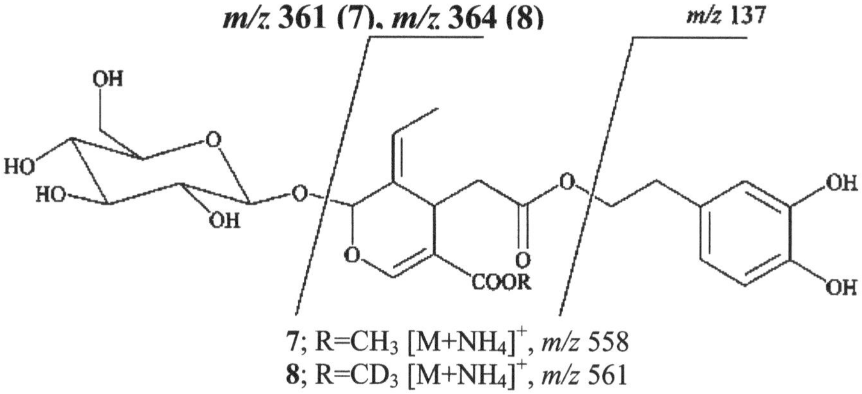

Chart 14.3 Oleuropein (**7**) and its d_3 labelled isomer (**8**). The origin of two diagnostic MS fragments are indicated

reliable and affordable procedure and does not depend either on the complexity of the instrument used or on the selected ionization method.

Years ago, The procedure has been applied to determine the amount of Oleuropein (OLP, Chart 14.3, **7**), a nutraceutical whose health benefits have been widely documented, in (i) food integrators extracted from olive leaves, (ii) in table-olives, and (iii) in extra virgin olive oils [5].

The MS hardware was represented by a simple triple quadrupole mass spectrometer equipped with an APCI source. The selection of a specific labeled internal standard is a consequence of the gas-phase chemistry undergone by the analyte. In the case of **7** and **8** the reported fragments (Chart 14.3) are deeply affected, in terms of *m/z* values, by the selected position within the molecule of the labeled site.

Many other antioxidants have been identified in olive tissues by mass spectrometry [4, 6, 7] a break through was, however, represented by the discovery that an already known molecule present in olive oil had anti-inflammatory properties similar to ibuprofen and many other healthy properties have been associated to this active principles [8]. They represent now markers of quality enhancing the commercial value of olive oil. It was, therefore, compelling to set-up reliable analytical methods to identify and assay these active molecules in the aliment. Once again this was and is a job for mass spectrometrists.

The dialdehyde **9** and **10** are the metabolites of the natural components of olive tissues, ligstroside and oleuropein, respectively. They were originally discovered by Montedoro et coll [9]. Their identification and assay, by soft ionization processes in extracts from olive tissues, poses the problems of the instability of the dialdehydic moiety which undergoes rearrangements presiding the formation of the $[M+H]^+$ species (Scheme 14.1) [10]. The consequence is that the ESI-MS spectra show the presence of all the derivative and does not allow a clear identification and assay of this active species (Scheme 14.1).

The problem has been solved, once more, by means of the *classical*, now, method of ESI-MS/MS supported, for the quantitation, by the creation of labelled internal standard. The dialdehydic active principles are *frozen* in solution by their derivatization to the bis-dioximate derivatives (Scheme 14.2)

Hydroxytyrosol derivative of the oleopentadialdehyde (**10**, Chart 14.4) and the corresponding d_6 labeled analogue provided the calibration curve reported in Fig. 14.3.

The sample was prepared from 100 mg of olive oil mixed with 900 μL of a 1.5 M solution of O-methylhydroxylammonium chloride, followed, after a conventional treatment, by the addition of 100 μL of a 10 mg/L concentration of the d_6 internal standard (Scheme 14.2) dissolved in CH3OH/H2O (70/30). Mass spectrometry was performed on triple-stage quadrupole using a C_{18} reversed-phase analytical column [11].

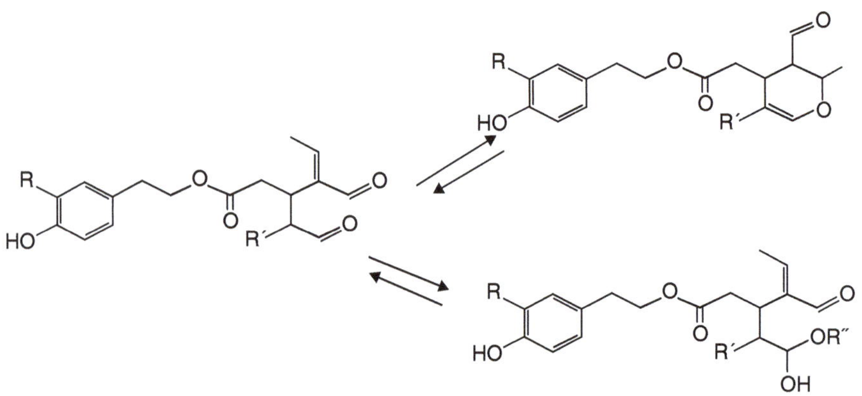

Scheme 14.1 Modification of the dialdehyde in polar media

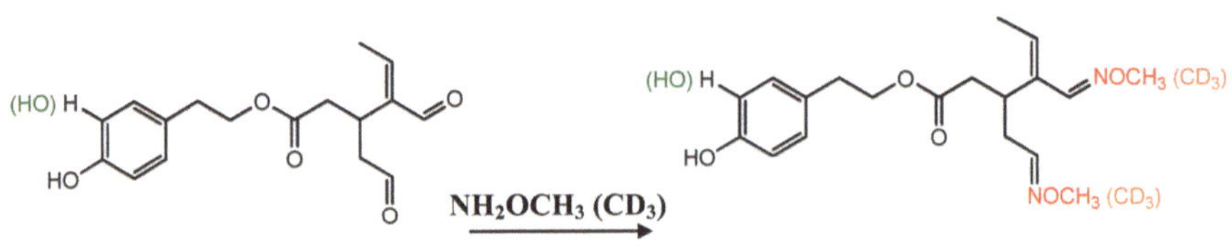

Scheme 14.2 *Freezing* of the dihaldehyde moiety as bis-dioximate

Chart 14.4 Anti-inflammatory molecules (**9**, **10**) extracted from olive drupes

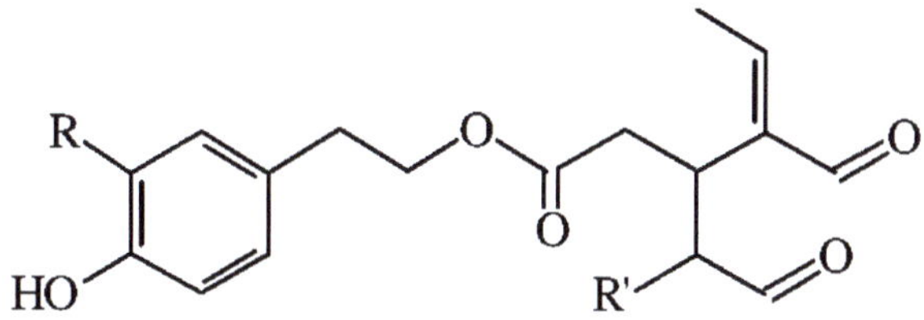

9; R=H; R'=CH3COO; Mr =362
10; R=OH; R'=CH3COO; Mr= 378

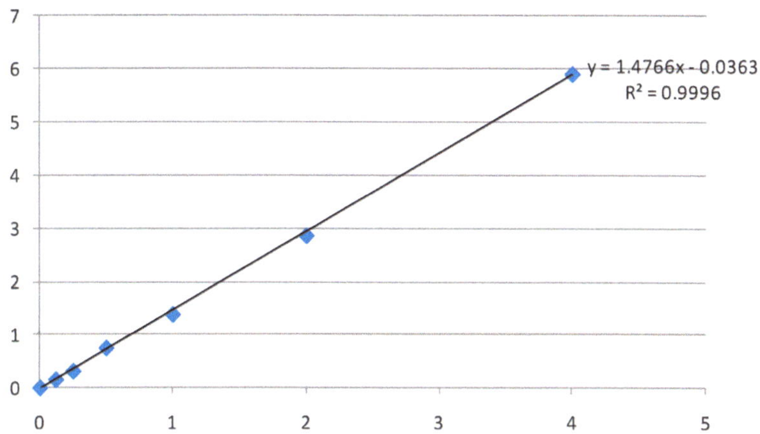

Fig. 14.3 Calibration curve of six standard solutions of HTyr-OLPD and internal standard d_6-HTyr-OLPD

The multiple reaction monitoring (MRM) mode was used for quantitative analysis. The assay of oleopentanedialdehydes (OLPD) was performed following two transitions per compound, the first one for quantitation and the second for confirmation [11].

It has been shown, very recently, that the so called Ambient Mass Spectrometry possesses the ability to record mass spectra of ordinary samples, in their native environment, without sample preparation or pre-separation [12, 13]. Many applications are present in the recent literature of mass spectrometry with reference to quality determination of foods [14, 15]. Paper spray (PS/MS), one of the new ambient approach of mass spectrometry, was applied in the identification of the inflammatory active principles **9** and **10** (Scheme 14.1) directly from different samples of commercial olive oil (Fig. 14.4).

Five hundred milligrams of olive oil were added of a 50 μL solution of benzaldehyde (100 ppm) and then vortexed for 3 min. The mixture thus obtained was then preloaded (15 μL) onto a paper triangle and dried at room temperature for 2 min, chemical derivatization reaction was carried out using 15 μL of 1.5 M methoxyamine hydrochloride methanol/water (1:1, v/v) solution; the latter was spotted onto

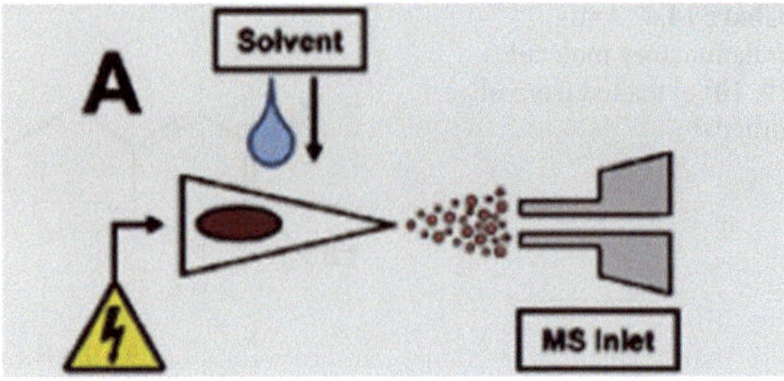

Fig. 14.4 Schematic of a Paper Spray (*PS*) device. Samples are directly loaded on a *triangular shaped* paper sheet which is located in a high electric field in front of the MS inlet

Table 14.2 Amount of analyte 10 founds in virgin olive oils, and repeatability value (RSD%)

Amount of analytes 10 (mg/kg)		
Olive oil Samples		RSD %
1	85 ± 12	14.1
2	92 ± 15	16.3
3	131 ± 19	14.5
4	75 ± 9	12.0
5	68 ± 6	8.8

a paper triangle preloaded with 15 μL volume of olive oil. After drying, the paper triangle was positioned in front of the ion transfer tube of the mass spectrometer and the high voltage was applied. Then, 3 × 15 μL volume of the methoxyamine solution were spotted onto the paper to allow 2 min acquisition for the precursor ion scan mode experiment. Table 14.2 summarizes the data obtained in the assay of **10** (Chart 14.4) [16].

Tandem mass spectrometry associated to the use of labeled internal standard is, therefore, a reference verification methodology in food quality evaluation, in general, and, in particular, when possible fraudulently modified aliments are introduced into the market. This is an important step in the identification of frauds, especially when food claims are often based on the amplification of their nutraceutical properties based on the presence of active principles which could be absent or negligible. Many applications of the method to worldwide distributed aliments different from olive oil have been reported.

The wine food chain is an important one. Starting from the early 1990s the *French Paradox*, based on the effect of Resveratrol (**11**) present in red wine which might prevent coronary heart diseases, has flooded the international wine market. The analytical method discussed thoroughly in this contribution was therefore applied to the identification and quantitation of **12** in the presence of the deuterated internal standard **12** (Chart 14.5).

The measurements were carried out in the selected ion monitoring (SIM) or multiple reaction monitoring (MRM) mode; the latter was unsuitable when positive ions were sampled because of the extensive isotopomerization of the d_4 internal standard [17]. Different red wines were checked (Table 14.3) and the amount of **1** ranged from 2.02 to 20.39 ppm for a Moroccan to a southern Italian wine.

Many other applications are available including those were the methodology has been applied to the amino acid assay in beverages [18], and to the identification of

Chart 14.5 Resveratrol and its d_4 labeled isomer

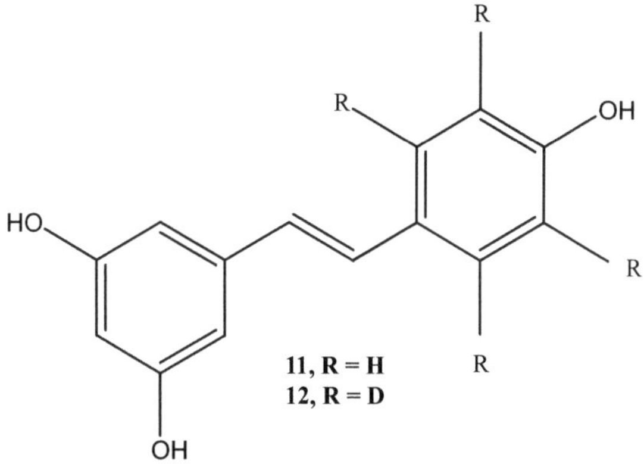

Table 14.3 Amount of resveratrol found in red wines by ESI-SIM

Wine	*trans*-resveratrol (ppm)	RSD %	*cis*-resveratrol (ppm)	RSD %
Dragone Lento	6.79 ± 0.49	7.25	1.43 ± 0.05	3.25
Federico II Lento	4.82 ± 0.41	8.46	1.64 ± 0.07	4.19
Riserva Lento	14.46 ± 0.32	2.19	2.46 ± 0.07	2.66
Gaglioppo Statti	13.08 ± 0.90	6.89	2.86 ± 0.07	2.39
Arvino Statti	10.78 ± 0.91	8.47	3.15 ± 0.19	6.14
I Gelsi Statti	20.39 ± 1.45	7.10	4.07 ± 0.23	5.72
Ceppereto	3.26 ± 0.16	4.87	1.09 ± 0.02	1.88
Cirò Classico	2.66 ± 0.03	0.98		
Elios	8.48 ± 0.42	4.96	1.20 ± 0.03	2.59
Chianti	2.70 ± 0.20	7.50		
Chateau Beni Chougrane (Algeria)	9.08 ± 0.54	5.98	1.53 ± 0.08	5.21
Sidi Brahim (Morocco)	2.02 ± 0.06	3.22	0.56 ± 0.02	3.68

flavonoids in citrus juices and beverages [19] and leeks [20]. The identification of unknown compound from its full scan mass spectrum only is, as expected, impossible. Tandem mass spectrometry provides, since its original applications in the 1970s of last century, a wealth of information which are determinant for a structure identification of a new species. In the case of unknowns other high tech analytical methodologies, such as NMR, should be applied.

A case study can be represented by the discovery of the presence of 3-hydroxy-3-methylglutarate (HMG) flavonoids **13** and **14** (Chart 14.6) in different tissues of bergamot fruit [21] (Chart 14.6). HMG moiety represents the active site which inhibits the reductase enzyme in the, in vivo, cholesterol biosynthesis [22].

The HMG moiety, present in the natural substrate HMG- CoA **15**, plays a fundamental role since it is recognized by the enzyme [23], but not metabolised by the NADH reducing agent present in the enzyme active site which cannot hydrolyze the ester bond linking HMG to flavonoids, whereas it interacts vey effectively with the thioester moiety of substrate **15**.

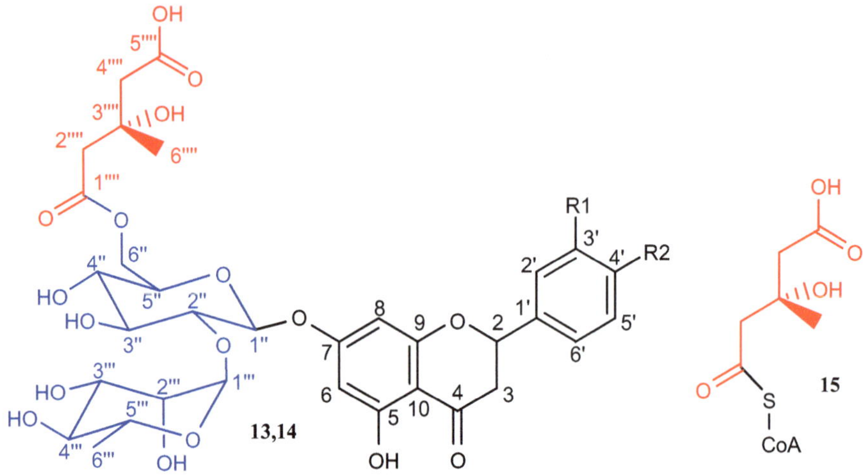

13 Brutieridin, $\mathbf{R_1 = OH, R_2 = OCH_3}$
14 Melitidin, $\mathbf{R_1 = H, \quad R_2 = OH}$

Chart 14.6 The natural statins HMG-Flavonoids (**13, 14**) share the same active species (HMG, *in red*) of the HMG-SCoA (**15**) the natural substrate for cholesterol synthesis

The active species **13** and **14** confer either to juices and fruit bergamot albedo [24] the role of nutraceuticals, thus adding value to natural products which are still considered wastes of those industries producing the bergamot oil for cosmetic applications.

The important step in the structure recognition of **13** and **14** was the identification of the HMG moiety, the two species were present in the juice since ever but never considered as important healthy molecules. Some funny interpretation can be found in the literature whereby HMG was confidently replaced by an isobaric di-oxalate moiety [25]! The real structure was determined using MS/MS assisted by the classic approach of analytical organic chemistry, i.e. the HMG species was isolated by hydrolysis and identified by MS, and the structure of the isolated molecule **13** and **14** confirmed by nuclear magnetic resonance (NMR) spectroscopy.

14.2.2 Food Safety

It is a complex issue which include frauds deliberately planned to introduce into the market low quality foods adulterated, occasionally, with poisoning substances, which could mask the nutritional and healthy properties of the aliment. The simplest case is represented by the evaluation of phytodrug residues when the operators do not follow the official directives or when superficial rules have been set-up by national or international bodies and this was the case of, a naturally occurring insecticide, Rotenone, whose use was allowed for more than 70 years in lakes, rivers and

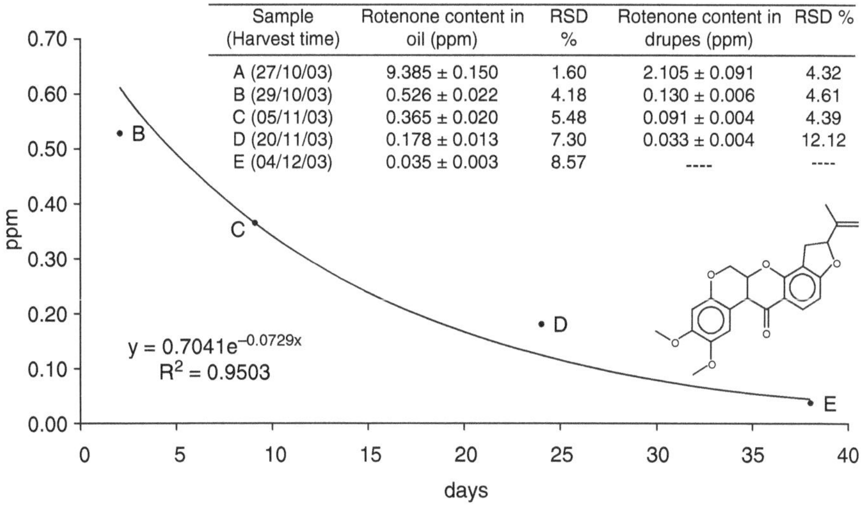

Sample (Harvest time)	Rotenone content in oil (ppm)	RSD %	Rotenone content in drupes (ppm)	RSD %
A (27/10/03)	9.385 ± 0.150	1.60	2.105 ± 0.091	4.32
B (29/10/03)	0.526 ± 0.022	4.18	0.130 ± 0.006	4.61
C (05/11/03)	0.365 ± 0.020	5.48	0.091 ± 0.004	4.39
D (20/11/03)	0.178 ± 0.013	7.30	0.033 ± 0.004	12.12
E (04/12/03)	0.035 ± 0.003	8.57	----	----

Fig. 14.5 Decay of the amount of rotenone with harvesting time

ponds as piscicide, and, last but not least, in the growing of olive trees whose drupes were used for the preparation of organic olive oil! The molecule has fairly recently been classified as highly toxic to humans, and its lethal effect on fish was known to American and Australian aborigens [26] (Fig. 14.5).

Rotenone was persistent in olive drupes for nearly 40 days after the original treatment. It was preserving the fruit from olive fly attack but it was transferred to the oil thus obtained.

Another threat to consumers was represented by the azo dyes deliberately added to hot chili pepper powder to preserve its reddish color. Those colorants belong to the family of Sudan dyes and they were found in any kind of foods where pepper powder was used. Studies exist on the dangerous exposure of humans to Sudan dyes [27], which have been classified as carcinogens by the International Agency for Research on Cancer. An MS/MS method was set-up for the identification of the different species of Sudan family dyes in foodstuff, which relies on the use of labeled internal standards. The application of the methodology required the synthesis of the labeled compounds (Chart 14.7) [28], not available in the market, and a fully investigation of their gas-phase chemistry (Fig. 14.6) [29]. The fraudulent addition of sudan dyes to foodstuff have been banned by EC directives published in the *Official Journal of the European Union* in the period 2003–2005.

The results presented in Fig. 14.6 show how tandem mass spectrometry in MRM mode with the use of specific labelled internal standards for each analyte allows the identification and assay of all the Sudan dyes present in a given foodstuff.

The use of ambient ionization devices such as PS has allowed recently the determination of "Sudan dyes in seconds". This was the enthusiastic comment [31] to the scientific results obtained by PS-MS/MS just spotting a tiny volume of 10–20 μL of solution on a wetted triangular shaped piece of paper [32] (Table 14.4).

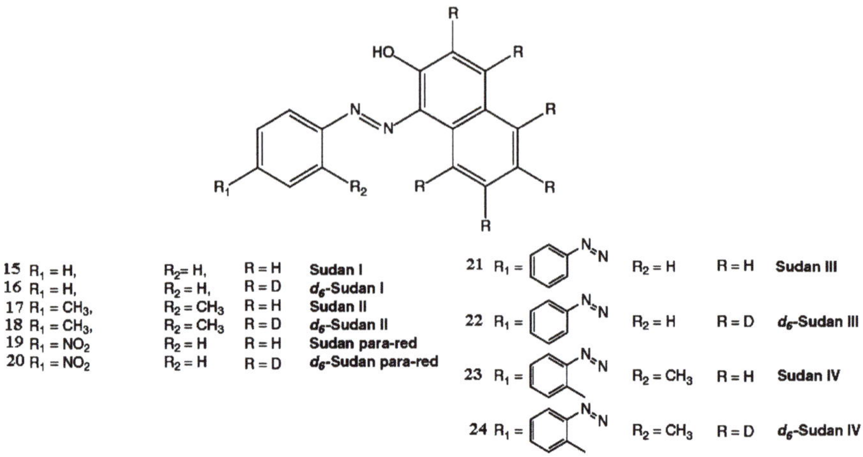

Chart 14.7 Investigated Sudan dyes and their corresponding labeled isomers

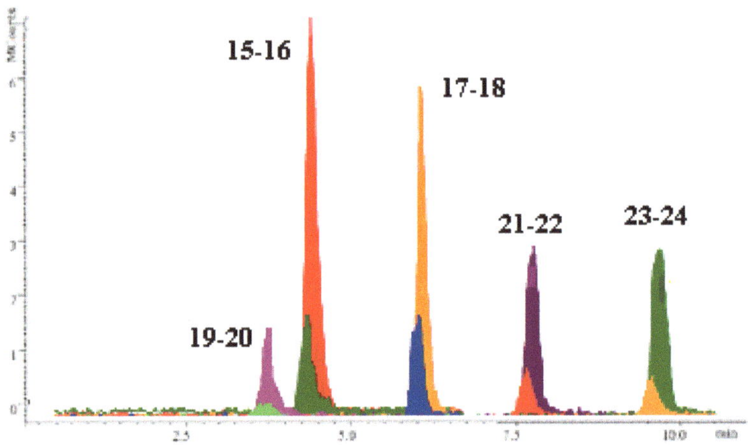

Fig. 14.6 MRM chromatogram of standard solution at 200 ppb of Sudan dyes and 50 ppb of the corresponding deuterium labeled isomers

14.2.3 Origin of Foods

The origin of particular aliment is often associated to some of its, supposed, healthy and/or quality properties. The methods worldwide used to trace the origin of foodstuff lack of scientific significance.

Trace element analysis by ICP-MS provides traceability markers of absolute value in agricultural and zootechnical food chains [32] since they relay the "breeding" of either the plant or the animal to what they "eat." Nothing new if we recall the Savarin's aphorism *tell me what you eat, and I will tell you what you are.*

Table 14.4 Accuracy and RSD values obtained in the quantitation Sudan dyes added to hot chili pepper at different concentration

Dyes	Spiked samples mg/L	Calculated amount mg/L	RSD %	Accuracy %
Sudan I	20	26.63 ± 2.1	7.89	133.1
	100	94.35 ± 9.85	10.44	94.4
	200	201.32 ± 15.36	7.63	100.7
Sudan II	20	19.24 ± 1.81	9.41	96.2
	100	112.95 ± 8.56	7.58	113.1
	200	206.92 ± 11.32	5.47	103.5
Sudan III	20	23.11 ± 1.95	8.44	115.5
	100	101.46 ± 8.48	8.36	101.5
	200	204.95 ± 10.23	4.99	102.5
Sudan IV	20	23.37 ± 1.83	7.83	116.9
	100	100.13 ± 7.56	7.55	100.1
	200	203.04 ± 9.84	4.85	101.5
Sudan Para Red	20	23.71 ± 2.12	8.94	118.5
	100	103.7 ± 13.23	0.13	103.7
	200	199.87 ± 15.18	0.08	99.9

Table 14.5 Number of PGI "Clementine of Calabria" and non PGI samples

Corigliano Calabro	Calabria	18	18
Lamezia Terme	Calabria	9	9
Pizzo Calabro	Calabria	9	9
Rosarno	Calabria	18	18
Algiers	Algeria	10	10
Blida	Algeria	4	4
Valencia	Spain	8	8
	Tunisia	12	12

A case study can be represented by the evaluation of origin of Clementine (*Citrus clementina Hort. ex Tan.*) which is one of the most important cultivated variety of citrus mandarins in the Mediterranean basin [33]. It is the result of a cross between mandarin and bitter orange achieved in Algeria in the early twentieth century. The fruit produced in Calabria, a southern region of Italy, can be labeled with the protected geographical indication (PGI) "Clementine of Calabria." The traceability of the latter was successfully determined from the concentration range of the elements Ce, Dy, Er, Eu, Gd, Ho, La, Lu, Nd, Pr, Sc, Sm, Tb, Th, Tm, U, Y, Yb, Ag, Al, As, Ba, Be, Bi, Ca, Cd, Co, Cr, Cs, Cu, Fe, Ga, In, K, Li, Mg, Mn, Na, Ni, Pb, Rb, Se, Sr, Tl, V and Zn evaluated by a simple and rapid method based on a mineralization process assisted by microwaves and a subsequent ICP-MS analysis of the digested samples (Table 14.5).

The statistical evaluation of the data thus obtained was based on the evaluation of three pattern recognition chemometric models. The results of chemometric analysis expressed in terms of prediction ability show that all of the statistical

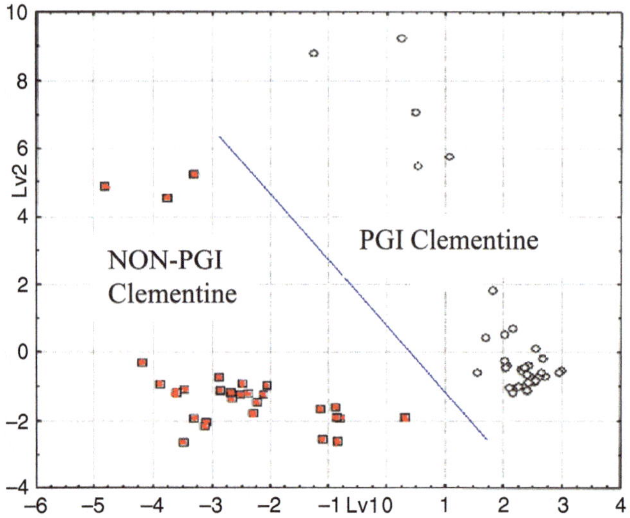

Fig. 14.7 PLS-DA plot LV1 versus LV2 for clementine peel samples

techniques involved (S-LDA, SIMCA, and PLS-DA) can be successfully employed for traceability purposes (Fig. 14.7).

The analytical approach presented for evaluation of origin of this peculiar citrus fruit [34] can be extended either to the same fruit produced in other areas or to other agrifoods. It seems important to underline that traceability based on the application of scientifically qualified methods is the only way to protected consumer from frauds.

14.3 Conclusions

Identification of Fraudulently Modified Foods can be confidently performed by means of all the methodologies developed in the last 40 years in the field of mass spectrometry. This powerful analytical approach has the unique property that to each single analyte present in a complex mixture can be attributed a molecular formula and a molecular weight thus smoothing all the steps for the description of a given foodstuff in terms of its minor components and their relative amounts.

References

1. Wu QQ, Fan KX, Sha W, Ruan HQ, Zeng R (2009) Highly sensitive detection of melamine based on reversed phase liquid chromatography mass spectrometry. Chin Sci Bull 54:732–737

2. Ciafardini G, Zullo BA, Iride A (2006) Lipase production by yeasts from extra virgin olive oil. Food Microbiol 15:2360
3. Attya M, Russo A, Perri E, Sindona G (2012) Endogenous lipase catalyzed transesterification of olive oil fats. The formation of isomeric and oligomeric triacyleglycerols. J Mass Spectrom 47:1247–1253
4. Mazzotti F, Benabdelkamel H, Di Donna L, Napoli A, Sindona G (2012) Assay of tyrosol and hydroxytyrosol in olive oil by tandem mass spectrometry and isotope dilution method. Food Chem 135:1006–1010
5. De Nino A, Di Donna L, Mazzotti F, Muzzalupo E, Perri E, Sindona G, Tagarelli A (2005) Absolute method for the assay of oleuropein in olive oils by atmospheric pressure chemical ionization tandem mass spectrometry. Anal Chem 77:5961
6. De Nino A, Lombardo A, Perri E, Procopio A, Sindona G (1997) Direct identification of phenolic glucosides from olive leaves extracts by ion spray ionisation and MS/MS analysis. J Mass Spectrom 32:533, e
7. Di Donna L, Mazzotti F, Napoli A, Salerno R, Sajjad A, Sindona G (2007) Secondary metabolism of olive secoiridoids. New microcomponents detected in drupes by electrospray ionization and high-resolution tandem mass spectrometry. Rapid Commun Mass Spectrom 21:273–278
8. Beauchamp GK, Keast RS, Morel D, Lin J, Pika J, Han Q, Lee CH, Smith AB, Breslin PA (2005) Ibuprofen-like activity in extra-virgin olive oil. Nature 437:45–46
9. Montedoro G, Servili M, Baldioli M, Selvaggini R, Miniati E, Macchioni A (1993) Simple and hydrolyzable phenolic compounds in virgin olive oil. 3. Spectroscopic characterization of the secoiridoid derivatives. J Agric Food Chem 41:2228–2234
10. De Nino A, Mazzotti F, Perri E, Procopio A, Raffaelli A, Sindona G (2000) Virtual freezing of the hemiacetal-aldehyde equilibrium of the aglycones of oluropein and ligstroside present in olive oils from carolea and coratina cultivars by ion-spray ionisation tandem mass spectrometry. J Mass Spectrom 35:461
11. Di Donna L, Benabdelkamel H, Mazzotti F, Napoli A, Nardi M, Sindona G (2011) High-throughput assay of oleopentane dialdheydes in extra virgin olive oil by the UHPLC-ESI-MS/MS and isotope dilution methods. Anal Chem 83:1990–1995
12. Cooks RG, Ouyang Z, Takats Z, Wiseman JM (2006) Ambient mass spectrometry. Science 311:1566–1570
13. Monge ME, Harris GA, Dwivedi P, Fernández FM (2013) Recent advances in direct open air surface sampling/ionization. Chem Rev 113:2269–2308
14. Malaj N, Ouyang Z, Sindona G, Cooks RG (2012) Analysis of pesticide residues by leaf spray mass spectrometry. Anal Methods 4:1913–1919
15. Malaj N, Gallucci G, Romano E, Sindona (2013) GHMG-flavonoids in bergamot fruits: identification and assay of natural statins by recently developed mass spectrometric methods. In: G Dugo, I Bonaccorsi (eds) Citrus bergamia-bergamot and its derivatives. Taylor & Francis
16. Mazzotti F, Di Donna L, Taverna D, Nardi M, Aiello D, Napoli A, Sindona G (2013) Evaluation of dialdehyde anti-inflammatory active principles in extra-virgin olive oil by reactive paper spray mass spectrometry. Int J Mass Spectrom 352:87–91
17. Di Donna L, Mazzotti F, Benabdelkamel H, Gabriele B, Plastina P, Sindona G (2009) Effect of H/D isotopomerization in the assay of resveratrol by tandem mass spectrometry and isotope dilution method. Anal Chem 81:8603–8609
18. Mazzotti F, Benabdelkamel H, Di Donna L, Athanassopoulos CM, Napoli A, Sindona G (2012) Light and heavy dansyl reporter groups in food chemistry: amino acid assay in beverages. J Mass Spectrom 47:932–939
19. Di Donna L, Taverna D, Mazzotti F, Benabdelkamel H, Attya M, Napoli A, Sindona G (2013) Comprehensive assay of flavanones in citrus juices and beverages by UHPLC-ESI-MS/MS and derivatization chemistry. Food Chem 141:2328–2333
20. Di Donna L, Mazzotti F, Taverna D, Napoli A, Sindona G (2013) Structural characterisation of Malonyl Flavonols in Leek (Allium porrum L.) using high-performance liquid chromatography and mass spectrometry. Phytochem Anal. doi:10.1002/pca.2493

21. Di Donna L, De Luca G, Mazzotti F, Napoli A, Salerno R, Taverna D, Sindona G (2009) Statin-like principles of bergamot fruit (Citrus bergamia): isolation of 3-hydroxymethylglutaryl flavonoid glycosides. J Nat Prod 72:1352–1354
22. Di Donna L, Dolce V, Sindona G patent nr. CS2008A00019
23. Leopoldini M, Malaj N, Toscano M, Sindona G, Russo N (2010) On the inhibitor effects of bergamot juice flavonoids binding to the 3-Hydroxy-3-methylglutaryl-CoA reductase (HMGR) enzyme. J Agric Food Chem 58:10768–10773
24. Di Donna L, Gallucci G, Malaj N, Romano E, Tagarelli A, Sindona G (2011) Recycling of industrial essential oil waste: brutieridin and melitidin, two anticholesterolaemic active principles from bergamot albedo. Food Chem 125:438–441
25. Gardana C, Nalin F, Simonetti P (2008) Evaluation of flavonoids and furanocoumarins from citrus bergamia (Bergamot) juice and identification of new compounds. Molecules 13:2220–2228
26. Caboni P, Sherer T, Zhang N, Taylor G, Na HM, Greenamyre J (2004) Rotenone, deguelin, their metabolites, and the rat model of Parkinson's disease. J Chem Res Toxicol 17:1540–1548
27. Stiborova M, Martinek V, Rydlova H, Koblas T, Hodek P (2005 (a) Expression of cytochrome P450 1A1 and its contribution to oxidation of a potential human carcinogen 1-phenylazo-2-naphthol (Sudan I) in human livers. Cancer Lett 220:145 (b) Xu H, Heinze T M, Paine DD, Cerniglia CE, Chen H (2010) Sudan azo dyes and Para Red degradation by prevalent bacteria of the human gastrointestinal tract. Anaerobe 16:114–119
28. De Nino A, Di Donna L, Sindona G, Maiuolo L, Mazzotti F (2008) The synthesis of deuterium labelled azo-dyes of the Sudan Family. Synthesis 3:459–463
29. Di Donna L, De Nino A, Maiuolo L, Mazzotti F, Napoli A, Salerno R, Sindona G (2007) High-throughput mass spectrometry: the mechanism of sudan azo dye fragmentation by ESI tandem mass spectrometry and extensive deuterium labeling experiments. J Mass Spectrom 42:1057–1061, e
30. Mazzotti F, Di Donna L, Maiuolo L, Napoli A, Salerno R, Sajjad A, Sindona G (2008) Assay of the set of all sudan azodye (I, II, III, IV, and Para-Red) contaminating agents by liquid chromatography-tandem mass spectrometry and isotope dilution methodology 56:63–67
31. Sudan dyes in seconds: paper spray mass spectrometry for chili powder tests. www.spectroscopynow.com. Jun 15, 2013. S. Down
32. Taverna D, Di Donna L, Mazzotti F, Policicchio B, Sindona G (2013) High-throughput determination of sudan azo-dyes within powdered chili pepper by paper spray mass spectrometry. J Mass Spectrom 48:544–547
33. Aiello D, De Luca D, Gionfriddo E, Naccarato A, Napoli A, Romano E, Russo A, Sindona G, Tagarelli A (2011) Multistage mass spectrometry in quality, safety and origin of foods. Eur J Mass Spectrom 17:1–31
34. Benabdelkamel H, Di Donna L, Mazzotti F, Sindona G, Tagarelli A, Taverna D (2012) Authenticity of PGI "Clementine of Calabria" by multielement fingerprint. J Agric Food Chem 11:3717–3726

Chapter 15
Tandem Mass Spectrometric Analysis of Novel Antineoplastic Curcumin Analogues

H. Awad, U. Das, J. Dimmock, and A. El-Aneed

Abstract The fragmentation behavior of two antineoplastic curcumin analogues with a structural backbone of 1,5-diaryl-3-oxo-1,4-pentadiene was studied using ESI-MS/MS. Tandem mass spectrometric analysis confirmed their molecular structure and identified their diagnostic product ions that will be used for qualitative and quantitative applications. Two curcumin analogues (designated as NC2067 and NC2081) have been studied using a triple quadrupole linear ion trap hybrid mass spectrometer (Qq-LIT-MS). It was observed that the two compounds behave in the same manner during MS/MS analysis. The fragmentation of the 3,5-bis(benzylidene)-4-piperidone was mainly based on cleavage within the 4-piperidone ring. However, the side chain fragmentation was mainly observed on the carbon-nitrogen and carbon-oxygen bonds. Based on this behavior, a general fragmentation pattern has been established.

Keywords Tandem mass spectrometry (MS/MS) • Multiple reaction monitoring (MRM) • Isotope dilution method • Ambient ionization methods • Food quality • Food safety • Food origin

15.1 Introduction

The design of new anticancer agents usually aims to improve potency, safety and selective toxicity towards malignant cells rather than normal cells. Several studies have been conducted to explore the anticancer properties of curcumin which shows anti-inflammatory, antioxidant, antiproliferative and antiangiogenic activities [1]. However, the poor bioavailability of curcumin has hampered its wide use as an anticancer agent [2]. Several approaches have been adopted to improve curcumin potency, targeting and bioavailability [3, 4]; one of which is the synthesis of structurally-related compounds, namely curcumin analogues [4].

H. Awad (✉) • U. Das • J. Dimmock • A. El-Aneed
College of Pharmacy and Nutrition, University of Saskatchewan,
Saskatoon, SK S7N 5E5, Canada
e-mail: anas.el-aneed@usask.ca

J. Banoub (ed.), *Detection of Chemical, Biological, Radiological and Nuclear Agents for the Prevention of Terrorism*, NATO Science for Peace and Security Series A: Chemistry and Biology, DOI 10.1007/978-94-017-9238-7_15

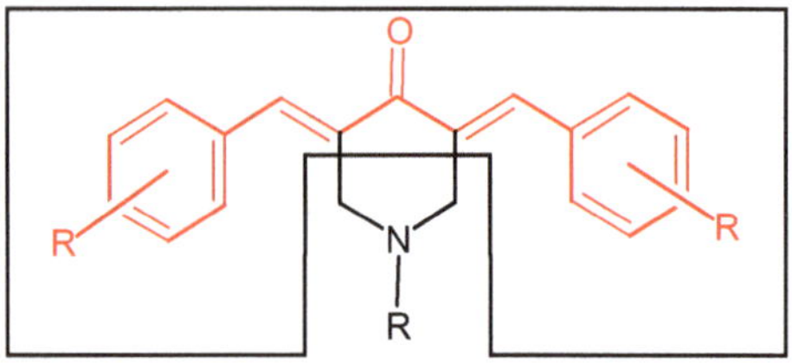

Scheme 15.1 3,5-bis (benzylidene)-4-piperidones (1,5-diaryl-3-oxo-1,4-pentadienyl pharmacophore *highlighted in box*)

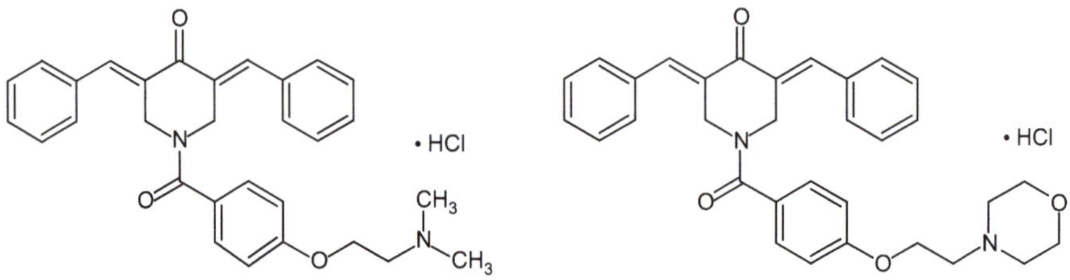

NC 2067
466.2256 Da

NC 2081
508.2362Da

Scheme 15.2 Structures and monoisotopic masses of the tested curcumin analogues (NC2067 and NC2081)

Curcumin analogues have been designed as 3,5-bis(benzylidene)-4-piperidones which contains the 1,5-diaryl-3-oxo-1,4-pentadienyl pharmacophore (Scheme 15.1) which is responsible for the selective antitumor effect of these compounds [5, 6]. Various newly designed curcumin analogues displayed a significant cytotoxic effect towards different cancer cell lines including leukemia, melanoma and colon cancer [5, 7, 8]. Sensitive and rapid screening of curcumin analogues during pharmacological and pharmacokinetic studies has become a demanding process that requires sensitive high throughput analytical tools.

Mass spectrometry (MS) plays a critical role in the qualitative and quantitative analysis of drug molecules based on its sensitivity and high throughput capability [9, 10]. For highly selective MS-quantification, Multiple Reaction Monitoring (MRM) is used where the ion of interest (precursor ion) is selected in the first mass analysis stage while a fragment ion (product ion) is monitored in the second stage after the fragmentation reaction of that analyte [11]. Therefore, the MS/MS fragmentation behavior of an analyte is important to be investigated.

In this study, two representative drugs of the 3,5-bis(benzylidene)-4-piperidone curcumin analogues family were chosen, designated as NC 2067 and NC 2081 (Scheme 15.2). These agents are currently being evaluated as anti-melanoma entities. A general ESI- collision induced dissociation (CID)-MS/MS fragmentation pattern for the two tested curcumin analogues was established using Qq-LIT-MS/MS. Their molecular structure was confirmed and unique product ions of each compound were also identified.

15.2 Experimental

15.2.1 Materials

Curcumin analogues (NC 2067 and NC 2081) (Scheme 15.2) were synthesized as described [6, 7] by the research team of Drs. J. Dimmock and U. Das, College of Pharmacy and Nutrition, University of Saskatchewan.

15.2.2 Methods

Samples were prepared by dissolving trace amounts of the curcumin analogue in water: methanol (1:1) and injected into an AB SCIEX 4000 QTRAP® instrument (Qq-LIT-MS) with a flow rate of 10 μL/min using a Harvard Syringe Pump. Low energy CID-MS/MS was operated in the positive ion mode with ion spray voltage 5,500 V at room temperature, DP 100 V and collision energy (CE) 25–35 V using nitrogen as a collision gas. MS/MS, MS^3 and neutral loss analysis were performed for each sample to establish the fragmentation pattern. Parameters of the MS/MS analysis were optimized to ensure the formation of product ions while maintaining the presence of the precursor ion.

15.3 Results and Discussion

The single stage ESI-MS analysis of curcumin analogues (NC2067 and NC2081) indicated the presence of the protonated molecules $[M+H]^+$ at *m/z* 267.2 and 509.2 respectively as shown in Fig. 15.1. The gas-phase fragmentation of the protonated molecules at *m/z* 267.2 and 509.2 of both NC2067 and NC2081 respectively was performed by low-energy collision dissociation tandem mass spectrometry (CID-MS/MS) analysis. The CID-fragmentation behavior for curcumin analogues (NC2067 and NC2081) appeared to be similar. Accordingly, the fragmentation of 3,5-bis(benzylidene)-4-piperidone was mainly based on the breaking of the 4-piperidone ring. However, the side chain fragmentation was targeted primarily on the amide and carbon-oxygen bonds as highlighted in Scheme 15.3.

15.3.1 3,5-bis(benzylidene)-4-piperidone Fragmentation

As shown in Scheme 15.4a, the CID-fragmentation of the precursor ions of both NC2067 and NC2081 at *m/z* 267.2 and 509.2 respectively occurs through the breakage of the 4-piperidone ring resulting in the formation of ten product ions

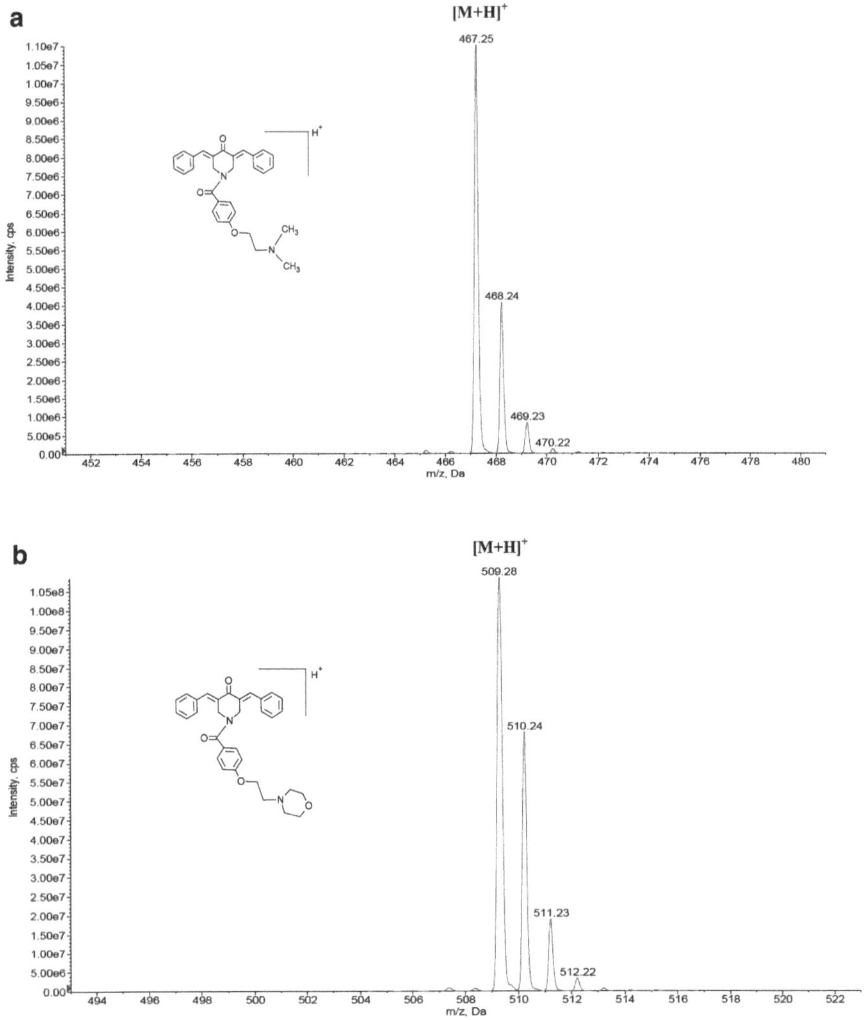

Fig. 15.1 The single stage ESI-MS spectra of NC2067 (**a**) and NC2081 (**b**) showing the protonated ion $[M+H]^+$ of each compound

(Figs. 15.2 and 15.3). Nine of these ions (observed at *m/z* 276, 259, 247, 231, 229, 219, 205, 143 and 117) were identical for the two compounds. However, one product ion showed a 42 Da difference (*m/z* 349 for NC2067 and 391 for NC2081) due to differences in the side chain (Scheme 15.2). This ion, with relatively good abundance, can be considered as a diagnostic product ion that could be used during the development of multiple reaction monitoring (MRM) quantification method of such compounds.

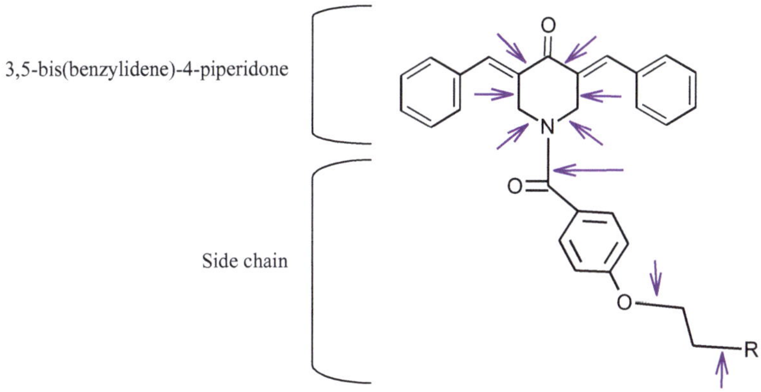

Scheme 15.3 Sites of fragmentation in the general structure of NC2067 and NC2081 curcumin analogues, **NC2067** (R= 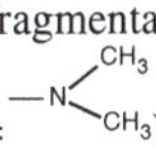), **NC2081** (R= 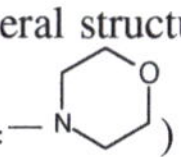)

To confirm the sequence of the proposed fragmentation, MS^3 analysis was performed for various MS/MS product ions of both NC2067 and NC2081 (namely product ions observed at *m/z* 349.1, 259.1, 247.1, 231.1 and 192.1) (Scheme 15.4).

Table 15.1 shows a summary for MS^3 experiments of NC2067; illustrating that one product ion could be formed from different sources. For example, the product ion observed at *m/z* 143 was formed from two other product ions (*m/z* 259 and 247) in addition to its direct formation from the precursor ion. On the other hand, a unique product ion has been formed during the MS/MS analysis of NC2067 (not shown in Scheme 15.4) with *m/z* value = 189. ESI-MS^3 confirmed that this ion is produced from another product ion of 349 Da via serial fragmentation.

15.3.2 Side Chain Fragmentation

The side chain fragmentation of the tested curcumin analogues (NC2067 and NC2081) was mainly based on breaking the carbon-nitrogen and carbon-oxygen bonds during CID-MS/MS analysis of their precursor ions at *m/z* 267.2 and 509.2 respectively (Scheme 15.4b). The most abundant product ion that was observed due to side chain fragmentation was a result of carbon-oxygen bond breakage producing an ion observed at *m/z* 72 for NC2067 and 114 for NC2081 (Figs. 15.2 and 15.3). Such product ions are structure-dependent and were among the most abundant ions; therefore, they could be used for the development of MRM-MS/MS quantification methods of NC 2067 and NC2081.

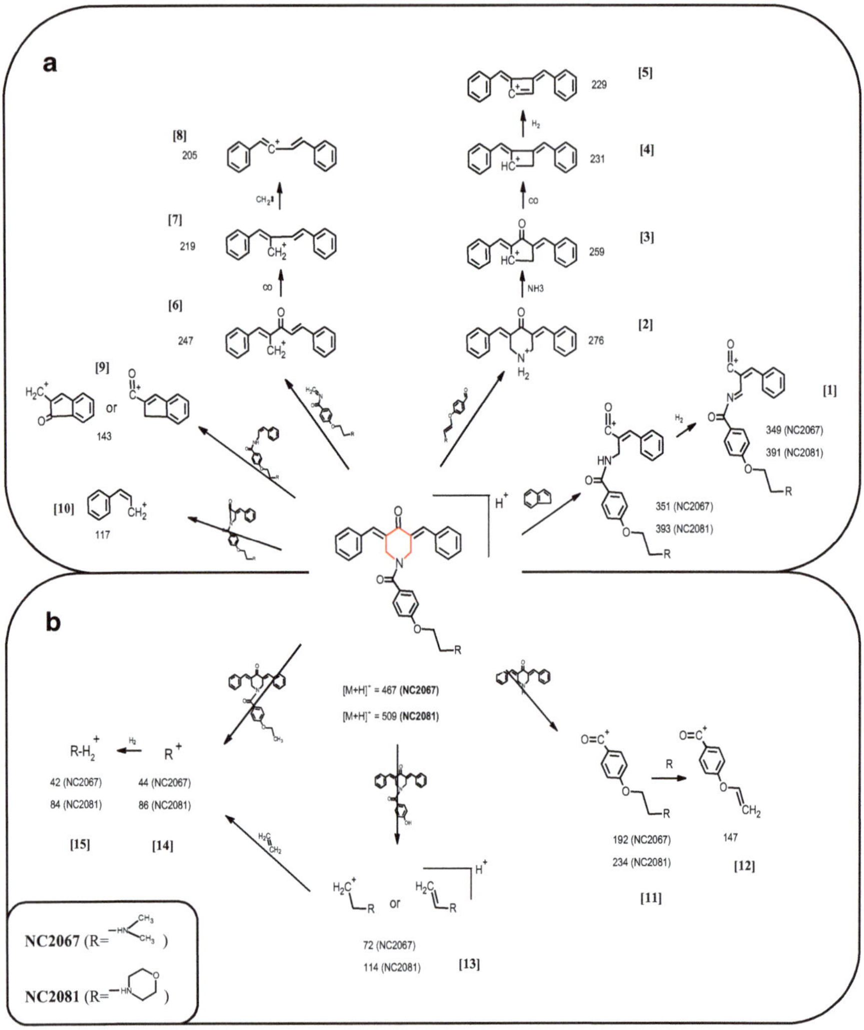

Scheme 15.4 A general fragmentation pattern of NC 2067 and NC2081 curcumin analogues. (**a**) *Fragmentation of 3,5-bis(benzylidene)-4-piperidone.* (**b**) Fragmentation of *the side chain*

In addition to MS^3 experiments, neutral loss analysis was also used to aid in establishing the fragmentation patterns of both protonated molecules obtained from both compounds. A neutral loss of 395 Da supports the proposed direct formation of the product ion *m/z* 72 for NC2067 and 114 for NC2081 from their precursor ions.

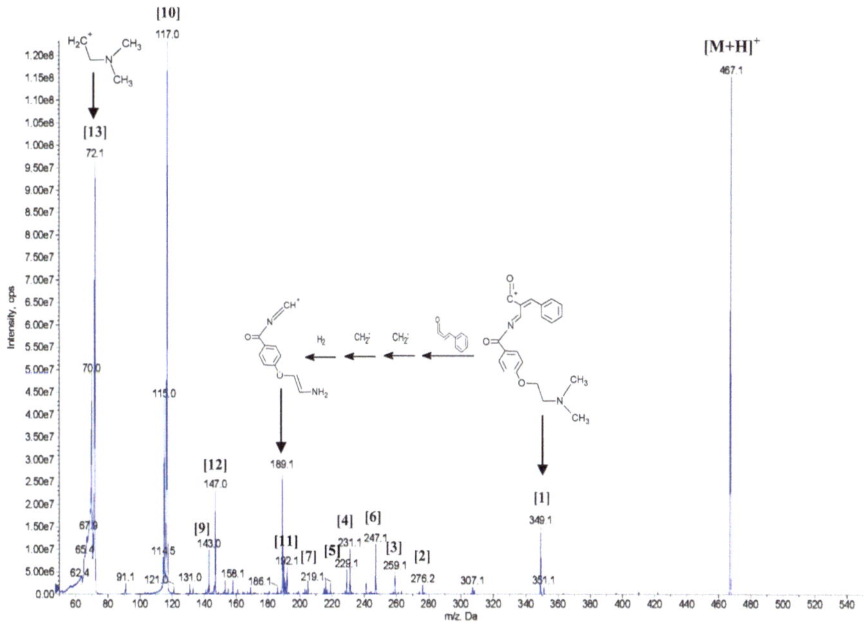

Fig. 15.2 ESI-MS/MS Spectrum of NC2067 showing the diagnostic product ions

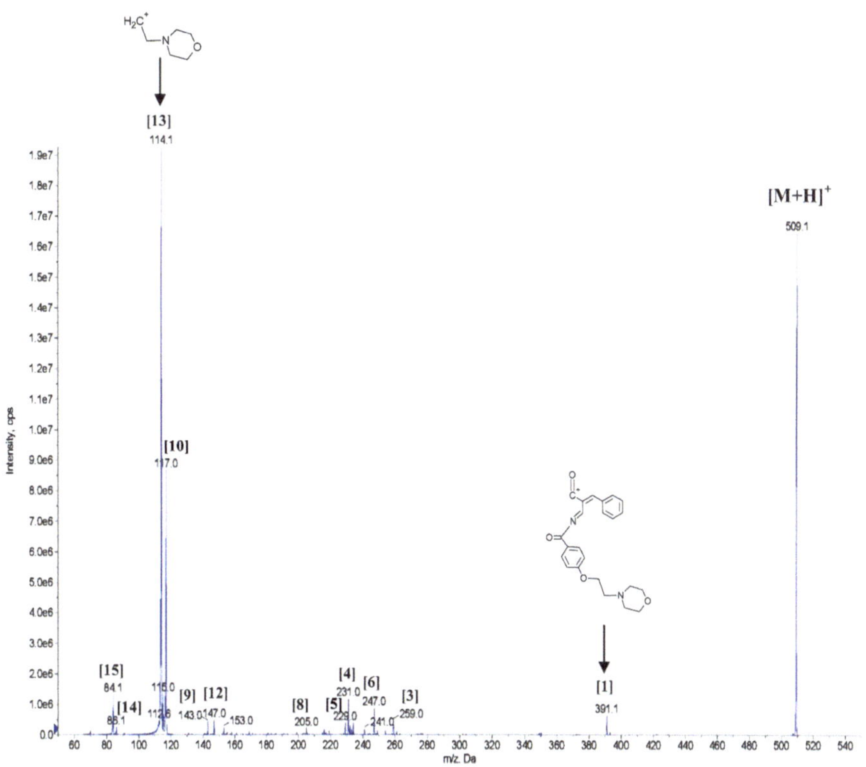

Fig. 15.3 ESI-MS/MS spectrum of NC2081 showing the diagnostic product ions

Table 15.1 Summary of MS^3 experiments for NC2067

Precursor ion I	Precursor ion II	MS/MS/MS product ions
467.1	349.1	189.1, 147.0
	259.1	231.1, 229.1, 143.0
	247.1	219.1, 205.0, 143.0
	231.1	229.1
	192.1	147.2

15.4 Conclusions

Two curcumin analogues that belong to 3,5-bis(benzylidene)-4-piperidone structural family were evaluated: NC2067 and NC2081. By comparing the ESI-fragmentation behavior of both compounds using CID-MS/MS, we can conclude that these compounds have similar fragmentation behavior and such behavior enables the establishing of a general fragmentation pattern for both compounds. Structural confirmation was achieved and diagnostic product ions were identified. We are currently evaluating 13 curcumin analogues using both ESI-MS/MS and matrix assisted laser desorption ionization (MALDI)-MS/MS. The goal is to establish a universal MS/MS fragmentation behavior that will be employed for the qualitative and quantitative analysis of these moieties.

References

1. Aggarwal BB, Kumar A, Bharti AC (2003) Anticancer potential of curcumin: preclinical and clinical studies. Anticancer Res 23(1A):363–398
2. Anand P, Kunnumakkara AB, Newman RA, Aggarwal BB (2007) Bioavailability of curcumin: problems and promises. Mol Pharm 4(6):807–818
3. Shaikh J, Ankola DD, Beniwal V, Singh D, Kumar MNV (2009) Nanoparticle encapsulation improves oral bioavailability of curcumin by at least 9-fold when compared to curcumin administered with piperine as absorption enhancer. Eur J Pharm Sci 37(3):223–230
4. Anand P, Thomas SG, Kunnumakkara AB, Sundaram C, Harikumar KB, Sung B, Tharakan ST, Misra K, Priyadarsini IK, Rajasekharan KN (2008) Biological activities of curcumin and its analogues (congeners) made by man and mother nature. Biochem Pharmacol 76(11): 1590–1611
5. Das U, Sharma RK, Dimmock JR (2009) 1, 5-diaryl-3-oxo-1,4-pentadienes: a case for antineoplastics with multiple targets. Curr Med Chem 16(16):2001
6. Das S, Das U, Selvakumar P, Sharma RK, Balzarini J, De Clercq E, Molnár J, Serly J, Baráth Z, Schatte G (2009) 3, 5-Bis (benzylidene)-4-oxo-1-phosphonopiperidines and related diethyl esters: potent cytotoxins with multi-drug-resistance reverting properties. ChemMedChem 4(11):1831–1840
7. Das U, Alcorn J, Shrivastav A, Sharma RK, De Clercq E, Balzarini J, Dimmock JR (2007) Design, synthesis and cytotoxic properties of novel 1-[4-(2-alkylaminoethoxy) phenylcarbonyl]-3, 5-bis (arylidene)-4-piperidones and related compounds. Eur J Med Chem 42(1): 71–80

8. Das U, Sakagami H, Chu Q, Wang Q, Kawase M, Selvakumar P, Sharma RK, Dimmock JR (2010) 3, 5-Bis (benzylidene)-1-[4-2-(morpholin-4-yl) ethoxyphenylcarbonyl]-4-piperidone hydrochloride: a lead tumor-specific cytotoxin which induces apoptosis and autophagy. Bioorg Med Chem Lett 20(3):912–917
9. Gillespie TA, Winger BE (2011) Mass spectrometry for small molecule pharmaceutical product development: a review. Mass Spectrom Rev 30(3):479–490
10. Korfmacher WA (ed) (2011) Mass spectrometry for drug discovery and drug development. Wiley, Hoboken
11. Gross JH (2011) Tandem mass spectrometry. Mass spectrometry. Springer, Berlin/Heidelberg, 415–478

Chapter 16
Glycoconjugate Vaccines Used for Prevention from Biological Agents: Tandem Mass Spectrometric Analysis

Farid Jahouh, Wael L.L. Demian, Rina Sakksena, Shu-jie Hou, Robert J. Brown, Pavol Kováč, René Roy, and Joseph Banoub

Abstract In this review, we present the determination of the various glycation sites of synthetic neoglycoconjugates formed by conjugation of the antigenic saccharide hapten to BSA using tandem mass spectrometry. The ratio of hapten: BSA was determined by the matrix-assisted laser desorption/ionization-TOF/TOF-MS analyses of the glycoconjugates. We also tentatively propose that all glycated residues are located mainly near the outer surface of the protein.

Keywords MALDI-TOF-MS • LC-ESI-QqTOF-MS/MS • Glycoconjugate vaccines • Glycation sites

---

F. Jahouh
Department of Chemistry, Memorial University of Newfoundland, 232 Elizabeth Avenue, A1B 3X7 St. John's, NL, Canada

W.L.L. Demian • R.J. Brown
Department of Biochemistry, Memorial University of Newfoundland, St. John's, NL A1B 3X9, Canada

R. Sakksena • S.-j. Hou • P. Kováč
NIDDK, LBC, National Institutes of Health, Bethesda, MD 20892-0815, USA

R. Roy
Department of Chemistry, University of Quebec at Montreal, CP 8888, Montreal, QC H3C 3P8, Canada

J. Banoub (✉)
Department of Chemistry, Memorial University of Newfoundland, 232 Elizabeth Avenue, A1B 3X7 St. John's, NL, Canada

Fisheries and Oceans Canada, Science Branch, Special Projects, St John's, NL A1C 5X1, Canada
e-mail: joe.banoub@dfo-mpo.gc.ca

J. Banoub (ed.), *Detection of Chemical, Biological, Radiological and Nuclear Agents for the Prevention of Terrorism*, NATO Science for Peace and Security Series A: Chemistry and Biology, DOI 10.1007/978-94-017-9238-7_16

16.1 Introduction

In 1796, Edward Jenner discovered that the inoculation with cowpox was able to protect against smallpox infection. Since then, different vaccines were developed to prevent infectious disease [1, 2]. It was found that the protection conferred by these vaccines was due to the adaptive immunity (cellular and/or humoral immunity) [1–4]. Unlike adaptive immunity, innate immunity does not recognize every possible antigen. Instead, it is designed to recognize the microbial molecules that are essential for survival of the pathogens. These unique microbial molecules are called pathogen-associated molecular patterns (PAMPS) [3, 4].

PAMPS include lipopolysaccharides (LPS, also called endotoxin) from gram-negative cell wall, peptidoglycan lipoteichoic acid from gram-positive cell wall, flagellin of bacterial flagella, the sugar mannose (a terminal sugar common in bacterial, viral or fungal glycolipid and glycoprotein), bacterial or viral unmethylated CpG DNA, double-stranded and single-stranded RNA from viruses and glucans from fungal cell wall [3, 4].

Several pathogens and tumor cells exhibit unique glycan structures on their cellular membrane surfaces (Fig. 16.1). For example, bacteria, microbes and viruses all possess a cell wall consisting of a plasma membrane and a capsule which were formed of glycoproteins or complex polysaccharides [5]. These carbohydrate antigens can be used as targets for the development of carbohydrate vaccines.

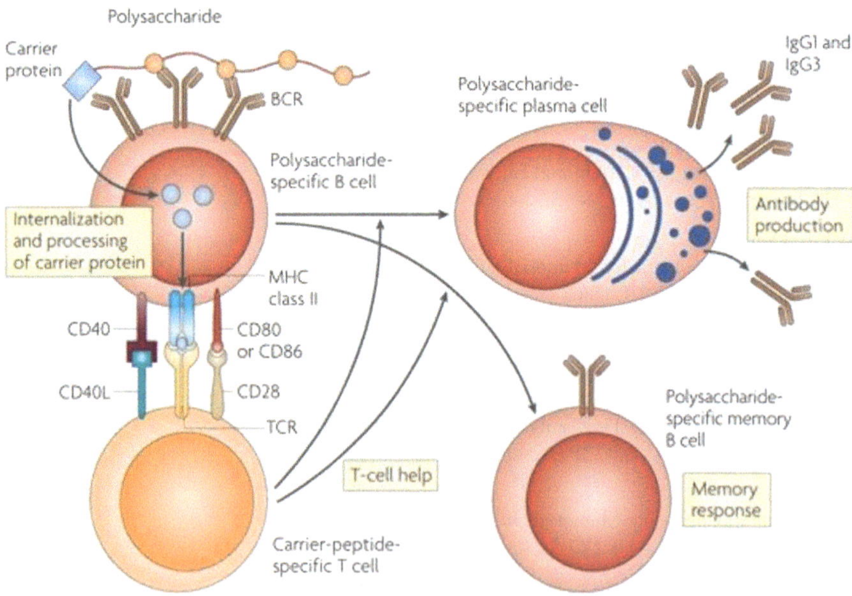

Fig. 16.1 Generation of the immune responses against oligosaccharide antigen-carrier protein conjugate vaccines [18]

In the case of pathogenic bacteria, one has to distinguish Gram positive to Gram negative bacteria that possess respectively, a either capsular polysaccharides (CPS) and/or a lipopolysaccharides (LPS) that are implicated in all virulence factors [6, 7]. The LPS are located on the outer membranes of Gram-negative bacterial cells. The study of the LPS revealed that it is composed of amphiphilic macromolecule, corresponding to an external core oligosaccharide (O-specific chain, or hydrophilic antigen) and an internal core oligosaccharide covalently linked to the lipid A [8]. Knowing that the oligosaccharide portion confers the immunological properties to the bacteria, different portions of the LPS have been tcstcd for vaccines development [9].

16.1.1 Lipopolysaccharide (LPS)

The lipopolysaccharides are located on the outer membranes of Gram-negative bacterial cells. Several studies on the isolation and structure and composition determination of LPS were reported [8, 10]. It was s observed that LPS are composed an external polysaccharide composed of repeating of identical sugar oligosaccharide units, called the *O*-specific chain, and an internal core oligosaccharide covalently attached to a lipid A. The lipid A is a glycolipid composed by a β-D-(1→6) GlcN disaccharide in which O-3, O-3′, O-4′, C_2-N and C_2-N′ are acylated with different C:12 and C:14 fatty acids. The lipid A confers the toxicity to the LPS. In addition, it has to be noted that the *O*-specific oligosaccharide that corresponds to a sequence of oligosaccharide units is unique for each bacterial serotype and provides an immunological property to the bacteria. As a result, several studies involving different moieties of lipopolysaccharides were carried out for the development of vaccines and drugs [9, 11, 12].

16.1.2 LPS-Derived Vaccines or LPS-Protein Neoglycoconjugates

Different studies were carried out on the use of LPS derivatives as vaccines. It was found that the immune reaction relies mainly on the LPS, and it has been efficiently utilized for vaccination. However, the use of the LPS alone for vaccination is not efficient, as, because of its small size, it is not recognized by the immune system as it has resistance to non-specific host immunity such as complement system and resistance to specific host immunity (poor antibody response) [13, 14].

However, when conjugated to a protein carrier, the LPS is able to induce an immunological reaction and an extended immunity. Landsteiner's group were the first to utilize carbohydrate-protein conjugates as immunogens [15, 16], and lately, it was discovered that they can induce strong antibody reaction [17]. Landsteiner's group were also first to refer to the carbohydrate-protein conjugate as a hapten [15, 16].

Figure 16.1 is representing the immune response generated by oligosaccharide antigen-carrier protein vaccines [18]. The interaction of the glycoconjugate vaccine with the B cell receptor (BCR) stimulates the lymphocytes and results in activation of plasma B cells to secrete immunoglobulins, while the neoglycoconjugate vaccine is recognized by the polysaccharide specific B cell receptors. The carrier protein will enter the B cell by phagocytosis. Then, the degradation of the protein carrier in B cells will be done by lysosomal enzymes into short peptides called epitopes. At that time these epitopes will be recruited by a special protein called major histocompatibility protein II (MHC II) that are presented at the cell surface of antigen presenting cell (APC) or B cell and interact with the carrier-peptide-specific T cells (helper T cell) which sends a signal to produce polysaccharide specific plasma cells and polysaccharide specific memory B cells [18].

The synthesis of efficient glycoconjugate vaccines has been challenging, since their efficacy relies on different factors, such as the saccharide size, the average number of saccharide chains per conjugate molecule, the nature of the carrier and the distance between the saccharide and the protein in the formed glycoconjugate [19–22].

Different methods have been used for the synthesis of carbohydrate antigens-protein neoglycoconjugates. One of these methods consisted on the use of the squaric acid chemistry for the single-point attachment of carbohydrates to proteins [23–26]. Tietze et al. used squaric acid diethyl esters for the conjugation [24], while it has also been reported that squaric acid dimethyl esters [27, 28], as well as didecyl squarate [29] were used for the single point attachment of carbohydrates to proteins. In addition, Kamath et al. used squaric acid amide ethyl esters for the conjugation of oligosaccharides to protein and monitored the conjugation using matrix assisted laser desorption ionization time of flight mass spectrometry (MALDI-TOF-MS) [30].

More recently, the group of Kováč also utilized the squaric acid chemistry to conjugate different carbohydrate antigens to a protein carrier [31–33]. They used this strategy to conjugate the synthetic tetrasaccharide side chain of the *Bacillus anthracis* exosporium to the bovine serum albumin (BSA) protein [32].

16.1.3 Mass Spectrometry Methods for the Characterization of Carbohydrate Vaccines

Mass spectrometry has emerged as a powerful technique for the characterization of different biomolecules ranging from small molecules to larger molecules. Thus, mass spectrometry is extensively used in proteomics [34], glycomics [35], metabolomics [36], lipidomics [37], and in oligonucleotides [38] analysis. Initially, the exploring of MALDI-TOF-MS of the different hapten-BSA glycoconjugate vaccines allowed us to determine the hapten-to-BSA ratios. Then the glycoconjugate vaccine samples were then digested and analyzed by MALDI-TOF/TOF-MS/MS and LC-ESI-QqTOF-MS/MS for the determination of glycation sites. The digestion was done by two different enzymes; trypsin which will not be able to digest or react with the glycated lysines of the protein and the other enzyme was GluC V8 endoproteinase which is known to digest proteins at C-terminus of the aspartic acid and glutamic acid residues. Finally the MS/MS spectra will be submitted to MASCOT library to get the matched and non-matched peptides.

16.1.4 Molecular Weight and Carbohydrate-to-Protein Ratio Determination

The aim of determining the molecular weight and the carbohydrate-to-protein ratio of a carbohydrate-protein neoglycoconjugate is to define the number of carbohydrates that are incorporated in the protein carrier, as a result of the conjugation. Two main methods are currently used for the molecular weight determination carbohydrate-protein glycoconjugates: matrix-assisted-laser-ionization time-of-flight mass spectrometry (MALDI-TOF-MS) [39, 40] and surface enhanced laser desorption ionization time-of-flight mass spectrometry (SELDI-TOF-MS) [31–33, 41–43].

Both of these methods allow determining the carbohydrate-to-protein ratio of the neoglycoconjugates by comparing the molecular weight of the protein before and after the conjugation to that of the neoglycoconjugate.

16.1.4.1 MALDI-TOF-MS

Matrix assisted laser desorption ionization mass spectrometry has been successfully used for the determination of the molecular weight of biomolecules, such as proteins, oligosaccharides and glycoproteins [39–41].

Kamath et al. used MALDI-TOF-MS to characterize neoglycoconjugates formed by the conjugation of oligosaccharide amines to carrier proteins by the aim of diethyl squarate [41]. The MALDI-MS was recorded in linear mode and with positive ion detection. Figure 16.2 displays the MALDI-TOF-MS analysis of BSA (A) and the following oligosaccharide-BSA glycoconjugates: (B) GlcNAc-BSA (carbohydrate-BSA ratio (n) = 8.4); (C) lactose-BSA (n = 3.5); (D) lactose-BSA (n = 8.2); (E) Fucal-2Galfll-3[Fucal-4]GlcNAc- BSA (n = 11); (F) Fucal-2Fucal-3GalNAc-BSA (n = 13). The MALDI-TOF-MS spectrum of BSA was used a calibration standard. Then the analysis allowed to reveal that the average carbohydrate-BSA ratios for the hapten-BSA neoglycoconjugates are (Fig. 16.2).

16.1.4.2 SELDI-TOF-MS

Surface enhanced laser desorption ionization time-of-flight mass spectrometry (SELDI-TOF-MS) has been extensively used for biomarkers discovery [42], and cancer diagnosis [43]. This technique consists in using a modified target for the immuno affinity purification of proteins before analysis.

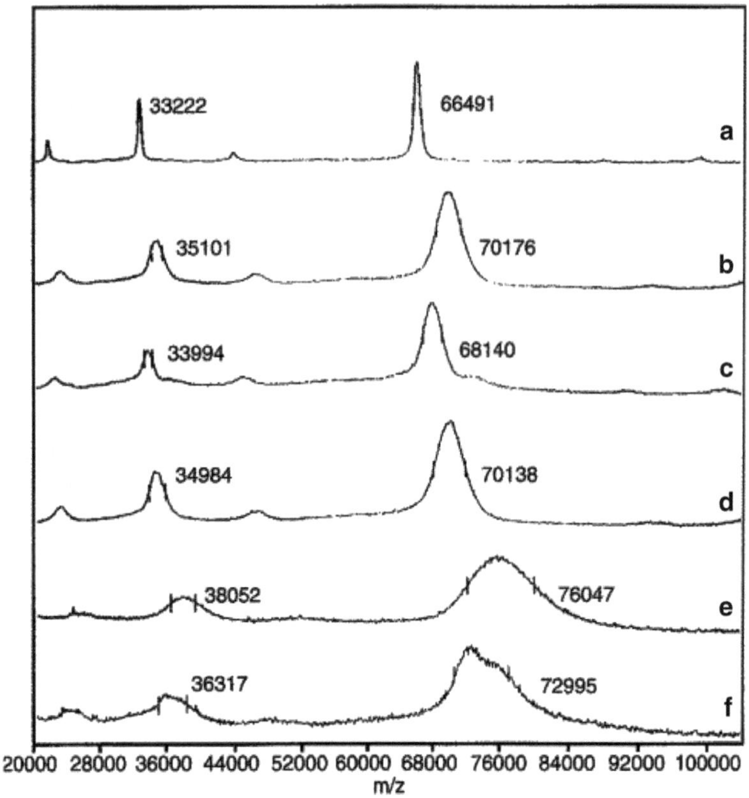

Fig. 16.2 MALDI-TOF spectra of: (**a**) BSA calibration standard; (**b**) GlcNAc-BSA (n=8.4); (**c**) lactose-BSA (n=3.5); (**d**) lactose-BSA (n=8.2); (**e**) Fucal-2Galfll-3[Fucal-4]GlcNAc- BSA (n=11); (**f**) Fucal-2Fucal-3GalNAc-BSA (n=13) [41]

The group of Kováč successfully used this analytical method to monitor the conjugation of synthetic carbohydrate antigens to different protein carriers [31–33, 44]. Figure 16.3 shows the SELDI-TOF-MS analysis with a ProteinChip® System of a neoglycoconjugate prepared by the dialkyl squarate chemistry attachment of the hexasaccharide of *Vibrio cholerae* O:1 to the BSA [44]. This technology allowed the analysis of neoglycoconjugates at different reaction time by taking an aliquot of the reaction mixture, in the picomolar concentration range, and analyzing it without any purification step. In addition, this method was found to be fast and only a small amount of sample is used for analysis.

The same group followed the progress of conjugation of the hexasaccharide (exact mass: 1,780.79 Da) to the BSA protein (molecular mass: 66,430 Da) at different reaction times [44]. It was observed that the conjugation rate increased with the reaction time. In addition, the SELDI-TOF-MS also allowed observing that after a reaction time of 7 h, the fine structure the peak of the neoglycoconjugate (Fig. 16.4) shows the polydispersity of the neoglycoconjugate formed from the hexasaccharide and BSA.

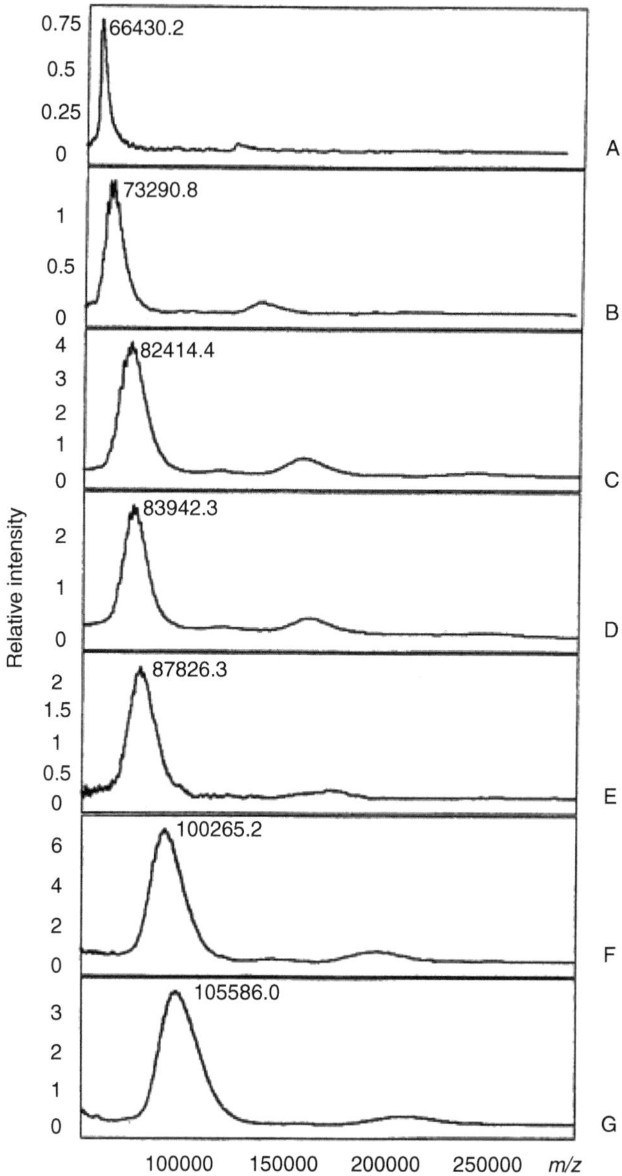

Fig. 16.3 The progress of conjugation of the hexasaccharide of Vibrio cholerae serotype Ogawa (exact mass, 1,780.79 Da) and BSA (molecular mass 66,430 Da) as revealed by monitoring the reaction by SELDI-TOF MS. Spectrum A was taken at the onset of the reaction (t-0); spectra B-G were taken at 1, 3, 7, 9, 27, and 54 h, respectively [44]

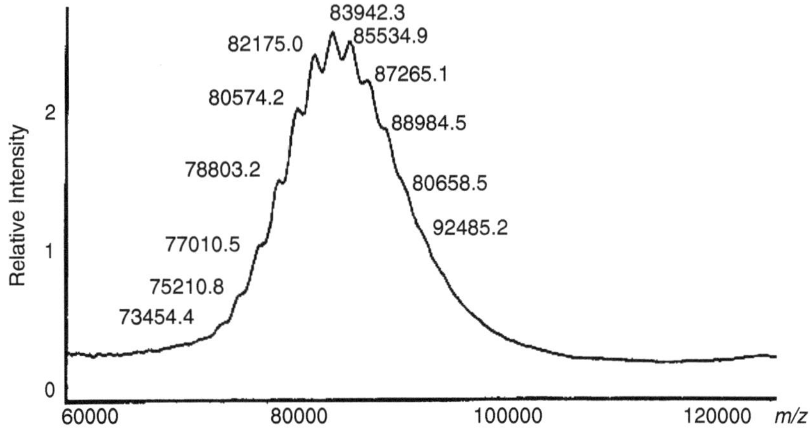

Fig. 16.4 The fine structure of the peak D (Fig. 16.3) showing the polydispersity of the neoglycoconjugate formed from the hexasaccharide 4 and BSA after 7 h of reaction time. For further details, see text [44]

Moreover, after the conjugation is complete, the excess oligosaccharide can be recovered for further use, allowing an economy of labor and time for the preparation of synthetic oligosaccharides and their conjugation [44].

16.1.5 Glycation Sites Determination

The glycation sites determination of carbohydrates-protein neoglycoconjugates is usually carried out by first digesting the neoglycoconjugate using a protease, such as trypsin or GluC V8 endoproteinase, followed by MALDI-TOF-MS/MS or liquid chromatography tandem mass spectrometry [45–50].

It has to be noted that during the tandem mass spectrometry analysis of the glycoconjugate digests, the identification of the glycated peptides are confirmed by the presence of diagnostic product ions of the carbohydrate in the mass spectrum. In addition, the tandem mass spectrometry analysis also reveals the sequence of the peptide through diagnostic product ions of the peptide moiety of the glycated peptide. The combined information allows the unambiguous characterization of the carbohydrate-peptide and the glycation site identification. It has to be noted that during the tandem mass spectrometry analyses of the glycated peptides, the product ions corresponding to the fragmentation of the peptide portion were identified using the nomenclature established by Roepstorff et al. and lately modified by Johnson and coworkers [51, 52], and the product ions resulting from the fragmentation of the carbohydrate moiety were assigned using the nomenclature introduced by Domon and Costello, as: A, B, C, X, Y and Z [53].

Different examples of the mass spectrometry characterization of neoglycoconjugate vaccines are discussed in the next section.

16.2 Examples of the Mass Spectrometry Characterization of Neoglycoconjugate Vaccines Against Bacterial Biological Agents

16.2.1 Neoglycoconjugate Vaccines from the Terminal Monosaccharide Antigen of the O-PS of **Vibrio cholerae** *O1, Serotype Ogawa, and BSA [45, 46]*

Vibrio cholera is a gram-negative bacterium that infects the intestine and causes cholera. The *Vibrionaceae* bacteria family encompasses the *Vibrio cholerae* strain and is composed of 200 serogroups [54]. Cholera is an acute diarrhoeal infection caused by ingestion of water and food contaminated with the bacterium V. cholerae [55]. Different studies have been carried out for the development of conjugate vaccines for cholera [56, 57]. The group of Kováč used synthetic fragments of the *O*-PS of *Vibrio cholerae* O:1 for the design of synthetic carbohydrate-protein glycoconjugate vaccines, by means of the single point attachment of spacer-equipped synthetic oligosaccharides similar to the *O*-PS of the two main strains of *Vibrio cholerae* serogroup O1, serotypes Ogawa and Inaba, to protein carriers, using the squaric acid chemistry [15, 22, 57]. It is important to indicate that the *O*-PS of Ogawa and Inaba strains are similar and consist in an oligosaccharide chain of 15–20 repeating units composed of α-(1→2)-linked monomers of 4-amino-4,6-dideoxy-D-mannopyranosyl residues (D-perosamine) in which the amino group is acetylated with 3-deoxy-L-*glycero*-tetronic acid [58, 59]. The only difference between the *O-PS* of the two strains is that the upstream terminal perosamine residue of the *O*-PS of the Ogawa strain is methylated [60–62].

In a recent study, we reported the mass spectrometry characterization of neoglycoconjugate vaccines formed by the conjugation of the terminal monosaccharide antigen of the O-PS of *Vibrio cholerae* O1, serotype Ogawa, and BSA [45, 46]. The neoglycoconugates were prepared by the group of Kováč, using the dialkyl squarate chemistry [63].

A schematic representation of the terminal monosaccharide of the O-specific polysaccharide (O-PS) of *Vibrio cholerae* serotype Ogawa antigen-BSA vaccine is displayed in Fig. 16.5a.

16.2.1.1 Carbohydrate:BSA Ratio

The MALDI-TOF-MS analysis of three selected carbohydrate-BSA vaccines allowed to measure their carbohydrate-to-protein ratio. Figure 16.5b shows the MALDI-TOF-MS analysis of the BSA that was used as a standard calibration, while Fig. 16.5c–e displays the mass spectrum of three hapten-BSA glycoconjugate vaccines obtained with a reaction time of 45 min, 1.5 h and 6 h, respectively. Table 16.1 displays the molecular masses of these synthetic neoglycoconjugates.

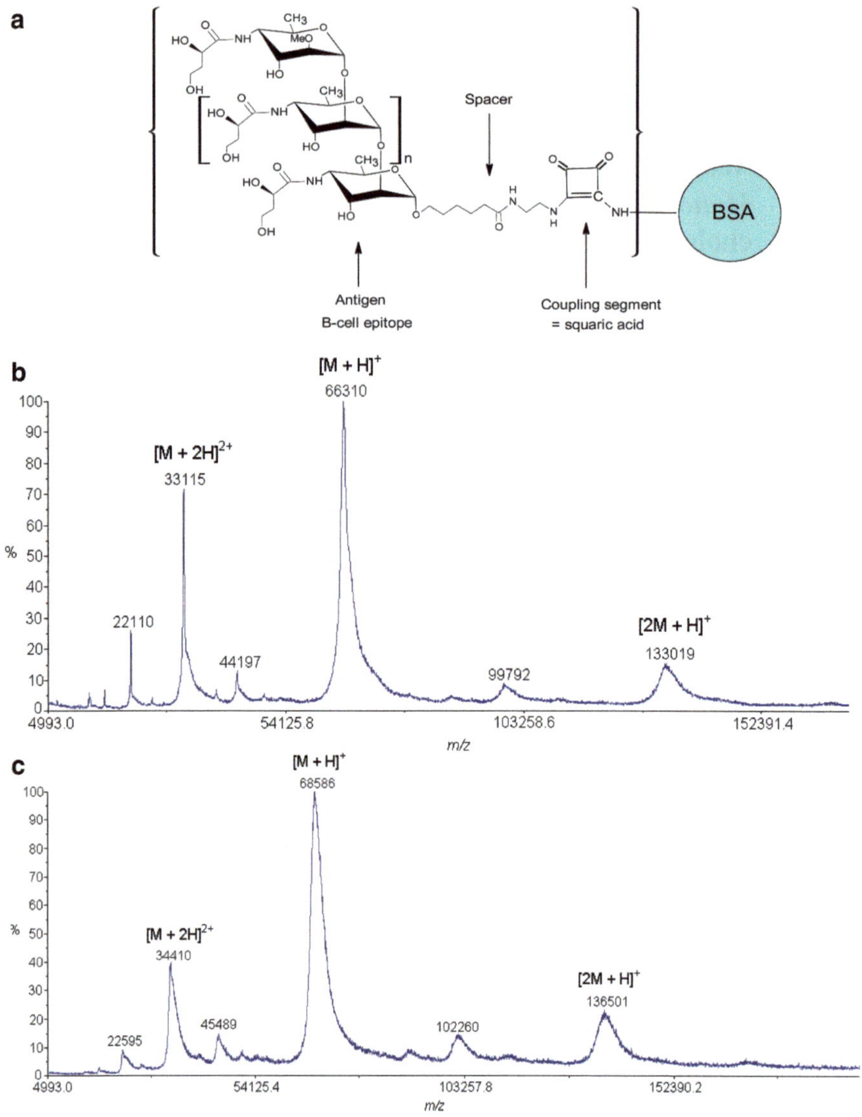

Fig. 16.5 Schematic representation of the general structure of carbohydrate-spacer-squaric acid-protein constructs from oligosaccharide fragments of the *O*-polysaccharide of the *Vibrio cholerae* O1, serotype Ogawa (**a**) and MALDI-TOF/TOF-MS analysis of native BSA (**b**), hapten-BSA neoglycoconjugates with a hapten:protein ratio of 45 min (**c**), 1.5 h (**d**) and 6 h (**e**) [45]

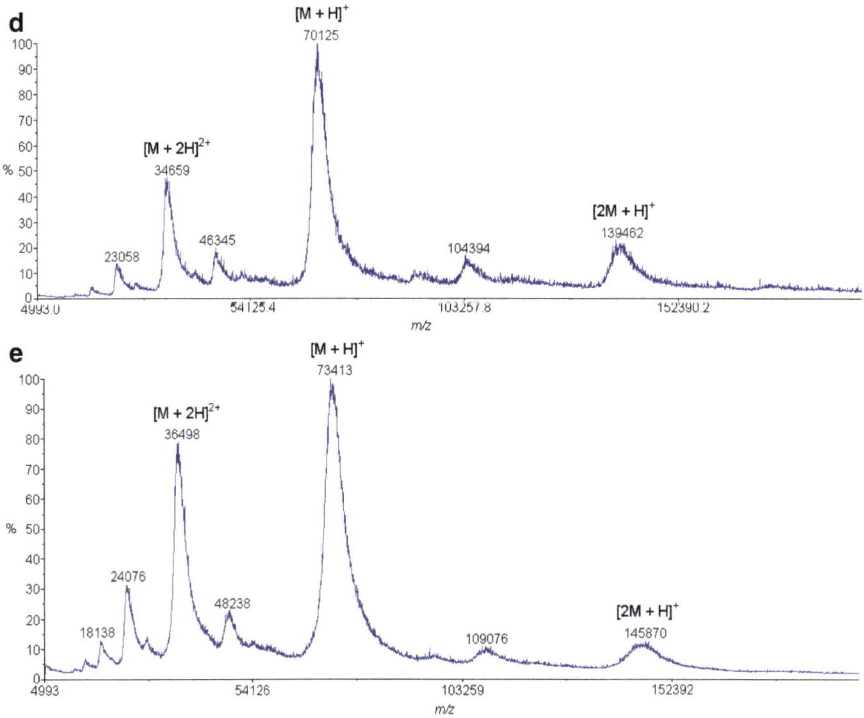

Fig. 16.5 (continued)

Table 16.1 Molecular masses and calculated hapten-BSA ratios of BSA and hapten-BSA conjugates [45]

	$[M + 2H]^{2+}$	$[M + H]^{+}$	$[2 M + H]^{+}$	Remeasured Hapten-BSA ratio
Standard BSA	33,115	66,310	133,019	–
Hapten-BSA 4:1	34,410	68,586	136,501	4.3
Hapten-BSA 7:1	34,659	70,125	139,462	6.6
Hapten-BSA 13:1	36,498	73,413	145,870	13.2

It was shown that the conjugation using the squaric acid chemistry covalently links the carbohydrate squaric acid ethyl ester to the ε-amino function of the lysine residues in BSA with the loss of EtOH. In addition, knowing that the molecular mass of the monosaccharide-squaric acid portion was 513.23 Da, we calculated the hapten-BSA ratio to be 4.3, 6.6 and 13.2, for a reaction time of 45 min, 1.5 h and 6 h, respectively. This was in accordance to the measured hapten-BSA ratios of this neoglycoconjugates, by the group of Kováč [63, 64].

16.2.1.2 Glycation Sites Determination

MALDI-MS/MS

The glycation sites determination in these neoglycoconujugates was carried out as described in the previous section. First, the neoglycoconjugate vaccine was digested using the trypsin proteinase [45–49].

The MALDI-TOF/TOF-MS analysis of the carbohydrate-BSA trypic digests were compared to the tryptic digests of BSA, to determine the potentially glycated peptides. In addition, tandem mass spectrometry analysis was carried out on these peptides by high-energy collision dissociation CID-TOF/TOF-MS/MS, and the MS/MS were manually sequenced to identify the sequence of the peptides. Figure 16.6 shows the MALDI-TOF/TOF-MS analysis of the tryptic digests of BSA (a) and the hapten-BSA glycoconjugates with a carbohydrate:protein ratio of 4.3 (b), 6.6 (c) and 13.2 (d). Table 16.2 displays the identified peptides resulting from the MALDI-TOF/TOF-MS/MS analysis of the tryptic digests of the hapten-BSA glycoconjugate with an average carbohydrate:protein ratio of 4.3. Consequently, the MALDI-TOF/TOF-MS analysis of the tryptic digests revealed that the following molecular ions corresponded to the exact addition of the glycan mass increment of 513.23 Da to well characterized BSA tryptic digests: *m/z* 830.41, *m/z* 912.48, *m/z* 1,069.60, *m/z* 1,330.72, *m/z* 1,405.73, *m/z* 1,514.82, *m/z* 1,662.70, *m/z* 1,691.79, *m/z* 2,153.16 and *m/z* 3,025.37 (Fig. 16.6).

High energy collision induced dissociation tandem mass spectrometry was performed on these precursor ions. Figure 16.7 shows the product ion scan of the precursor ions at *m/z* 1,330.72 (a), *m/z* 1,514.82 (b) and *m/z* 2,153.16 (c), that were assigned to the following glycated peptides: SLGK*VGTR, ALK*AWSVAR and K*VPQVSTPTLVEVSR, respectively (star = glycation site). It has to be noted that the glycation sites were exclusively located on lysine residues. The CID MS/MS not only allowed to establish the sequence of the glycated pepetide, but also revealed product ions corresponding to a sequence of the peptide attached to a portion of the carbohydrate were also observed: $[M + H - C_2H_5O]^+$ at *m/z* 2,108.15 and $[M + H - C_{11}H_{19}NO_6]^+$ at *m/z* 1,892.01. As expected, the product ion scan of the glycated peptide also contained the following series of a- and b-product ions: $[b_2 - C_{23}H_{37}N_3O_{10}]^+$ at *m/z* 226.15, $[a_3 - C_{23}H_{35}N_3O_{10}]^+$ at *m/z* 297.21, $[b_3 - C_{23}H_{35}N_3O_{10}]^+$ at *m/z* 325.23, $[b_4 - C_{23}H_{36}N_3O_{10}{}^{\bullet}]^{+\bullet}$ at *m/z* 452.31, $[b_2 - C_{11}H_{19}NO_6]^+$ at *m/z* 480.33, $[b_4 - C_{11}H_{19}NO_6]^+$ at *m/z* 705.45 and $[b_5 - C_7H_{13}NO_3]^+$ at *m/z* 804.51.

Thus, MALDI-CID-MS/MS allowed to identify only three glycation sites on the hapten-BSA glycoconjugates, on the following lysine residues: Lys235, Lys437 and Lys455.

LC-MS/MS

As the glycation sites were located exclusively on lysine residues, we presumed that the glycated lysine residues were not able to be digested by the trypsin protease. Thus, we used the GluC V8 endoproteinase for the digestion of the vaccines.

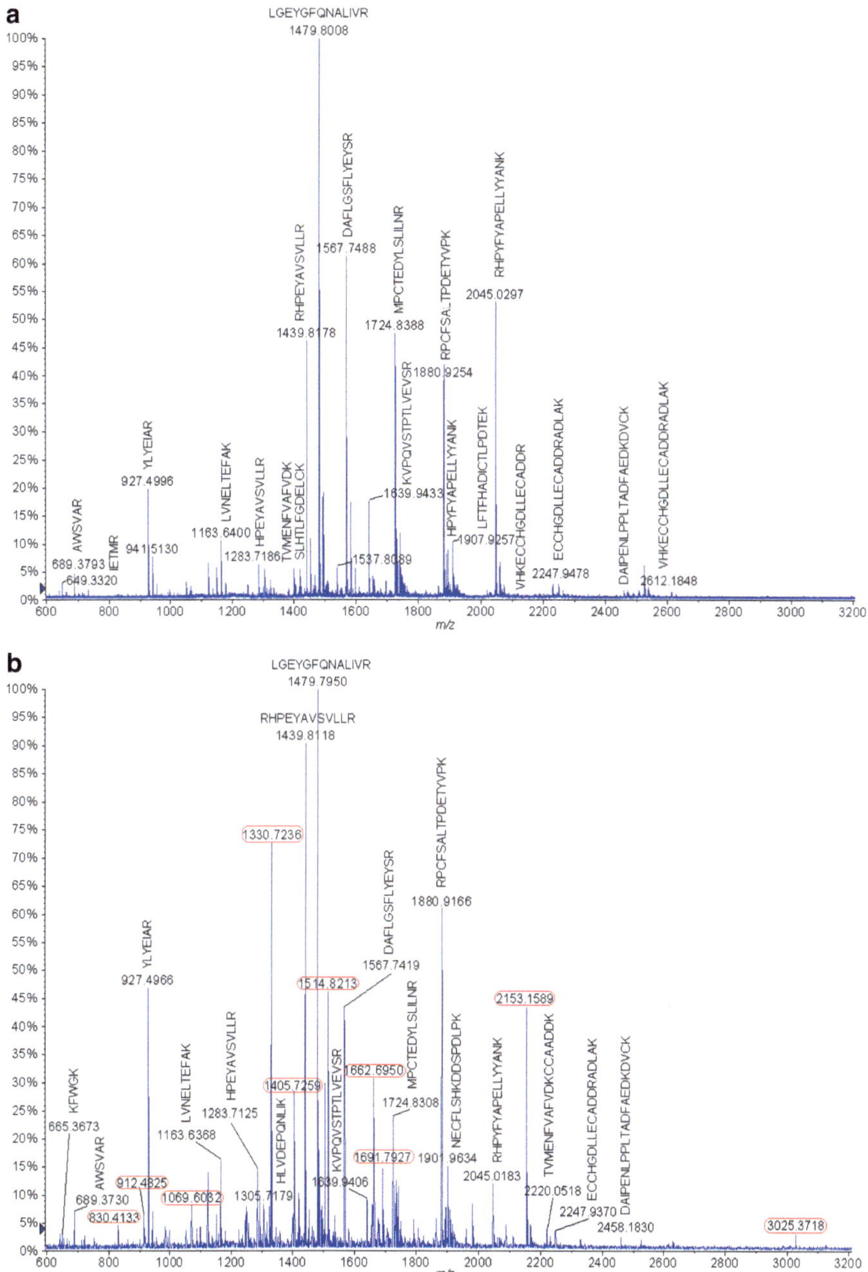

Fig. 16.6 MALDI-TOF/TOF-MS spectra of the digested (**a**) BSA, and hapten-BSA conjugates with hapten:protein ratios of (**b**) 4.3, (**c**) 6.6 and (**d**) 13.2 [45]

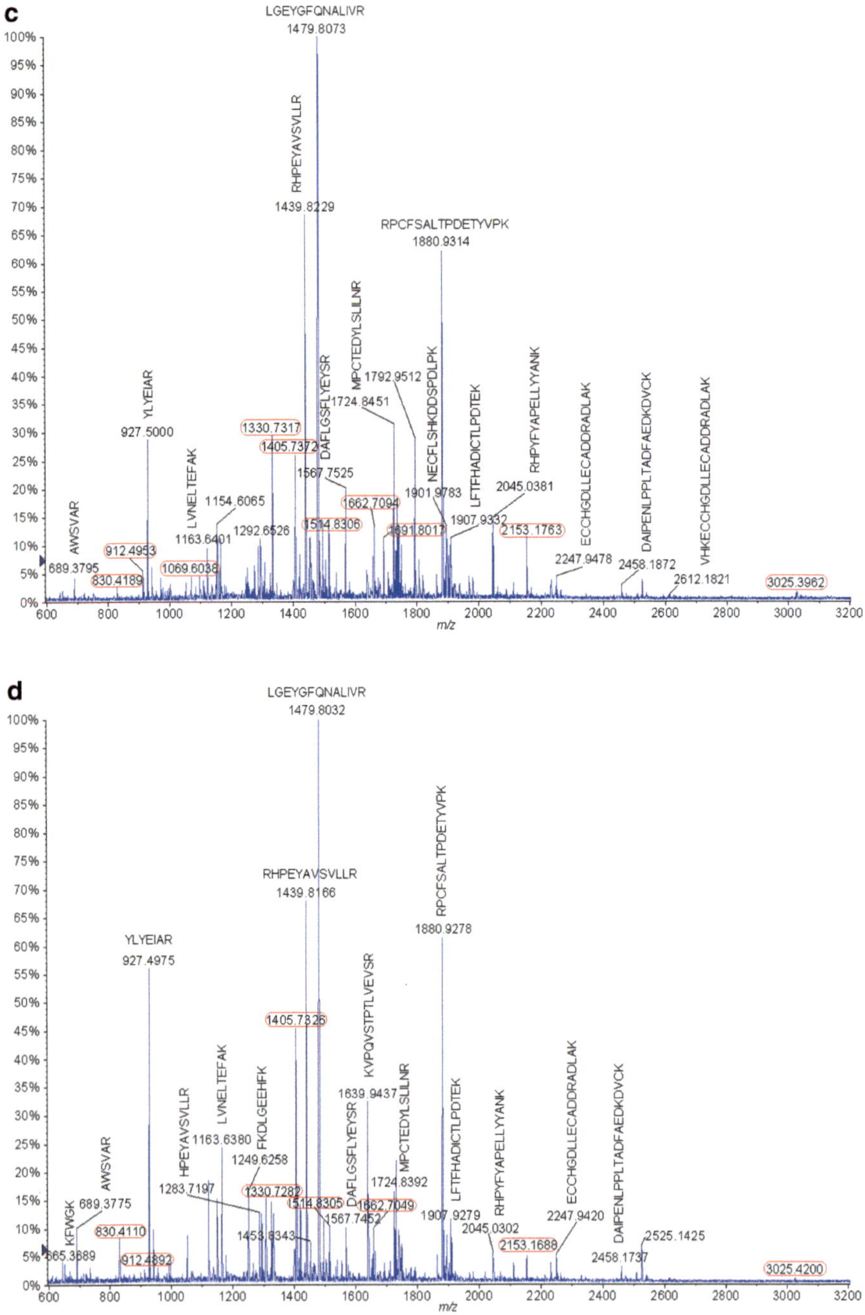

Fig. 16.6 (continued)

Table 16.2 MALDI-TOF/TOF-MS peptide mapping of the digested neoglycoconjugate with a hapten:protein ratio of 4.3 (after recalibration) [45]

Peptide		Observed	Calculated	Accuracy	Missed	
#	Location	*m/z*	*m/z*	ppm	Cleavage	Sequence
T46–T48	310–340	3,511.6660	3,511.6725	2	2	SHCIAEVEKDAIPENLPPLTADFAEDKDVCK
T47–T48	319–340	2,458.1830	2,458.1812	−1	1	DAIPENLPPLTADFAEDKDVCK
T41–T42	267–285	2,247.9370	2,247.9433	3	2	ECCHGDLLECADDRADLAK
T78–T79	569–587	2,220.0518	2,219.9775	−33	1	TVMENFVAFVDKCCAADDK
T61–T62	**437–451**	**2,153.1589**	**2,153.1700**	**5**	**0**	**K*VPQVSTPTLVEVSR**
T23–T24	168–183	2,045.0183	2,045.0285	5	0	RHPYFYAPELLYYANK
T16–T17	123–138	1,901.9634	1,901.8703	−49	2	NECFLSHKDDSPDLPK
T71	508–523	1,880.9166	1,880.9216	3	0	RPCFSALTPDETYVPK
T66	469–482	1,724.8308	1,724.8351	2	0	MPCTEDYLSLILNR
T61–T62	437–451	1,639.9406	1,639.9383	−1	1	KVPQVSTPTLVEVSR
T50	347–359	1,567.7419	1,567.7433	1	0	DAFLGSFLYEYSR
T34–T35	**233–241**	**1,514.8213**	**1,514.8213**	**0**	**1**	**ALK*AWSVAR**
T59	421–433	1,479.7950	1,479.7960	1	0	LGEYGFQNALIVR
T51–T52	360–371	1,439.8118	1,439.8123	0	1	RHPEYAVSVLLR
T12	89–100	1,419.7020	1,419.6942	−5	0	SLHTLFGDELCK
T63–T64	**452–459**	**1,330.7236**	**1,330.7213**	**−2**	**1**	**SLGK*VGTR**
T57	402–412	1,305.7179	1,305.7167	−1	0	HLVDEPQNLIK
T52	361–371	1,283.7125	1,283.7112	−1	0	HPEYAVSVLLR
T7–T8	35–44	1,249.6237	1,249.6217	−2	1	FKDLGEEHFK
T10	66–75	1,163.6368	1,163.6312	−5	0	LVNELTEFAK
T22	161–167	927.4966	927.4940	−3	0	YLYEIAR
T35	236–241	689.3730	689.3735	1	0	AWSVAR
T20–T21	156–160	665.3673	665.3775	15	1	KFWGK
T13	101–105	545.3493	545.3411	−15	0	VASLR

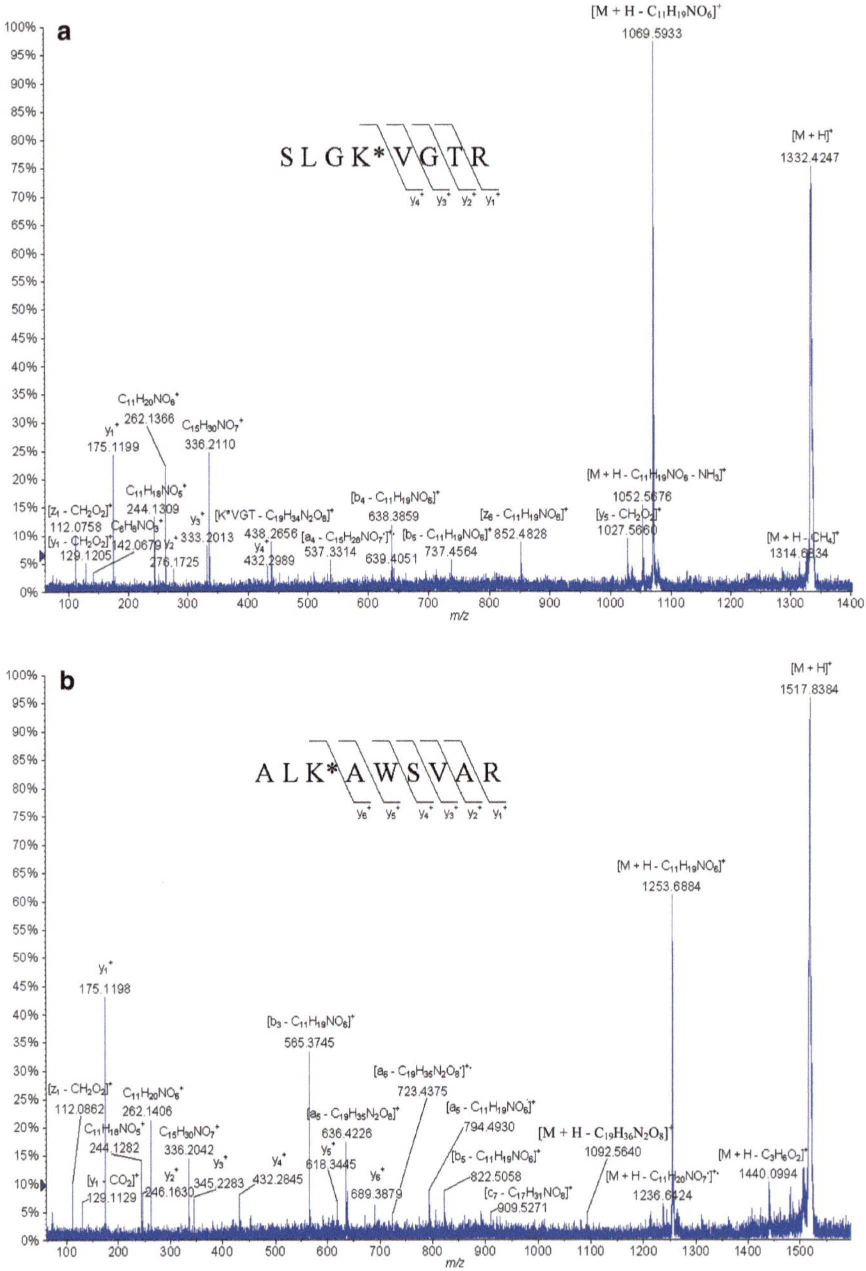

Fig. 16.7 MALDI-CID-TOF/TOF-MS/MS spectra of the glycated peptide (**a**) SLGK*VGTR at *m/z* 1,330.72, (**b**) ALK*AWSVAR at *m/z* 1,514.82 and (**c**) K*VPQVSTPTLVEVSR at *m/z* 2,153.16 [45]

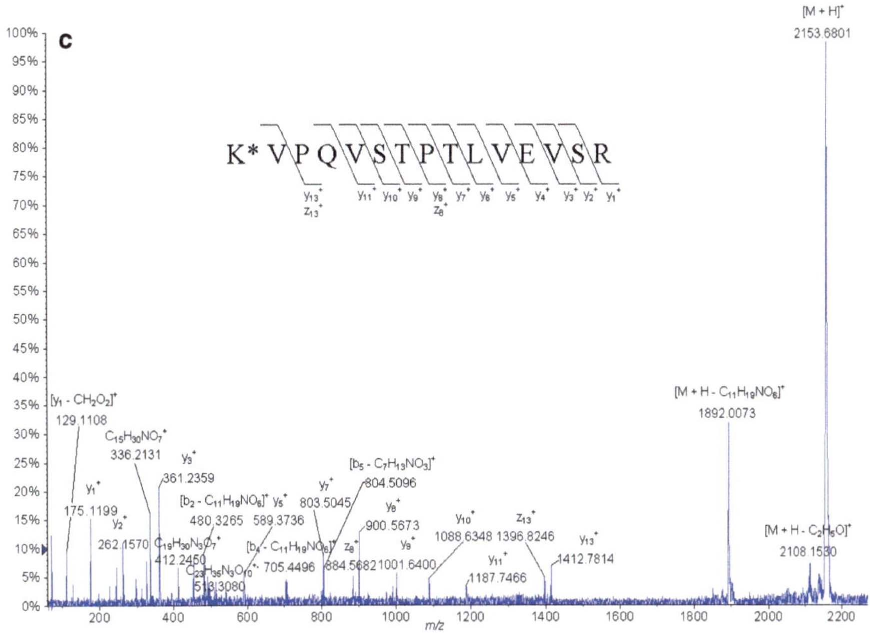

Fig. 16.7 (continued)

Liquid chromatography coupled to tandem mass spectrometry (LC-MS/MS) was then used for the analysis of tryptic and GluC V8 digests of the same hapten-BSA glycoconjugates [46].

Figure 16.8a displays the MS/MS spectrum of the identified glycopeptide EK*VLTSSAR (Lys 211) at *m/z* 752.4233 (+2) obtained during the LC-MS/MS analysis of tryptic digests of the glycoconjugate vaccine with a hapten:BSA ratio of 4.3:1. The product ion scan of the glycopeptide EK*VLTSSAR (Lys 211) at *m/z* 752.4233 (+2) allowed to observe diagnostic product ions of the carbohydrate moiety: B^+ at *m/z* 262.1265, $[B - H_2O]^+$ at *m/z* 244.1142, $[B - 2H_2O]^+$ at *m/z* 226.1047, $[^{2,5}A - 2H_2O]^+$ at *m/z* 152.0663, $[B - C_4H_6O_3]^+$ at *m/z* 142.0815 and $[^{2,5}A - CH_4O_3]^+$ at *m/z* 124.0708 (Fig. 16.9). Moreover, the following product ions were assigned to the linker portion attached to a lysine residue: $C_{17}H_{26}N_3O_4^+$ at *m/z* 336.1918 and $[C_{11}H_{16}N_3O_2]^+$ at *m/z* 222.1187. Product ions corresponding to the entire peptide attached to the linker portion were also identified: Y^+ at *m/z* 1,242.6794 and Z^+ at *m/z* 1,224.6785. In addition, as observed during the MALDI-MS/MS analysis of the glycated peptides, the LC-MS/MS analyses also revealed product ions consisting in peptide product ions attached to the linker portion that lost the carbohydrate moiety, allowing to cover the peptide sequence: $[y_8 - B]^+$ at *m/z* 1,113.6384, $[b_8 - B]^+$ at *m/z* 1,068.5932, $[b_7 - B]^+$ at *m/z* 997.5436, $[b_6 - B]^+$ at *m/z* 910.5089, $[b_5 - B]^+$ at

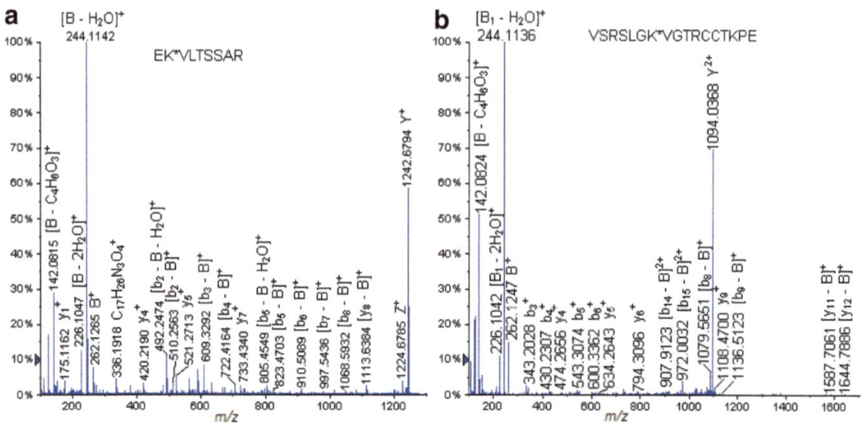

Fig. 16.8 LC-ESI-QqTOF-MS/MS spectra of (**a**) tryptic glycopeptide EK*VLTSSAR (Lys 211) at *m/z* 752.4233 (+2) and (**b**) the GluC V8 endoproteinase digest glycopeptide VSRSLGK*VGTRCCTKPE (Lys 455) at *m/z* 816.7646 (+3) [46]

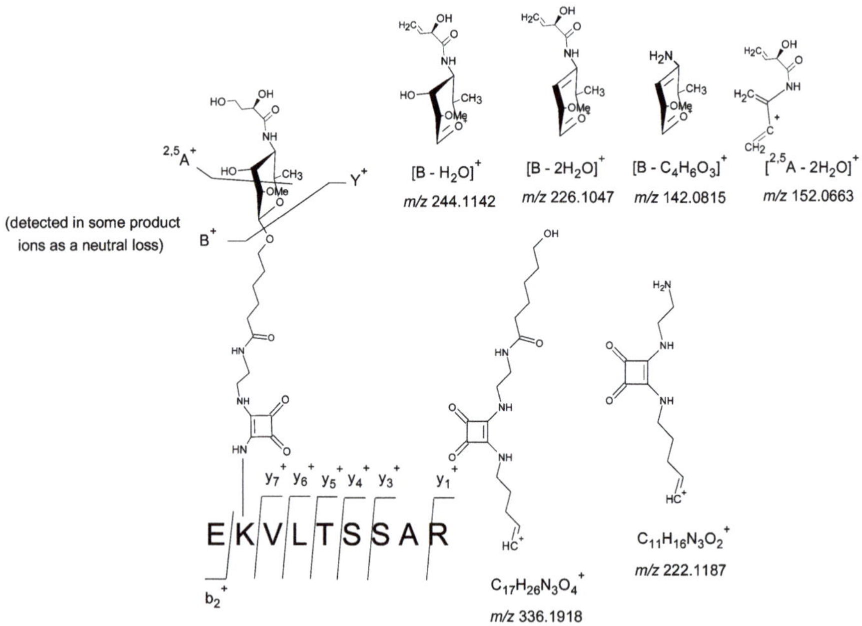

Fig. 16.9 Different product ions involving the fragmentation of the carbohydrate hapten observed during the LC-ESI-QqTOF-MS/MS analysis of the tryptic glycopeptide EK*VLTSSAR (Lys 211) at *m/z* 752.4233 (+2) [46]

m/z 823.4703, $[b_5 - B - H_2O]^+$ at *m/z* 805.4549, $[b_4 - B]^+$ at *m/z* 722.4164, $[b_4 - B - H_2O]^+$ at *m/z* 704.4081, $[b_3 - B]^+$ at *m/z* 609.3292, $[b_3 - B - H_2O]^+$ at *m/z* 591.3171, $[b_2 - B]^+$ at *m/z* 510.2563 and $[b_2 - B - H_2O]^+$ at *m/z* 492.2474. The identified glycated peptides from the LC-MS/MS analysis of the tryptic digests of the neoglycoconjugate vaccines are reported in Table 16.3. It was observed that the LC-MS/MS analyses allowed discovering a higher number of glycation sites.

The analysis of the tryptic digests allowed to identify 11, 16 and 17 glycation sites for the neoglycoconjugate vaccines with hapten:BSA ratios of 4.3:1, 6.6:1 and 13.2:1, respectively, displayed in Fig. 16.10.

An example of the product ion scan of a glycopeptide identified during the LC-MS/MS analysis of the GluC V8 digests is displayed in Fig. 16.8b and corresponds to the glycated peptide VSRSLGK*VGTRCCTKPE (Lys 455) at *m/z* 816.7646 (+3). The fragmentation route of this glycopeptide was the same as the one described previously for the glycopeptide EK*VLTSSAR (Lys 211) at *m/z* 752.4233 (+2). Thus, the CID-MS/MS analysis of the glycated peptide VSRSLGK*VGTRCCTKPE (Lys 455) at *m/z* 816.7646 (+3) allowed to observe diagnostic product ions of the carbohydrate moiety: B^+ at *m/z* 262.1247, $[B - H_2O]^+$ at *m/z* 244.1136, $[B - 2H_2O]^+$ at *m/z* 226.1042, $[^{2,5}A - 2H_2O]^+$ at *m/z* 152.0669, $[B - C_4H_6O_3]^+$ at *m/z* 142.0824 and $[^{2,5}A_1 - CH_4O_3]^+$ at *m/z* 124.0714. Moreover, a product ion that corresponds to the entire glycated peptide that lost the carbohydrate portion was detected: Y^{2+} at *m/z* 1,094.0368.

Finally, the peptide sequence was covered by the observation of the following product ions consisting in glycated peptide fragment ions that lost the carbohydrate portion: $[y_{12} - B]^+$ at *m/z* 1,644.7886, $[y_{11} - B]^+$ at *m/z* 1,587.7061, $[b_9 - B]^+$ at *m/z* 1,136.5123, $[b_8 - B]^+$ at *m/z* 1,079.5651, $[b_{16} - B]^{2+}$ at *m/z* 1,020.5277, $[b_7 - B]^+$ at *m/z* 980.5699, $[b_{15} - B]^{2+}$ at *m/z* 972.0032, $[b_{14} - B]^{2+}$ at *m/z* 907.9123 and $[GK^* - B]^+$ at *m/z* 438.2288. Table 16.4 displays the identified glycated peptides from the LC-MS/MS analysis of the GluC V8 digests of the different neoglycoconjugate vaccines. In addition, the GluC V8 digests revealed 20, 27 and 33 glycation sites for the vaccines with hapten:BSA ratios of 4.3:1, 6.6:1 and 13.2:1 (Fig. 16.10).

The combination of the identified glycated peptides during the LC-MS/MS analyses of the tryptic and GluC V8 digests allowed to record a total number of glycation sites of 20 (protein sequence coverage: 89 %), 27 (protein sequence coverage: 90 %) and 33 (protein sequence coverage: 88 %) for the hapten-BSA glycoconjugates with a carbohydrate:hapten ratio of 4.3:1, 6.6:1 and 13.2:1, respectively (Fig. 16.10, Tables 16.3 and 16.4).

The 3-D representation of the BSA protein was displayed using the Swiss-Pdb Viewer software (Fig. 16.11) [65, 66]. The glycated lysine residues for the different neoglycoconjugate vaccines (hapten:BSA ratios of 4.1:1, 6.6:1 and 13.2:1) were highlighted in red. We were thus able to observe that all the glycated lysine residues are located at the outer surface of the protein.

Table 16.3 Tryptic glycopeptides identified during LC-ESI-QqTOF-MS/MS analysis of the hapten-BSA glycoconjugate with a hapten-BSA ratio of 4.3:1, 6.6:1 and 13.2:1 [46]

Peptide			Hapten-BSA ratio 4.3:1		Hapten-BSA ratio 6.6:1		Hapten-BSA ratio 13.2:1	
	Calculated	Missed	Observed	Deviation	Observed	Deviation	Observed	Deviation
Sequence (asterisk = glycation site)	*m/z* (charge)	Cleavage	*m/z* (charge)	(Da)	*m/z* (charge)	(Da)	*m/z* (charge)	(Da)
SLGK*VGTR (Lys 455)	665.8643 (+2)	1	665.8810 (+2)	0.0168	665.8694 (+2)	0.0052	665.8699 (+2)	0.0057
EK*VLTSSAR (Lys 211)	752.3987 (+2)	1	752.4233 (+2)	0.0246	752.4114 (+2)	0.0127	752.3971 (+2)	–0.0016
ALK*AWSVAR (Lys 235)	757.9143 (+2)	1	757.9261 (+2)	0.0118	757.9189 (+2)	0.0046	757.9248 (+2)	0.0105
K*QTALVELLK (Lys 548)	828.4770 (+2)	1	828.4992 (+2)	0.0222	828.4872 (+2)	0.0103	828.4857 (+2)	0.0087
CCTK*PESER (Lys 463)	840.3662 (+2)	1	840.3949 (+2)	0.0287	840.3788 (+2)	0.0126	840.3769 (+2)	0.0107
LK*PDPNTLCDEFKADEK (Lys 140)	845.0720 (+3)	3	845.0908 (+3)	0.0188	845.0839 (+3)	0.0119	845.0911 (+3)	0.0191
DTHK*SEIAHR (Lys 28)	853.9209 (+2)	1	853.9445 (+2)	0.0236	853.9366 (+2)	0.0158	–	–
GACLLPK*IETMR (Lys 204)	951.4892 (+2)	1	951.5091 (+2)	0.0199	951.5014 (+2)	0.0122	951.5006 (+2)	0.0114
K*VPQVSTPTLVEVSR (Lys 437)	1,077.0887 (+2)	0	1,077.1129 (+2)	0.0243	1,077.0985 (+2)	0.0099	1,077.0975 (+2)	0.0089
LAK*EYEATLEECCAK (Lys 374)	1,164.5347 (+2)	2	1,164.5375 (+2)	0.0028	1,164.5613 (+2)	0.0266	1,164.5540 (+2)	0.0193
QNCDQFEK*LGEYGFQNALIVR (Lys 420)	1,014.8220 (+3)	1	1,014.8462 (+3)	0.0242	1,014.8390 (+3)	0.0170	1,014.8534 (+3)	0.0314
HKPK*ATEEQLK (Lys 561)	607.9779 (+3)	3	–	–	607.9966 (+3)	0.0187	608.0033 (+3)	0.0254
LSQK*FPK (Lys 245)	680.8716 (+2)	3	–	–	680.8722 (+2)	0.0007	680.8755 (+2)	0.0040
LCVLHEK*TPVSEK (Lys 489)	685.0222 (+3)	2	–	–	685.0256 (+3)	0.0034	685.0310 (+3)	0.0088
CASIQK*FGER (Lys 228)	854.9142 (+2)	1	–	–	854.9284 (+2)	0.0143	854.9290 (+2)	0.0149
VTK*CCTESLVNR (Lys 498)	990.4743 (+2)	1	–	–	990.4813 (+2)	0.0071	990.4930 (+2)	0.0188
FK*DLGEEHFK (Lys 36)	881.9304 (+2)	2	–	–	–	–	881.9491 (+2)	0.0188
LKECCDK*PLLEK (Lys 304)	1,023.5103 (+2)	3	–	–	–	–	1,023.5348 (+2)	0.0245

```
a   1 MKWVTFISLL LLFSSAYSRG VFRRDTHK*SE IAHRFKDLGE EHFKGLVLIA
   51 FSQYLQQCPF DEHVKLVNEL TEFAKTCVAD ESHAGCEK*SL HTLFGDELCK
  101 VASLRETYGD MADCCAK*QEP ERNECFLSHK DDSPDLPKLK* PDPNTLCDEF
  151 KADEKK*FWGK YLYEIARRHP YFYAPELLYY ANK*YNGVFQE CCQAEDKGAC
  201 LLPK*IETMRE K*VLTSSARQR LRCASIQKFG ERALK*AWSVA RLSQKFPKAE
  251 FVEVTKLVTD LTKVHK*ECCH GDLLECADDR ADLAKYICDN QDTISSKLKE
  301 CCDKPLLEK*S HCIAEVEKDA IPENLPPLTA DFAEDKDVCK NYQEAK*DAFL
  351 GSFLYEYSRR HPEYAVSVLL RLAK*EYEATL EECCAKDDPH ACYSTVFDKL
  401 KHLVDEPQNL IKQNCDQFEK* LGEYGFQNEL IVRYTRK*VPQ VSTPTLVEVS
  451 RSLGK*VGTRC CTK*PESERMP CAEDYLSLIL NRLCVLHEKT PVSEK*VTKCC
  501 TESLVNRRPC FSALTPDETY VPKAFDEKLF TFHADICTLP DTEKQIKK*QT
  551 ALVELLKHKP KATEEQLKTV MENFVAFVGK CCAADDKEAC FAVEGPKLVV
  601 STQTALA

b   1 MKWVTFISLL LLFSSAYSRG VFRRDTHK*SE IAHRFKDLGE EHFKGLVLIA
   51 FSQYLQQCPF DEHVKLVNEL TEFAKTCVAD ESHAGCEK*SL HTLFGDELCK
  101 VASLRETYGD MADCCAK*QEP ERNECFLSHK DDSPDLPKLK* PDPNTLCDEF
  151 KADEKK*FWGK YLYEIARRHP YFYAPELLYY ANK*YNGVFQE CCQAEDKGAC
  201 LLPK*IETMRE K*VLTSSARQR LRCASIQK*FG ERALK*AWSVA RLSQK*FPKAE
  251 FVEVTKLVTD LTKVHK*ECCH GDLLECADDR ADLAKYICDN QDTISSKLKE
  301 CCDKPLLEK*S HCIAEVEKDA IPENLPPLTA DFAEDKDVCK NYQEAK*DAFL
  351 GSFLYEYSRR HPEYAVSVLL RLAK*EYEATL EECCAKDDPH ACYSTVFDKL
  401 KHLVDEPQNL IKQNCDQFEK* LGEYGFQNEL IVRYTRK*VPQ VSTPTLVEVS
  451 RSLGK*VGTRC CTK*PESERMP CAEDYLSLIL NRLCVLHEK*T PVSEK*VTK*CC
  501 TESLVNRRPC FSALTPDETY VPKAFDEKLF TFHADICTLP DTEK*QIKK*QT
  551 ALVELLKHKP K*ATEEQLKTV MENFVAFVGK CCAADDKEAC FAVEGPK*LVV
  601 STQTALA

c   1 MKWVTFISLL LLFSSAYSRG VFRRDTHKSE IAHRFK*DLGE EHFKGLVLIA
   51 FSQYLQQCPF DEHVK*LVNEL TEFAKTCVAD ESHAGCEK*SL HTLFGDELCK*
  101 VASLRETYGD MADCCAK*QEP ERNECFLSHK DDSPDLPKLK* PDPNTLCDEF
  151 KADEKK*FWGK* YLYEIARRHP YFYAPELLYY ANK*YNGVFQE CCQAEDKGAC
  201 LLPK*IETMRE K*VLTSSARQR LRCASIQK*FG ERALK*AWSVA RLSQK*FPKAE
  251 FVEVTK*LVTD LTKVHK*ECCH GDLLECADDR ADLAKYICDN QDTISSKLKE
  301 CCDK*PLLEK*S HCIAEVEK*DA IPENLPPLTA DFAEDKDVCK NYQEAK*DAFL
  351 GSFLYEYSRR HPEYAVSVLL RLAK*EYEATL EECCAKDDPH ACYSTVFDKL
  401 KHLVDEPQNL IKQNCDQFEK* LGEYGFQNEL IVRYTRK*VPQ VSTPTLVEVS
  451 RSLGK*VGTRC CTK*PESERMP CAEDYLSLIL NRLCVLHEK*T PVSEK*VTK*CC
  501 TESLVNRRPC FSALTPDETY VPKAFDEKLF TFHADICTLP DTEK*QIKK*QT
  551 ALVELLKHKP K*ATEEQLKTV MENFVAFVGK* CCAADDKEAC FAVEGPK*LVV
  601 STQTALA
```

Fig. 16.10 Neoglycoconjugate model sequences where the glycation sites are indicated by an asterisk (*red* = identified on tryptic digests, *blue* = identified on GluC V8 digests and *red* and *underlined* = identified on both tryptic and GluC V8 digests) for the neoglycoconjugates with a hpaten-BSA ratio of (**a**) 4.3:1. (**b**) 6.6:1 and (**c**) 13.2:1 [46]

16.2.2 Neoglycoconjugate Vaccine of Biological Agent **Bacillus anthracis**, *the Etiological Agent of Anthrax [47]*

Bacillus anthracis is a Gram-positive bacterium that causes anthrax to both humans and animals [67]. The formation of endospores [68] at the maturity stage of the bacterium allows a protection against severe conditions, such as extreme temperatures, radiations, physical damages and chemicals [69]. *Bacillus anthracis* is

Table 16.4 GluC V8 glycopeptide digests identified during LC-ESI-QqTOF-MS/MS analysis of the hapten-BSA glycoconjugate with a hapten-BSA ratio of 4.3:1, 6.6:1 and 13.2:1 [46]

Peptide			Hapten-BSA ratio 4.3:1		Hapten-BSA ratio 6.6:1		Hapten-BSA ratio 13.2:1	
Sequence (asterisk = glycation site)	Calculated m/z (charge)	Missed Cleavage	Observed m/z (charge)	Deviation (Da)	Observed m/z (charge)	Deviation (Da)	Observed m/z (charge)	Deviation (Da)
K*SHCIAE (Lys 309)	679.3189 (+2)	0	679.3320 (+2)	0.0131	679.3343 (+2)	0.0154	679.3284 (+2)	0.0095
LTKVHK*E (Lys 266)	684.3745 (+2)	0	684.3867 (+2)	0.0122	684.3851 (+2)	0.0106	684.3795 (+2)	0.0050
K*VTKCCTE (Lys 495)	769.8575 (+2)	0	769.8677 (+2)	0.0102	769.8720 (+2)	0.0145	769.8668 (+2)	0.0093
K*QEPERNE (Lys 117)	771.8678 (+2)	2	771.8852 (+2)	0.0174	771.8795 (+2)	0.0117	771.8883 (+2)	0.0205
K*SLHTLFGDE (Lys 88)	830.4093 (+2)	1	830.4932 (+2)	0.0839	830.4246 (+2)	0.0153	830.4196 (+2)	0.0103
DKGACLLPK*IE (Lys 204)	878.9556 (+2)	0	878.9783 (+2)	0.0227	878.9669 (+2)	0.0113	878.9643 (+2)	0.0087
AK*DAFLGSFLYE (Lys 346)	937.4590 (+2)	1	937.4871 (+2)	0.0281	937.4759 (+2)	0.0169	937.4798 (+2)	0.0208
KK*FWGKYLYE (Lys 156)	937.9824 (+2)	0	937.9877 (+2)	0.0053	937.9862 (+2)	0.0038	937.9648 (+2)	–0.0176
LLYYANK*YNGVFQE (Lys 183)	1,118.0465 (+2)	0	1,118.0711 (+2)	0.0246	1,118.0587 (+2)	0.0122	1,118.0511 (+2)	0.0046
VSRSLGK*VGTRCCTKPE (Lys 455)	816.7456 (+3)	0	816.7646 (+3)	0.0190	816.7583 (+3)	0.0127	816.7483 (+3)	0.0027
K*TPVSE (Lys 489)	587.2979 (+2)	0	–	–	587.3103 (+2)	0.0124	587.3041 (+2)	0.0062
GPK*LVVSTQTALA (Lys 597)	899.4959 (+2)	0	–	–	899.5080 (+2)	0.0121	899.5092 (+2)	0.0133
LLKHKPK*ATEE (Lys 561)	903.9960 (+2)	1	–	–	904.0190 (+2)	0.0230	903.9916 (+2)	–0.0044
K*QIKKQTALVE (Lys 544)	900.0117 (+2)	0	–	–	900.0234 (+2)	0.0117	900.0165 (+2)	0.0048
VTK*LVTD (Lys 256)	644.8478 (+2)	0	–	–	–	–	644.8498 (+2)	0.0020
HVK*LVNE (Lys 65)	676.3588 (+2)	0	–	–	–	–	676.3647 (+2)	0.0059
VEK*DAIPE (Lys 318)	707.3534 (+2)	2	–	–	–	–	707.3578 (+2)	0.0044
CCDK*PLLE (Lys 304)	774.3520 (+2)	1	–	–	–	–	774.3662 (+2)	0.0142
LCK*VASLRE (Lys 100)	794.9162 (+2)	0	–	–	–	–	794.9163 (+2)	0.0001
VSRSLGK*VGTRCCTK*PE (Lys 455, Lys 463)	987.8230 (+3)	0	–	–	–	–	987.8313 (+3)	0.0083
KK*FWGK*YLYE (Lys 156, 160)	1,194.5985 (+2)	0	–	–	–	–	1,194.5626 (+2)	–0.0359
NFVAFVDK*CCAADDKE (Lys 580)	1,201.5300 (+2)	3	–	–	–	–	1,201.5350 (+2)	0.0050

Fig. 16.11 3D-structure of the neoglycoconjugate model vaccines. The glycated lysine residues are highlighted in *red* (Swiss-Pdb Viewer software) for the neoglycoconjugates with a hapten-BSA ratio of (**a**) 4.3:1, (**b**) 6.6:1 and (**c**) 13.2:1 [46]

a

b

c

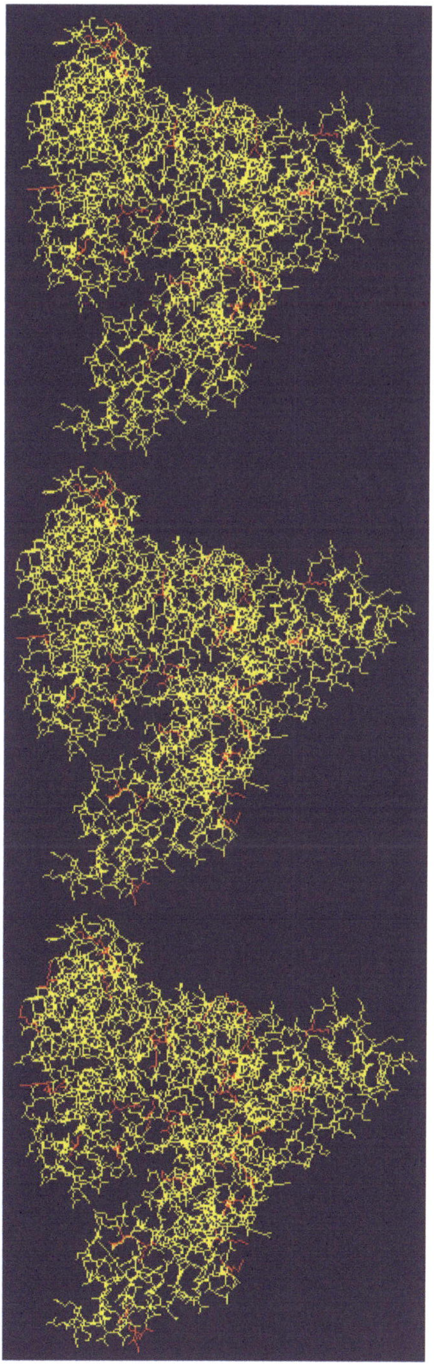

the etiologic agent of anthrax that can be used as a biological weapon [70, 71]. Indeed, *Bacillus anthracis* a pathogen that is lethal in most cases for both humans and animals. There are different *Bacillus anthracis* strains, among those, 89 of them were identified, such as the Sterne strain [72]. the Vollum strain [73] the Ames strain [71, 74] and the H9401 strain [75]. The last one has also been studied for the development of anthrax vaccine [75]. In addition, *Bacillus anthracis* has been extensively studied in the aim of understanding its pathogenesis, identifying new biomarkers and vaccines design [76]. The capsular polypeptide (polyglutamic acid) of the *Bacillus anthracis* has been targeted for the development of synthetic vaccines [77].

Daubenspeck et al. reported the structure of the tetrasaccharide side chain of the collagen like region of the major glycoprotein of the *B. anthracis* exosporium [78]. Their findings was that the upstream terminal of the tetrasaccharide corresponds to the sugar anthrose [4,6-dideoxy-4-(3-hydroxy-3-methylbutyramido)-2-O-methyl-D-glucopyranose].

The group of Kováč prepared a vaccine composed of a synthesized tetrasaccharide side chain of the collagen like region of the major glycoprotein of the *B. anthracis* exosporium (MW = 950.43 Da) attached to the BSA using the squaric acid chemistry [32]. The conjugation leaded to the formation of vaccines with different carbohydrate:BSA ratios.

16.2.2.1 Carbohydrate:BSA Ratio

MALDI-TOF-MS analysis of one of these synthetic vaccines allowed to observe the following protonated molecular ions: $[M + H]^+$ at *m/z* 71,448.36 and $[M + 2H]^{2+}$ at *m/z* 35,730.08. The molecular weight of this vaccine was thus found to be 71,447.36 Da. As the molecular weight of the synthetic tetrasaccharide side chain of the collagen like region being 950.43 Da, the carbohydrate:protein ratio of the synthetic vaccine was determined to be 5.4:1.

16.2.2.2 Glycation Sites Determination

The determination of the glycation sites of the hapten-BSA vaccine neoglycoconjugate was carried out by an enzymatic digestion with the trypsin and/or GluC V8 proteases, followed by the MALDI-MS/MS and LC-MS/MS analysis of the digests. The enzymatic digestion of the hapten-BSA glycoconjugate and MALDI-MS/MS as well as LC-MS/MS analyses were carried out as previously described [45–49].

MALDI-MS/MS

The spectra obtained during the ***MALDI-MS/MS*** analysis of the tryptic and GluC V8 digests were submitted to the MASCOT library to identify by PMF the peptides matching to the BSA (Fig. 16.12). Two Serum Albumin protein isoforms from the *Bos taurus* species were identified for the tryptic and GluC V8 digests: the serum

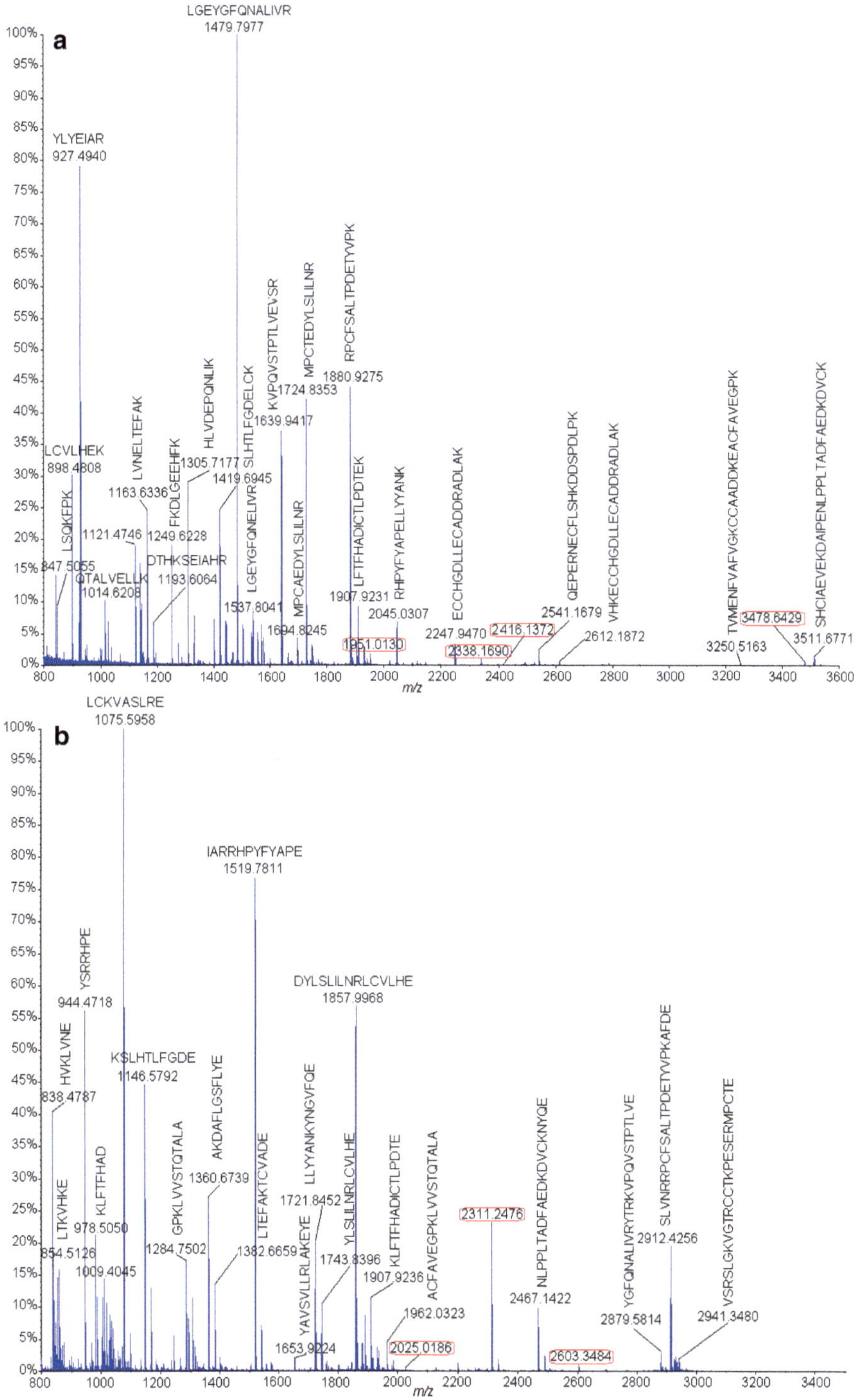

Fig. 16.12 MALDI-MS analysis of the glycoconjugate trypsin digests (**a**) and GluC V8 endoproteinase digests (**b**) [47]

albumin precursor (gi|1351907) and the serum albumin (gi|74267962). The detected peptides that were not identified in the database were analyzed by tandem mass spectrometry. The MALDI-MS/MS analysis of the tryptic digests allowed to identify three glycated peptides (Fig. 16.12) corresponding to BSA peptides with an increment of 950 Da to their original molecular mass, namely: ALK*AWSVAR at *m/z* 1,951.0130, VTK*CCTESLVNR at *m/z* 2,416.1372, and QNCDQFEK*LGEYGFQNALIVR at *m/z* 3,478.6429 (glycation represented with an asterisk on the lysine residue).

The high-energy CID-MS/MS analysis of the glycopeptide ALK*AWSVAR (Fig. 16.13) at *m/z* 1,951.0130 afforded a series of product ions corresponding to the entire precursor ion that losses different carbohydrate portions, the monosaccharide B_1 (−259 Da), disaccharide B_2 (−405 Da), trisaccharide B_3 (−551 Da) and/or the tetrasaccharide B_4 (−697 Da), leading respectively to the formation of the following product ions: Y_3^+ at *m/z* 1,691.8829, Y_2^+ at *m/z* 1,545.8376, Y_1^+ at *m/z* 1,399.7768 and Y_0^+ at *m/z* 1,253.7010 (Fig. 16.13). The mass difference between Y_3^+ and Y_2^+, Y_2^+ and Y_1^+, Y_1^+ and Y_0^+ was found to correspond to one α-L-rhamnopyranosyl unit (146 Da). Moreover, product ions resulting from the fragmentation of the same tetrasaccharide were observed: $C_{17}H_{26}N_3O_4^+$ at *m/z* 336.1242, B_1^+ at *m/z* 260.0739, $[B_1 - H_2O]^+$ at *m/z* 242.0709, $^{2,5}A_1^+$ at *m/z* 230.0719, $C_{11}H_{16}N_3O_2^+$ at *m/z* 222.0670 and $[C_1 - C_5H_{11}NO_2]^+$ at *m/z* 159.0486 (Fig. 16.13). Accordingly, the fragmentation of the carbohydrate portion allowed to confirm its structure and contributed to establish a diagnostic fragmentation signature of the synthetic tetrasaccharide. In addition, the fragmentation of the peptide portion leaded to the formation of b- and y- product ions that allowed the determination of the sequence of the glycated peptide. However, some peptide product ions had the particularity to have lost the carbohydrate portion but were still attached to the spacer-squaric acid chain: $[b_3 - B_4]^+$ at *m/z* 565.2661, $[b_4 - B_4]^+$ at *m/z* 636.2715 and $[y_7 - B_2]^+$ at *m/z* 1,361.7265. Thus, the identified glycated peptides ALK*AWSVAR at *m/z* 1,951.0130, VTK*CCTESLVNR at *m/z* 2,416.1372, and QNCDQFEK*LGEYGFQNALIVR at *m/z* 3,478.6429 allowed to determine the glycation site on the following lysine residues: Lys 235, Lys 498 and Lys 420, respectively.

Similarly, the MALDI-MS/MS analysis of the GluC V8 digests afforded different glycated peptides: LCK*VASLRE at *m/z* 2,025.0186, YAVSVLLRLAK*E at *m/z* 2,311.2476 and YAVSVLLRLAK*EYE at *m/z* 2,603.3484, allowing to identify the following glycation sites on lysine residues: Lys 100 and Lys 374. To sum it up, only five glycation sites were identified during the MALDI-MS and MS/MS analyses of the tryptic and GluC V8 digests of the hapten-BSA glycoconjugate: Lys 100, Lys 235, Lys 374, Lys 420 and Lys 498.

LC-MS/MS

The second approach for the determination of the glycation sites of the vaccine neoglycoconjugate was the LC-MS/MS analysis of the tryptic and GluC V8 digests (Fig. 16.12). It has to be noted that the LC-MS/MS analysis of peptides has the advantage of minimizing the ionization suppression effect comparing to the MALDI-MS/MS analysis [79, 80].

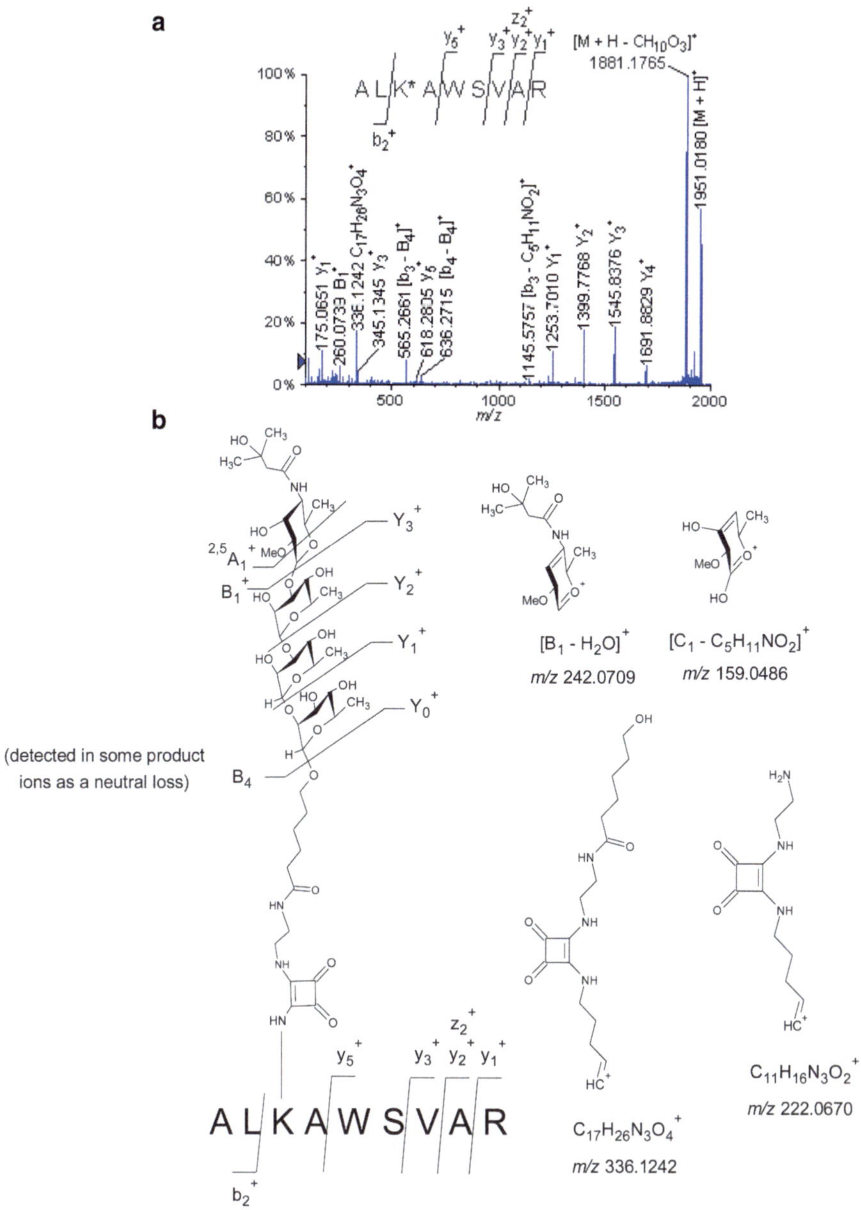

Fig. 16.13 (**a**) MALDI-TOF/TOF-MS/MS spectra of the glycated peptide ALK*AWSVAR (Lys 235) at *m/z* 1,951.0130. (**b**) Different product ions involving the fragmentation of the carbohydrate hapten observed during the MALDI-TOF/TOF-MS/MS analysis of the glycated peptide ALK*AWSVAR (Lys 235) at *m/z* 1,951.0130 [47]

The data of the LC-MS/MS analysis of the tryptic and GluC V8 digests were submitted to the MASCOT library and matched two serum albumin protein isoforms: the serum albumin precursor (gi|1351907) and serum albumin (gi|74267962) from *Bos taurus*.

For the tryptic digests, the BSA sequence coverage was found to be 57 % for the serum albumin precursor from *Bos taurus* (gi|1351907) and 58 % for serum albumin protein from *Bos taurus* (gi|74267962). The LC-MS/MS analysis of the hapten-BSA tryptic digests allowed the identification of 18 glycated peptides, reported in Table 16.5. The low-energy CID-MS/MS analysis of the extracted precursor ions of the glycated peptides allowed to localize the glycation sites on the following 18 lysine residues: Lys 140, Lys 155, Lys 156, Lys 204, Lys 211, Lys 228, Lys 235, Lys 304, Lys 374, Lys 401, Lys 420, Lys 437, Lys 455, Lys 463, Lys 495, Lys 498, Lys 547 and Lys 559.

The LC-MS/MS analysis of the hapten-BSA vaccine GluC V8 digest allowed the identification of the serum albumin from *Bos taurus* (gi|74267962) with a sequence coverage of 42 % and the precursor serum albumin from *Bos taurus* (gi|1351907) with a sequence coverage of 45 %, in the MASCOT database.

Table 16.6 displays the identified glycated peptides during the LC-MS/MS analysis of the GluC V8 digests. Thus, the CID-MS/MS analysis of theses glycated peptides allowed to discover 17 glycation sites, localized on the following lysine residues: Lys 65, Lys 75, Lys 88, Lys 100, Lys 117, Lys 151, Lys 183, Lys 197, Lys 256, Lys 266, Lys 304, Lys 309, Lys 336, Lys 374, Lys 420, Lys 455 and Lys 495.

In summary, the LC-MS/MS analysis of both tryptic and GluC V8 digests allowed the identification of a total of 30 glycation sites on the lysine residues (Fig. 16.14a). Mapping these glycation sites on the 3D representation of the BSA (Fig. 16.14b, lysines highlighted in red) permitted to observed that they correspond to lysine residues located at the outer surface of the BSA. In addition, the number of the identified glycation sites (30 lysine residues) being higher than the determined (hapten:BSA ratio 5.4:1) of the tetrasaccharide-BSA neoglycoconjugate, it was concluded that the vaccine is composed of a mixture of glycoforms.

16.3 Atypical Example of the Mass Spectrometry Characterization of Anti-cancer Vaccine

16.3.1 Anti-tumor Thomsen-Friedenreich Neoglycoconjugate Vaccine Prepared by Michael Addition

Tumor-associated carbohydrate antigens (TACAs) are recognized in a conventional class I MHC-restricted fashion. In Class I, the MHC proteins are found in the membrane of all nucleated cells and they are continually synthesized and transported to the cell membrane [81–87]. During their synthetic process, they pick up protein fragments from within the cytoplasm and carry them to the cell surface. If the cell is healthy and the protein fragments are normal, the T cells will ignore them.

Table 16.5 Tryptic glycopeptides identified of the bovine serum albumin protein by LC-ESI-QqTOF-MS/MS analysis of the hapten-BSA glycoconjugate [47]

Precursor ion *m/z* (Charge)	Mr (expt)	Mr (calc)	Deviation Da	Missed Cleavage	Peptide
729.8826 (+2)	1,457.7506	1,457.7389	0.0117	1	KHK*P (Lys 559)
733.3170 (+2)	1,464.6194	1,464.7698	−0.1504	1	QIK*K (Lys 547)
770.3757 (+2)	1,538.7369	1,538.7338	0.0031	1	ADEK*K (Lys 155)
807.9138 (+2)	1,613.8129	1,613.7964	0.0165	0	K*FWGK (Lys 156)
828.0811 (+3)	2,481.2213	2,481.2005	0.0208	2	LKECCDK*PLLEK (Lys 304)
832.7770 (+3)	2,495.3092	2,495.3146	−0.0054	1	LK*HLVDEPQNLIK (Lys 401)
863.7965 (+3)	2,588.3677	2,588.3572	0.0105	0	K*VPQVSTPTLVEVSR (Lys 437)
883.9655 (+2)	1,765.9164	1,765.9085	0.0079	1	SLGK*VGTR (Lys 455)
922.0941 (+3)	2,763.2605	2,763.2493	0.0112	1	LAK*EYEATLEECCAK (Lys 374)
969.5086 (+2)	1,937.0026	1,936.9868	0.0158	1	TPVSEK*VTK (Lys 495)
970.4900 (+2)	1,938.9738	1,938.9772	−0.0034	1	EK*VLTSSAR (Lys 211)
976.0173 (+2)	1,950.0201	1,950.0085	0.0116	1	ALK*AWSVAR (Lys 235)
990.4710 (+3)	2,968.3911	2,968.3886	0.0025	2	LK*PDPNTLCDEFKADEK (Lys 140)
1,073.0144 (+2)	2,144.0143	2,144.0082	0.0061	1	CASIQK*FGER (Lys 228)
1,058.4652 (+2)	2,114.9159	2,114.9123	0.0036	1	CCTK*PESER (Lys 463)
1,208.5857 (+2)	2,415.1569	2,415.1284	0.0285	1	VTK*CCTESLVNR (Lys 498)
1,160.2172 (+3)	3,477.6297	3,477.6385	−0.0088	1	QNCDQFEK*LGEYGFQNALIVR (Lys 420)
1,169.5978 (+2)	2,337.1810	2,337.1583	0.0227	1	GACLLPK*IETMR (Lys 204)

Table 16.6 Glycopeptides identified in the bovine serum albumin protein by LC-ESI-QqTOF-MS/MS analysis of the hapten-BSA glycoconjugate digested with the endoproteinase GluC V8 [47]

Precursor ion *m/z* (Charge)	Mr (expt)	Mr (calc)	Deviation Da	Missed Cleavage	Peptide
698.3598 (+2)	1,394.7051	1,394.6803	0.0248	0	K*LGE (Lys 420)
790.3774 (+2)	1,578.7403	1,578.7288	0.0115	1	K*QEPE (Lys 117)
844.3721 (+2)	1,686.7297	1,686.7499	–0.0202	1	EFK*ADE (Lys 151)
894.4637 (+2)	1,786.9128	1,786.8976	0.0152	0	HVK*LVNE (Lys 65)
862.9539 (+2)	1,723.8933	1,723.8754	0.01179	1	VTK*LVTD (Lys 256)
897.4200 (+2)	1,792.8255	1,792.8176	0.0079	0	K*SHCIAE (Lys 309)
902.4801 (+2)	1,802.9456	1,802.9289	0.0167	0	LTKVHK*E (Lys 266)
962.1552 (+2)	2,883.4439	2,883.4093	0.0346	0	VSRSLGK*VGTRCCTKPE (Lys 455)
987.9658 (+2)	1,973.9171	1,973.8949	0.0222	0	K*VTKCCTE (Lys 495)
989.9697 (+2)	1,977.9248	1,977.9154	0.0094	2	K*QEPERNE (Lys 117)
992.4568 (+2)	1,982.8989	1,982.884	0.0149	1	CCDK*PLLE (Lys 304)
995.4702 (+2)	1,988.9258	1,988.8912	0.0346	1	FAK*TCVADE (Lys 75)
1,013.0142 (+2)	2,024.0138	2,024.0123	0.0015	0	LCK*VASLRE (Lys 100)
1,048.5225 (+2)	2,095.0305	2,094.9984	0.0321	1	K*SLHTLFGDE (Lys 88)
1,097.0595 (+2)	2,192.1045	2,192.0909	0.0136	0	DK*GACLLPKIE (Lys 197)
1,124.5018 (+2)	2,246.9890	2,246.9876	0.0014	1	DK*DVCKNYQE(Lys 336)
1,336.1446 (+2)	2,670.2746	2,670.2728	0.0018	0	LLYYANK*YNGVFQE (Lys 183)
1,156.132 (+2)	2,310.2495	2,310.2345	0.0150	0	YAVSVLLRLAK*E (Lys 374)

a

```
  1 MKWVTFISLL LLFSSAYSRG VFRRDTHKSE IAHRFKDLGE EHFKGLVLIA
 51 FSQYLQQCPF DEHVK*LVNEL TEFAK*TCVAD ESHAGCEK*SL HTLFGDELCK*
101 VASLRETYGD MADCCAK*QEP ERNECFLSHK DDSPDLPKLK* PDPNTLCDEF
151 K*ADEK*K*FWGK YLYEIARRHP YFYAPELLYY ANK*YNGVFQE CCQAEDK*GAC
201 LLPK*IETMRE K*VLTSSARQR LRCASIQK*FG ERALK*AWSVA RLSQKFPKAE
251 FVEVTK*LVTD LTKVHK*ECCH GDLLECADDR ADLAKYICDN QDTISSKLKE
301 CCDK*PLLEK*S HCIAEVEKDA IPENLPPLTA DFAEDK*DVCK NYQEAKDAFL
351 GSFLYEYSRR HPEYAVSVLL RLAK*EYEATL EECCAKDDPH ACYSTVFDKL
401 K*HLVDEPQNL IKQNCDQFEK* LGEYGFQNEL IVRYTRK*VPQ VSTPTLVEVS
451 RSLGK*VGTRC CTK*PESERMP CAEDYLSLIL NRLCVLHEKT PVSEK*VTK*CC
501 TESLVNRRPC FSALTPDETY VPKAFDEKLF TFHADICTLP DTEKQIK*KQT
551 ALVELLKHK*P KATEEQLKTV MENFVAFVGK CCAADDKEAC FAVEGPKLVV
601 STQTALA
```

b

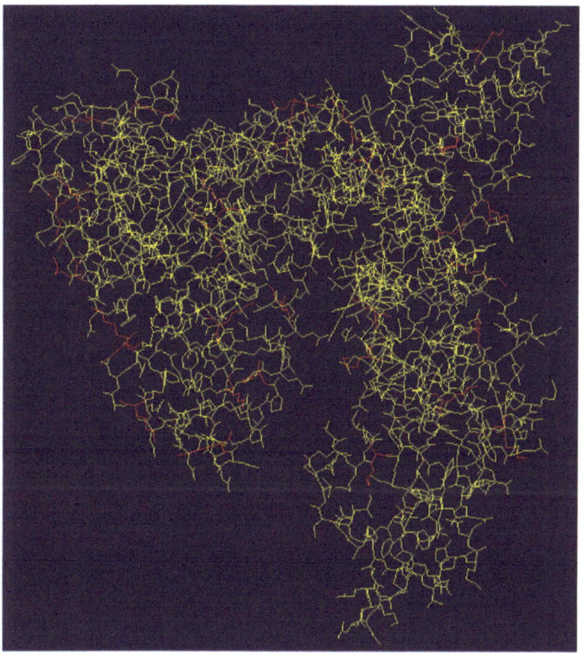

Fig. 16.14 (**a**) BSA sequence where the glycation sites are indicated by an *asterisk* (*red* = identified on tryptic digests, *blue* = identified on GluC V8 digests and *red* and *underlined* = identified on both tryptic and GluC V8 digests) and (**b**) 3D structure of the BSA. The glycated lysine residues are highlighted in *red* (Swiss-Pdb Viewer software) [47]

If the cytoplasm contains abnormal protein fragments derived from viral proteins and/or alien proteins made by a cancer cell, the CD8+ T cells (Cytotoxic T-Lymphocytes) will recognize the cell as being infected and ultimately destroy it [81, 87]. Actually, cancer cells are subjected to many observed changes in their behavior (changes in shape and size of the cancer cell which became invading for the normal cell) and their surrounding environment [88, 89]. Among the changes observed during human cancers, cell surface glycoprotein modifications results from incomplete or aberrant glycosylation by unambiguous phenomenon [88–90]. Accordingly, in some forms, the resulting increase in the numbers of glycosylation sites are reflected by the presence of the so-called TACAs such as; Tn (GalNAcα1 → Ser/Thr), sialyl-Tn (Neu5Acα2 → 6GalNAc) and Thomsen-Friedenreich (TF: Galβ1 → 3GalNAcα1 → Ser/Thr) antigens [91–94]. Two different syntheses of TF have attempted based on different stoichiometry of the glycosyl donor (TF-antigen) and acceptor carrier protein (BSA) of 2:1 and 8:1, respectively. As expected, these conditions resulted in the formation of two different TF neoglycoconjugate vaccines. The best advantage of the synthetic tactic used for producing the TF-antigen:BSA conjugate over other existing approaches for protein glycosylation, is that it can be applied to cysteine containing proteins (free of disulphide bonds with neighbouring residues) in their native form without any additional chemical modification or activation.

16.3.2 Preparation of the TF-BSA Vaccine Conjugate

The TF-antigen was prepared according to a published procedure [95]. Briefly, the photocatalyzed reaction of the allyl glycoside of Thomsen–Friedenreich (TF) disaccharide [allyl 3-*O*-(β-D-galactopyranosyl)-2-acetamido-2-deoxy-α-D-glucopyranoside] (**1**) with cysteamine gave the TF glycoside derivative **2** with a terminal primary amino group in 83 % yield (Scheme 16.1, Fig. 16.15). Subsequently, compound **2** was treated with acryloyl chloride in methanol to yield the α,β-unsaturated amide derivative **3** via the nucleophilic displacement of the chloride with the amino function. This reaction was carried out in the presence a strong base anion resin to remove the formed HCl. Finally, the acrylamide-ending TF-antigen **3** was anchored to the BSA carrier protein by treating the α,β-unsaturated amide with BSA at room temperature and at 40 °C, respectively in a 0.2 M carbonate buffer for 3 days to afford the two desired TF-BSA anti-tumor vaccine conjugates having a low and a high TF-content. A colorimetric phenol-sulfuric acid test was performed onto the BSA-TF conjugates using compound **1** a standard [96]. The analysis showed the conjugates to contain 2 ± 1 and 8 ± 2 TF-antigens, respectively.

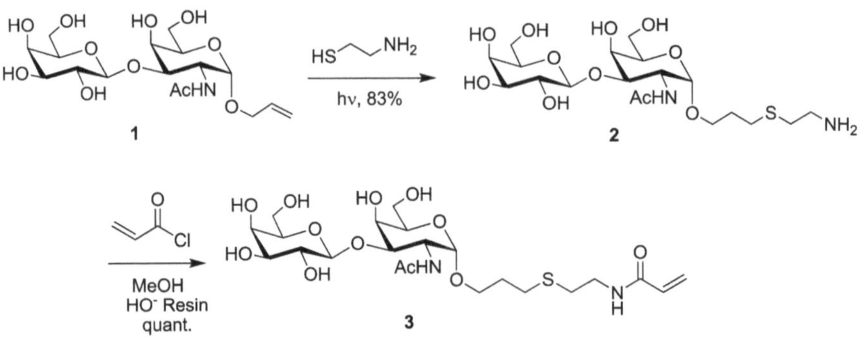

Scheme 16.1 Structure of the TF-antigen and its chemical transformation into a suitable Michael acceptor **3**

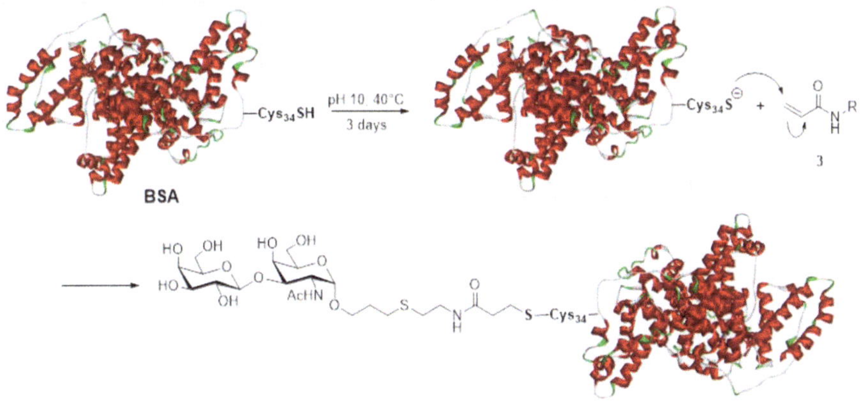

Fig. 16.15 Synthesis of Thomsen-Friedenreich antigen-BSA glycoconjugate vaccines using a Michael addition

16.3.3 MALDI-TOF/TOF-MS Analysis of the TF-BSA Glycoconjugates

For the determination of the average number of carbohydrate-spacer moieties linked to BSA (TF-antigen: BSA), the two glycoconjugates were analyzed by matrix assisted laser desorption/ionization- mass spectrometry (MALDI-TOF/TOF-MS). The MALDI-TOF/TOF-MS were characterized by the formation of a protonated molecular ions $[M+H]^+$ at *m/z* 67,599.2300 and m/z 70,904.7668 and their TF-BSA ratios were calculated to be individually 2:1 and 8:1. Please, note that in this work the molecular mass of the synthesized neoglycoconjugates are given with a standard error of the mean measurements (SEM). Henceforth, for the neoglycoconjugate prepared with hapten:BSA ratio of 2:1 gave the protonated molecule $[M+H]^+$ at *m/z* 67,599.2300, 67,599.2240, 67,599.2228 and 67,599.2271. Accordingly, the actual molecular weight is 67,598.2292 with a ± SEM = ± 0.0014 Da. Once again, the molecular weight for the neoglycoconjugate hapten:BSA ratio 8:1 gave the protonated molecule $[M + H]^+$ at *m/z* 70,904.7668, 70,904.7651, 70,904.7642, 70,904.7688, 70,904.7655 and thea standard error of the mean measurement was ± SEM ± 0.0007 Da.

The molecular mass calculations of these glycoconjugates (Hapten:BSA ratios 2:1 and 8:1) were based on the presumption that one molecule of the TF-antigen will covalently links by a Michael addition reaction on the free sulfhydryl (Cys-34) group forming a C-*S* glycoside bond. The remaining stoichiometric amounts of the TF-antigen, which is one remaining molecule for the glycoconjugate Hapten:BSA (ratio 2:1) and seven molecules of the glycoconjugate Hapten:BSA (ratio 8:1), respectively, will conjugate by direct targeting of the ε-amino groups of the various lysine residues of BSA, forming a *C-N* bond which occur by loss of ethanol molecules.

As mentioned previously the TF-antigen is composed of the disaccharide Galβ1 → 3GalNAcα1 → O-linked to spacer [-$(CH_2)_3S(CH_2)_2$NH-CO-CH = CH_2] which has a molecular mass of 555.22 Da.

16.3.4 LC-ESI-QqTOF-MS/MS Analyses of Tryptic Digest of the Glycoconjugate with a TF: BSA Ratio of 2:1 and 8:1

In this manuscript the use high-energy collision induced dissociation MALDI-MS/MS was circumvented in exploring the tryptic digests MS/MS data. Though, peptide identification was achieved by nano-LC-ESI-MS/MS analysis. This was completed in order to as improve the glycopeptides chromatographic separation and to hinder the ionization suppression effect observed during MALDI-TOF-MS analysis [79, 97]. Hereafter, nano-LC-ESI-QqTOF-MS/MS analysis was carried out on both tryptic digests of glycoconjugates with TF: protein ratio 2:1 and 8:1, and this was followed by *de novo* peptide sequencing using low-energy CID-QqTOF-MS/MS tandem mass spectrometry. The obtained low-energy CID-MS/MS were submitted to the Mascot library to identify the glycopeptides and peptides released from BSA.

The Mascot reports of LC-ESI-QqTOF-MS/MS data of the tryptic digests of both glycoconjugates demonstrated the identification of two serum albumin isoforms from *Bos taurus* species. The serum albumin precursor (gi|1351907) was identified and had the following sequence coverage: 46 % and 21 % both glycoconjugates formed with TF-antigen: BSA ratio of 2:1 and 8:1 respectively. In addition, serum albumin protein isoform (gi|229552) was also identified for the tryptic digest for both glycoconjugates and this allowed us to determine respectively. A sequence coverage of 42 % for the TF-antigen: BSA ratio of 2:1 and a sequence coverage of 19 % for the glycoconjugated having TF-antigen: BSA ratio of 8:1.

It is fundamental to remark that the tryptic glycopeptide *m/z* values of the precursor ions that were not matched to that of the predicted peptides, were calculated by the addition of the carbohydrate TF-antigen (Molecular Weight, 555.22 Da), to the *m/z* value of the correct peptide which actually form the complete analyzed glycopeptide.

As mentioned previously, de *novo* sequencing of the glycopeptides was achieved by measuring the precursor ion spectra. Hence, we were able to identify of three glycated peptides for glycoconjugate prepared with an antigen-TF:BSA ratio of 2:1 (Table 16.7); FK*DLGEEHFK (*C-N* glycation site is Lys 12) at 902.1155(+2), GLVLIAFSQYLQQC*PFDEHVK (*C-S* glycation site is Cyst 34) at 1,015.4871(+3) and HKPK*ATEEQLK (*C-N* glycation site is Lys 535) at 620.5411(+3). Additionally, 14 glycated peptides were also identified for the glycoconjugate prepared with an antigen-TF:BSA ratio of 8:1 respectively (Table 16.8): FK*DLGEEHFK (*C-N* glycation site is Lys 12) at *m/z* 902.1251(+2), LK*PDPNTLCDEFK (*C-N* glycation site is Lys 116) at m/z 1,065.5890(+2), K*QTALVELLK (*C-N* glycation site is Lys 522)

Table 16.7 Identified glycated peptides during the LC-MS/MS analysis of the trypsin digests of the TF-BSA vaccine with TF: BSA ratio 2:1

Glycated peptide sequence	Missed cleavage	TF:BSA ratio 2:1 Observed m/z (charge)	Deviation (Da)
FK*DLGEEHFK (Lys 12)	1	902.1155 (+2)	−0.0214
GLVLIAFSQYLQQC*PFDEHVK (Cyst 34)	0	1,015.4871 (+3)	0.0052
HKPK*ATEEQLK (Lys 535)	1	620.5411 (+3)	−0.0044

Table 16.8 Identified glycated peptides during the LC-MS/MS analysis of the trypsin digests of the TF-BSA vaccine with TF: BSA ratio 8:1

Glycated peptide sequence	Missed cleavage	TF: BSA ratio 8:1 Observed m/z (charge)	Deviation (Da)
FK*DLGEEHFK (Lys 12)	1	902.1251 (+2)	−0.0331
LK*PDPNTLCDEFK (Lys 116)	1	1,065.5890 (+2)	0.0762
K*QTALVELLK (Lys 522)	1	848.7161 (+2)	−0.0776
K*VPQVSTPTLVEVSR (Lys 412)	1	731.4685 (+3)	0.0317
GLVLIAFSQYLQQC*PFDEHVK (Cyst 34)	0	1,015.4614 (+3)	−0.0309
LKPDPNTLCDEFK*ADEK (Lys 127)	2	858.2895 (+3)	−0.0822
NECFLSHK*DDSPDLPK (Lys 106)	1	818.7910 (+3)	0.0198
DAIPENLPPLTADFAEDK*DVCK (Lys 312)	1	1,004.2688 (+3)	−0.0210
SLGK*VGTR (Lys 429)	1	685.2528 (+2)	0.0073
EK*VLTSSAR (Lys 186)	1	765.8557 (+2)	−0.0100
ALK*AWSVAR (Lys 210)	1	777.3762 (+2)	0.0036
DTHK*SEIAHR (Lys 4)	1	873.4264 (+2)	0.0033
HKPK*ATEEQLK (Lys 535)	1	620.5352 (+3)	0.0015
GACLLPK*IETMR (Lys 179)	1	942.6395 (+2)	−0.0094

at *m/z* 848.7161(+2), K*VPQVSTPTLVEVSR(*C-N* glycation site is Lys412) at *m/z* 731.4685(+3), GLVLIAFSQYLQQC*PFDEHVK(*C-S* glycation site is Cyst34) at *m/z* 1,015.3614(+3), LKPDPNTLCDEFK*ADEK (*C-N* glycation site is Lys 127) at *m/z* 858.2895(+3), NECFLSHK*DDSPDLPK (*C-N* glycation site is Lys 106) at *m/z* 818.7910(+3), DAIPENLPPLTADFAEDK*DVCK (*C-N* glycation site is Lys 312) *at m/z* 1,004.2688(+3), SLGK*VGTR (*C-N* glycation site is Lys 429) at m/z 685.2528 EK*VLTSSAR (*C-N* glycation site is Lys 186) at *m/z* 765.8557(+2), ALK*AWSVAR (*C-N* glycation site is Lys 210) at m/z 777.3762(+2), DTHK*SEIAHR (*C-N* glycation site is Lys 4) at m/z 873.4264(+2), HKPK*ATEEQLK (*C-N* glycation site is Lys 535) at *m/z* 620.5352(+3) and GACLLPK*IETMR (*C-N* glycation site is Lys 179) at m/z 942.6395(+2). All aforementioned glycation sites were displayed in Fig. 16.16.

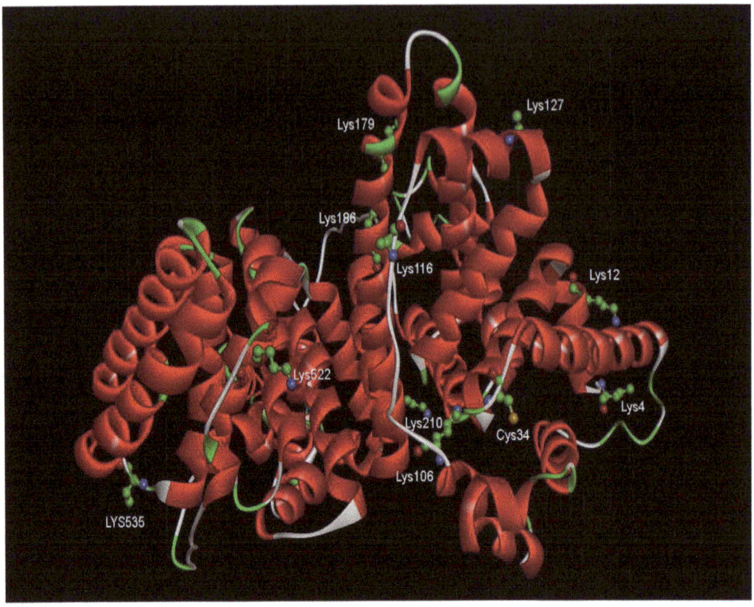

Fig. 16.16 3D structure of BSA generated from the X-ray data using the software Discovery Studio v 3.1

Consequently, the CID-fragmentation of the *S*-glycated peptide: GLVLIAFSQYLQQC*PFDEHVK(*C-S* glycation site located at Cyst34) at *m/z* 1,015.3614(+3) is shown in Table 16.9. This MS/MS afforded the following series of products ions. First, we noted the product ions formed only by the carbohydrate hapten: $[B_2 - Gal - CH_2CO - 2H_2O]^+$ at *m/z* 126.0578, $[B_2$ - Gal- H2O$]^+$ at *m/z* 186.0773, $[B_2 - Gal]^+$ at m/z 204.0845, B_2^+ at *m/z* 366.1487, Y_0^{2+} at 1,341.1178 and Y_1^{2+} 1,442.6371. Additionally, we also noted the peptide product ions resulting from the release of the carbohydrate portion from the glycopeptide $[-B_2^+]$ and these were the following: $[y_5 - B_2]^+$ at *m/z* 261.3074, $[b_7 - B_2]^+$ at *m/z* 348.4520, $[y_6 - B_2]^+$ at *m/z* 408.3712, $[b_8 - B_2]^+$ at *m/z* 435.4821, $[b_9 - B_2]^+$ at *m/z* 563.5424, $[y_8 - B_2]^+$ at *m/z* 665.4677, $[b_{10} - B_2]^+$ at *m/z* 726.6014, $[b_{11} - B_2]^+$ at *m/z* 839.7011, $[y_{10} - B_2]^+$ at *m/z* 921.5777, $[b_{13} - B_2]^+$ at *m/z* 1,095.8145, $[b_{14} - B_2]^+$ at *m/z* 1,255.8424 and $[b_{16} - B_2]^+$ at *m/z* 1,499.9621. Finally, we noted products ions created by the straight forward fragmentation of the peptide portion only, these are exemplified by the following product ions: b_2^+ at *m/z* 171.1100, b_5^+ at *m/z* 496.3487 and b_{17}^+ at *m/z* 1,980.9884. It is important to mention that for the CID-MS/MS of this described precursor ion at *m/z* 1,015.3614(+3) ($C_7H_{13}NO_2S^+$), no product ion, whatsoever, containing the spacer linker arm were noted.

Table 16.9 Identified product ions for the glycated peptide GLVLIAFSQYLQQC*PFDEHVK (Cyst34) at m/z 1,015.4614 (+3)

Product ion	Calculated m/z	Observed m/z	Deviation (Da)
b_{17}^+	1,980.9888	1,980.9884	0.0004
$[b_{16} - B_2]^+$	1,499.9619	1,499.9621	−0.0002
Y_1^{2+}	1,442.6635	1,442.6371	0.0264
Y_0^{2+}	1,341.1685	1,341.1178	0.0507
$[b_{14} - B_2]^+$	1,255.8407	1,255.8424	−0.0017
$[b_{13} - B_2]^+$	1,095.8100	1,095.8145	−0.0045
$[y_{10} - B_2]^+$	921.5786	921.5777	0.0009
$[b_{11} - B_2]^+$	839.7000	839.7011	−0.0011
$[b_{10} - B_2]^+$	726.6014	726.6004	0.0010
$[y_8 - B_2]^+$	665.4615	665.4677	−0.0062
$[b_9 - B_2]^+$	563.5455	563.5424	0.0031
b_5^+	496.3493	496.3487	0.0006
$[b_8 - B_2]^+$	435.4869	435.4821	0.0048
$[y_6 - B_2]^+$	408.3781	408.3712	0.0069
B_2^+	366.1400	366.1487	−0.0087
$[b_7 - B_2]^+$	348.4549	348.4520	0.0029
$[y_5 - B_2]^+$	261.3097	261.3074	0.0023
$[B_2 - Gal]^+$	204.0866	204.0845	0.0021
$[B_2 - Gal- H_2O]^+$	186.0761	186.0773	−0.0012
b_2^+	171.1128	171.1100	0.0028
$[B_2 - Gal - CH_2CO - 2H_2O]^+$	126.0550	126.0578	−0.0028

16.4 Conclusion

The systematic investigations presented herein constitute series of versatile examples for the identification of accurate quality control necessary in commercial production of glycoconjugate vaccines against infectious diseases. The glycopeptides isolated and fully characterized during this work may well represent useful reference compounds to be used in standardization analyses.

Moreover, although BSA usually serves a universal model carrier protein for novel conjugation chemistry, we found it perfectly legitimate as a vaccine in mouse experiments since the monoclonal antibodies isolated from the above conjugates were found to be anti-metastatic [23]. Most importantly, the identified glycopeptides may well form the basis for fully synthetic carbohydrate-based vaccines [2–5]. This would be particularly true when performed on other more immunogenic protein carriers such as tetanus toxoid and KLH.

References

1. Plotkin SA (2008) Vaccines: correlates of vaccine-induced immunity. Clin Infect Dis 47:401–409
2. Heidelberger M, Avery OT (1923) The soluble specific substance of pneumococcus. J Exp Med 38:73–79
3. Ausubel FM (2005) Are innate immune signaling pathways in plants and animals conserved? Nat Immunol 6:973–979
4. Rumbo M, Nempont C, Kraehenbuhl J-P, Sirard J-C (2006) Mucosal interplay among commensal and pathogenic bacteria: lessons from flagellin and Toll-like receptor 5. FEBS Lett 12:2976–2984
5. Shetty N, Aarons E, Andrews J (2009) Structure and functions of microbes. In: Shetty N, Tang JW, Andrews J (eds) Infectious disease: pathogenesis, prevention, and case studies. Wiley, London, p 15
6. Corbett D, Hudson T, Roberts IS (2010) Bacterial polysaccharide capsules. In: Konig H (ed) Prokaryotic cell wall compounds. Springer, Heidelberg, p 111
7. Monack DM, Mueller A, Falkow S (2004) Persistent bacterial infections: the interface of the pathogen and the host immune system. Nat Rev Microbiol 2:747–765
8. Westphal O, Liideritz O, Bister F (1952) Ueber die Extraktion von Bakterien mit Phenol/ Wasser. Z Naturforsch 7B:148–155
9. Pupo E, Aguila A, Santana H, Núnez JF, Castellanos-Serra L, Hardy E (1999) Mice immunization with gel electrophoresis-micropurified bacterial lipopolysaccharides. Electrophoresis 20:458–461
10. Davis MR Jr, Goldberg JB (2012) Purification and visualization of lipopolysaccharide from Gram-negative bacteria by hot aqueous-phenol extraction. J Vis Exp 28:e3916, 1–3
11. Nagy G, Pál T (2008) Lipopolysaccharide: a tool and target in enterobacterial vaccine development. Biol Chem 389:513–520
12. Reisser D, Pance A, Jeannin JF (2002) Mechanisms of the antitumoral effect of lipid A. Bioessays 24:284–289
13. Bowden RA, Cloeckaert A, Zygmunt MS, Dubray G (1995) Outer-membrane protein- and rough lipopolysaccharide-specific monoclonal antibodies protect mice against *Brucella ovis*. J Med Microbiol 43:344–347
14. Fulop M, Mastroeni P, Green M, Titball RW (2001) Role of antibody to lipopolysaccharide in protection against low- and high-virulence strains of *Francisella tularensis*. Vaccine 19:4465–4472
15. Ada G, Isaacs D (2003) Carbohydrate-protein conjugate vaccines. Clin Microbiol Infect 9:79–85
16. Landsteiner K (1945) The specificity of serological reactions. Harvard University Press, Cambridge
17. Avery OT, Goebel WF (1929) Chemo-immunological studies on conjugated carbohydrate-proteins. II. Immunological specificity of synthetic sugar-protein antigens. J Exp Med 50:533–550
18. Pollard AJ, Perrett KP, Beverley PC (2009) Maintaining protection against invasive bacteria with protein-polysaccharide conjugate vaccines. Nat Rev 9:213–220
19. Daum RS, Hogerman D, Rennels MB, Bewley K, Malinoski F, Rothstein E, Reisinger K, Block S, Keyserling H, Steinhoff M (1997) Infant immunization with pneumococcal CRM_{197} vaccines: effect of saccharide size on immunogenicity and interactions with simultaneously administrated vaccines. J Infect Dis 176:445–455
20. Lefeber DJ, Kamerling JP, Vliegenthart JFG (2001) Synthesis of Streptococcus pneumoniae type 3 neoglycoproteins varying in oligosaccharide chain length, loading and carrier protein. Chem Eur J 7:4411
21. Paoletti LC, Kasper DL, Michon F, DiFabio J, Jennings HJ, Tosteson TD, Wessels MR (1992) Effects of chain length on the immunogenicity in rabbits of group B Streptococcus type III oligosaccharide-tetanus toxoid conjugates. J Clin Invest 89:203

22. Chernyak A, Kondo S, Wade TK, Meeks MD, Alzari PM, Fournier JM, Taylor RK, Kováč P, Wade WF (2002) Induction of protective immunity by synthetic *Vibrio cholerae* hexasaccharide derived from *V. cholerae* O1 Ogawa lipopolysaccharide bound to a protein carrier. J Infect Dis 185:950–962
23. Dick WE Jr, Beurret M (1989) A survey and consideration of design and preparation factors. In: Cruse JM, Lewis RE Jr (eds) Glycoconjugates of bacterial carbohydrate antigens, vol 10. Krager, Basel, pp 48–114
24. Tietze LF, Arlt M, Beller M, Glüsenkamp KH, Jähde E, Rajewsky MF (1991) Anticancer agents, 15. Squaric acid diethyl ester: a new coupling reagent for the formation of drug biopolymer conjugates. Synthesis of squaric acid ester amides and diamides. Chem Ber 124:1215–1221
25. Glüsenkamp KH, Drosdziok W, Eberle G, Jähde E, Rajewsky MFZ (1991) Naturforsch C Biosci 46:498–501
26. Tietze LF, Schröter C, Gabius S, Brinck U, Goerlach-Graw A, Gabius HJ (1991) Conjugation of p-aminophenyl glycosides with squaric acid diesters to a carrier protein and the use of the neoglycoprotein in the histochemical detection of lectines. Bioconjug Chem 2:148–153
27. Cohen S, Cohen SG (1966) Preparation and reactions of derivatives of squaric acid. Alkoxy-, hydroxy-, and aminocyclobutenediones[1]. J Am Chem Soc 88:1533–1536
28. Grünefeld J, Bredhauer G, Zinner G (1985) Zur reaktion von quadratsäuredimethylester mit *N,N*-disubstituierten hydrazin-derivaten. Arch Pharm (Weinheim) 318:984–988
29. Bergh A, Magnusson BG, Ohlsson J, Wellmar U, Nilsson UJ (2001) Didecyl squarate – a practical amino-reactive cross-linking reagent for neoglycoconjugate synthesis. Glycoconj J 18:615–621
30. Kamath VP, Diedrich P, Hindsgaul O (1996) Use of diethyl squarate for the coupling of oligosaccharide amines to carrier proteins and characterization of the resulting neoglycoproteins by MALDI-TOF mass spectrometry. Glycoconj J 13:315–319
31. Hou S-J, Saksena R, Kováč P (2008) Preparation of glycoconjugates by dialkyl squarate chemistry revisited. Carbohydr Res 343:196–210
32. Saksena R, Adamo R, Kováč P (2007) Immunogens related to the synthetic tetrasaccharide side chain of the *Bacillus anthracis* exosporium. Bioorg Med Chem 15:4283–4310
33. Bongat AFG, Saksena R, Adamo R, Fujimoto Y, Shiokawa Z, Peterson DC, Fukase K, Vann WF, Kováč P (2010) Multimeric bivalent immunogens from recombinant tetanus toxin HC fragment, synthetic hexasaccharides and a glycopeptide adjuvant. Glycoconj J 27:69–77
34. Aebersold R, Mann M (2003) Mass spectrometry-based proteomics. Nature 422:198–207
35. Morelle W, Michalski JC (2005) Glycomics and mass spectrometry. Curr Pharm Des 11:2615–2645
36. Dettmer K, Aronov PA, Hammock BD (2007) Mass spectrometry-based metabolomics. Mass Spectrom Rev 26:51–78
37. Blanksby SJ, Mitchell TW (2010) Advances in mass spectrometry for lipidomics. Annu Rev Anal Chem 3:433–465
38. Banoub JH, Newton RP, Esmans E, Ewing DF, Mackenzie G (2005) Recent developments in mass spectrometry for the characterization of nucleosides, nucleotides, oligonucleotides, and nucleic acids. Chem Rev 105:1869–1915
39. Zhang Y, Go EP, Desaire H (2008) Maximizing coverage of glycosylation heterogeneity in MALDI-MS analysis of glycoproteins with up to 27 glycosylation sites. Anal Chem 80:3144–3158
40. Laštovičková M, Chmelik J, Bobalova J (2009) The combination of simple MALDI matrices for the improvement of intact glycoproteins and glycans analysis. Int J Mass Spectrom 281:82–88
41. Kamath VP, Diedrich P, Hindsgaul O (1996) Use of diethyl squarate for the coupling of oligosaccharide amines to carrier proteins and characterization of the resulting neoglycoproteins by MALDI-TOF mass spectrometry. Glycoconj J 13:315–319
42. Issaq HJ, Conrads TP, Prieto DA, Tirumalai R, Veenstra TD (2003) SELDI-TOF MS for diagnostic proteomics. Anal Chem 75:148A–155A

43. Liu C (2011) The application of SELDI-TOF-MS in clinical diagnosis of cancers. J Biomed Biotechnol 2011:6, Article ID 245821
44. Chernyak A, Karavanov A, Ogawa Y, Kováč P (2001) Conjugating oligosaccharides to proteins by squaric acid diester chemistry: rapid monitoring of the progress of conjugation, and recovery of the unused ligand. Carbohydr Res 330:479–486
45. Jahouh F, Saksena R, Aiello D, Napoli A, Sindona G, Kováč P, Banoub JH (2010) Glycation sites in neoglycoconjugates from the terminal monosaccharide antigen of the O-PS of *Vibrio cholerae* O1, serotype Ogawa, and BSA revealed by matrix-assisted laser desorption-ionization tandem mass spectrometry. J Mass Spectrom 10:1148–1159
46. Jahouh F, Saksena R, Kováč P, Banoub JH (2012) Revealing the glycation sites in synthetic neoglycoconjugates formed by conjugation of the antigenic monosaccharide hapten of *Vibrio cholerae* O1 serotype Ogawa with the BSA protein carrier using LC-ESI-QqTOF-MS/MS. J Mass Spectrom 47:890–900
47. Jahouh F, Hou SJ, Kováč P, Banoub JH (2011) Determination of the glycation sites of *Bacillus anthracis* neoglycoconjugate vaccine by MALDI-TOF/TOF-CID-MS/MS and LC-ESI-QqTOF-tandem mass spectrometry. J Mass Spectrom 46:993–1003
48. Jahouh F, Hou SJ, Kováč P, Banoub JH (2012) Determination of glycation sites by tandem mass spectrometry in a synthetic lactose-bovine serum albumin conjugate, a vaccine model prepared by dialkyl squarate chemistry. Rapid Commun Mass Spectrom 26:749–758
49. Jahouh F, Xu P, Vann WF, Kováč P, Banoub JH (2013) Mapping the glycation sites in the neoglycoconjugate from hexasaccharide antigen of *Vibrio cholerae*, serotype Ogawa and the recombinant tetanus toxin C-fragment carrier. J Mass Spectrom 48:1083–1090
50. McCarthy PC, Saksena R, Peterson DC, Lee CH, An Y, Cipollo JF, Vann WF (2013) Chemoenzymatic synthesis of immunogenic meningococcal group C polysialic acid-tetanus Hc fragment glycoconjugates. Glycoconj J 30:857–870
51. Roepstorff P, Fohlman J (1984) Proposal for a common nomenclature for sequence ions in mass spectra of peptides. Biol Mass Spectrom 11:601
52. Johnson RS, Martin SA, Biemann K, Stults JT, Watson JT (1987) Novel fragmentation process of peptides by collision-induced decomposition in a tandem mass spectrometer: differentiation of leucine and isoleucine. Anal Chem 59:2621–2625
53. Domon B, Costello C (1988) A systematic nomenclature for carbohydrate fragmentations in FAB-MS/MS spectra of glycoconjugates. Glycoconj J 5:397–409
54. Rietschel ET, Brade L, Lindner B, Zahringer U (1992) Biochemistry of lipopolysaccharides. In: Morrison DC, Ryan JL (eds) Bacterial endotoxic lipopolysaccharides, 1st edn. CRC Press, Boca Raton, p 3
55. Chatterjee S, Chaudhuri K (2003) Lipopolysaccharides of *Vibrio cholerae*. I. Physical and chemical characterization. Biochem Biophys Acta 1639:65–79
56. Kossaczka Z, Shiloach J, Johnson V, Taylor DN, Finkelstein RA, Robbins JB, Szu SC (2000) *Vibrio cholerae* O139 conjugate vaccines: synthesis and immunogenicity of *V. cholerae* O139 capsular polysaccharide conjugates with recombinant diphtheria toxin mutant in mice. Infect Immun 68:5037–5043
57. Xu P, Alam MM, Kalsy A, Charles RC, Calderwood SB, Qadri F, Ryan ET, Kováč P (2011) Simple, direct conjugation of bacterial O-SP-core antigens to proteins: development of cholera conjugate vaccines. Bioconjug Chem 22:2179–2185
58. Manning PA, Stroeher UH, Morona R (1994) In: Wachsmuth IK, Blake PA, Olsvik O (eds) Vibrio cholerae and cholera: molecular to global perspectives. American Society for Microbiology, Washington, DC, p 77
59. Dick WE Jr, Beurret M (1989) In: Cruse JM, Lewis RE Jr (eds) Conjugate vaccines, vol 10. Krager, Basel, p 48
60. McNaught AD (1997) International union of pure and applied chemistry and international union of biochemistry and molecular biology. Joint commission on biochemical nomenclature. Nomenclature of carbohydrates. Carbohydr Res 297:1–92

61. Kenne L, Lindberg B, Unger P, Gustafsson B, Holme T (1982) Structural studies of the *Vibrio cholerae* O-antigen. Carbohydr Res 100:341–349
62. Hisatsune K, Kondo S, Isshiki Y, Igushi T, Haishima Y (1993) Occurrence of 2-*O*Methyl-*N*-(3-Deoxy-*L*-*glycero*-tetronyl)-*D*-perosamine (4-amino-4,6-dideoxy-*D*-*manno*pyranose) in lipopolysaccharide from Ogawa but not from Inaba O forms of O1 *Vibrio cholerae*. Biochem Biophys Res Commun 190:302–307
63. Isshiki Y, Kondo S, Haishima Y, Iguchi T, Hisatsune K (1996) Identification of N-3-hydroxypropionyl-2-O-methyl-D-perosamine as a specific constituent of the lipopolysaccharide from *Vibrio* bio-serogroup 1875 which has Ogawa antigen factor B of *Vibrio cholerae* O1. J Endotoxin Res 3:143–149
64. Saksena R, Chernyak A, Karavanov A, Kovác P (2003) Conjugating low molecular mass carbohydrates to proteins. 1. Monitoring the progress of conjugation. Methods Enzymol 362:125–139
65. Arnold K, Bordoli L, Kopp J, Schwede T (2006) The SWISS-MODEL workspace: a web-based environment for protein structure homology modelling. Bioinformatics 22:195–201
66. Peitsch MC (1995) Protein modeling by E-mail. Biotechnology 13:658–660
67. Mock M, Fouet A (2001) Anthrax. Annu Rev Microbiol 55:647–671
68. Pries FG (1993) In: Sonenshein AL, Hoch JA, Losick R (eds) Bacillus subtilis and other gram-positive bacteria: biochemistry, physiology, and molecular biology. American Society for Microbiology, Washington, DC, p 3
69. Nicholson WL, Munakata N, Horneck G, Melosh HJ, Setlow P (2000) Resistance of *Bacillus* endospores to extreme terrestrial and extraterrestrial environments. Microbiol Mol Biol Rev 64:548–572
70. Boutiba-Ben Boubaker I, Ben Redjeb S (2001) *Bacillus anthracis*: causative agent of anthrax. Tunis Med 79:642–646
71. Read TD, Salzberg SL, Pop M, Shumway M, Umayam L, Jiang L, Holtzapple E, Busch JD, Smith KL, Schupp JM, Solomon D, Keim P, Fraser CM (2002) Comparative genome sequencing for discovery of novel polymorphisms in *Bacillus anthracis*. Science 296:2028–2033
72. Turnbull PCB (1999) Definitive identification of *Bacillus anthracis*-a review. J Appl Microbiol 87:237–240
73. Reed LJ, Muench H (1938) A simple method for estimating fifty percent endpoints. Am J Hyg 27:493–497
74. Hoffmaster AR, Fitzgerald CC, Ribot E, Mayer LW, Popovic T (2002) Molecular subtyping of *Bacillus anthracis* and the 2001 bioterrorism-associated anthrax outbreak, United States. Emerg Infect Dis 8:1111–1116
75. Chun J-H, Hong K-J, Cha SH, Cho M-H, Lee KJ, Jeong DH, Yoo C-K, Rhie G-e (2012) Complete genome sequence of *Bacillus anthracis* H9401, an isolate from a Korean patient with anthrax. J Bacteriol 194:4116–4117
76. Williams DD, Benedek O, Turnbough CL Jr (2003) Species-specific peptide ligands for the detection of *Bacillus anthracis* spores. Appl Environ Microbiol 69:6288–6293
77. Chabot DJ, Scorpio A, Tobery SA, Little SF, Norris SL, Friedlander AM (2004) Anthrax capsule vaccine protects against experimental infection. Vaccine 23:43–47
78. Daubenspeck JM, Zeng H, Chen P, Dong S, Steichen CT, Krishna NR, Pritchard DG Jr, Turnbough CL (2004) Novel oligosaccharide side chains of the collagen-like region of BclA, the major glycoprotein of the *Bacillus anthracis* exosporium. J Biol Chem 279:30945–30953
79. Burkitt WI, Giannakopulos AE, Sideridou F, Bashir S, Derrick PJ (2003) Discrimination effects in MALDI-MS of mixtures of peptides-analysis of the proteome. Aust J Chem 56:369–377
80. Kratzer R, Eckerskorn C, Karas M, Lottspeich F (1998) Suppression effects in enzymatic peptide ladder sequencing using ultraviolet – matrix assisted laser desorption/ionization – mass spectrometry. Electrophoresis 19:1910–1919

81. Gao GF, Jakobsen BK (2000) Molecular interactions of coreceptor CD+8 and MHC class I: the molecular basis for functional coordination with the T-cell receptor. Immunol Today 21:630
82. Roy R (2004) New trends in carbohydrate-based vaccines. Drug Discov Today Technol 1:327
83. Roy R, Shiao TC, Rittenhouse-Olson K (2013) Glycodendrimers: versatile tools for nanotechnology. Braz J Pharm Sci 49:85
84. Roy R, Shiao TC (2012) Glycodendrimers as functional antigens and antitumorales vaccines. New J Chem 36:324
85. Roy R, Shiao TC (2011) Organic chemistry and immunochemical strategies in the design of potent carbohydrate-based vaccines. Chimia 65:24
86. Icart LP, Fernandez-Santana V, Veloso RC, Carmenate T, Sirois S, Roy R, Verez Bencomo V (2007) T-cell immunity of carbohydrates. In: Roy R (ed) Carbohydrate-based vaccines. ACS Symposium Series, 989, p 1
87. Daniels MA, Jameson SC (2000) Critical role for CD+8 in T cell receptor binding and activation by peptide/major histocompatibility complex multimers. J Exp Med 191:335
88. Kim YS, Gum J, Brockhausen I (1996) Mucin glycoproteins in neoplasia. Glycoconj J 13:693
89. Kim YJ, Varki A (1997) Perspectives on the significance of altered glycosylation of glycoproteins in cancer. Glycoconj J 14:569
90. Ono M, Hakomori S (2004) Glycosylation defining cancer cell motility and invasiveness. Glycoconj J 20:71
91. Springer GF (1997) Immunoreactive T and Tn epitopes in cancer diagnosis, prognosis, and immunotherapy. J Mol Med 75:594
92. Campbell BJ, Finnie IA, Hounsell EF, Rhodes JM (1995) Direct demonstration of increased expression of Thomsen-Friedenreich (TF) antigen in colonic adenocarcinoma and ulcerative colitis mucin and its concealment in normal mucin. J Clin Invest 95:571
93. Springer GF (1984) T and Tn, general carcinoma autoantigens. Science 224:1198
94. Dippold W, Steinborn A, Büschenfelde KHM (1990) The role of the Thomsen-Friedenreich antigen as a tumor-associated molecule. Environ Health Perspect 88:255
95. Baek M-G, Roy R (2002) Glycodendrimers: novel glycotope isosteres unmasking sugar coding. Case study with T-antigen markers from breast cancer MUC1 glycoprotein. Rev Mol Biotechnol 90:291
96. Dubois M, Gilles KA, Hamilton JK, Rebers PA, Smith F (1956) Colorimetric method for determination of sugars and related substances. Anal Chem 28:350
97. Kartazer R, Eckerskorn C, Karas M, Lottspeich F (1998) Suppression effects in enzymatic peptide ladder sequencing using ultraviolet – matrix assisted laser desorption/ionization- mass spectrometry. Electrophoresis 19:1910

Chapter 17
Nano-structured Solids and Heterogeneous Catalysts for the Selective Decontamination of Chemical Warfare Agents

Matteo Guidotti, Claudio Evangelisti, Alessandra Rossodivita, and Massimo C. Ranghieri

Abstract The destruction of chemical hazardous agents can be required on the field, for decontamination after an accidental or deliberate release, as well as in laboratories, pilot plants and chemical agent destruction sites, for abatement of stockpiled chemical weapons. Nanostructured inorganic metal oxides and/or metal particles, in all forms and formulations, constitute a large class of materials that are suitable for such purposes. They are robust, rich in specific surface sorption sites, active in the degradation of hazardous compounds via catalytic or photocatalytic mechanisms and, in most cases, relatively cheap. Such nanosystems show promising performances in terms of activity and selectivity, even at very low catalyst to toxic agent ratios. It is thus possible to move from conventional stoichiometric destruction to catalytic chemical decontamination of hazardous compounds. However, the recent ever-increasing concerns about the consequences on human health and environment of nanosized inorganic systems must induce a careful investigation about their toxicological and pathogenic impact on living organisms.

Keywords Heterogeneous catalysis • Inorganic oxides • Microporous and mesoporous materials • Decontamination • Absorption • Chemical warfare agents

---

M. Guidotti, Ph.D. (✉) • C. Evangelisti, Ph.D.
CNR – Institute of Molecular Sciences and Technologies, Via C. Golgi 19, Milan, Italy
e-mail: m.guidotti@istm.cnr.it

A. Rossodivita, M.D.
"San Raffaele" Hospital and University, Via Olgettina 60, Milan, Italy

M.C. Ranghieri
1st Field Unit, ACISMOM Military Corps – Auxiliary of the Italian Army,
Via Saint Bon 7, Milan, Italy

J. Banoub (ed.), *Detection of Chemical, Biological, Radiological and Nuclear Agents for the Prevention of Terrorism*, NATO Science for Peace and Security Series A: Chemistry and Biology, DOI 10.1007/978-94-017-9238-7_17

17.1 Introduction

Decontamination techniques, aimed at eliminating the hazard of extremely toxic chemical warfare agents (CWA), started to be developed since the first cases of use of non-conventional weapons on European battlefields during World War I [1]. Since the beginning to the end of the Cold War in the mid-twentieth century, most of the research activity was focused on military-oriented on-field conditions. These crucial chemical and/or physical method activities depended on the speed and ease of application of the decontaminant for the rapid removal of the contaminant chemical agents from vehicles, equipment, personnel, weapons and facilities [2].

Later, after the release of the nerve agent *sarin* in the Tokyo subway system in 1995 by the fanatic organization Aum Shinrikyo, it appeared that terrorist organizations and sub-national groups do have the capabilities to produce or purchase toxic industrial chemicals and CWA to threaten the civilian population. Accordingly, efficient countermeasures for the protection and decontamination from threats became then a primordial priority for first responders (policemen, rescuers, fire-fighters, pre-hospital personnel) and for all workers active in non-conventional emergencies and threatening situations [3].

In addition to these considerations, the chemical research in the field of the abatement and decontamination witnessed a noteworthy improvement after 1993, when the Chemical Weapons Convention [4] came into force and the related Organization for the Prohibition of Chemical Weapons started its supranational control activity [5]. Such Convention prohibits not only the use of CWA (as the earlier Geneva Protocol, signed by most countries in 1925), but also the production, detention and stockpiling of highly toxic warfare agents. It became obligatory to impose the destruction and detoxification worldwide of the huge amounts of CWA produced since the early twentieth century (mainly in the Cold War times) and still stored by several Western and Eastern Countries.

Currently, the present unstable international situation (as the recent news about the use of CWA in Syrian civil war [6]) and the high level of concern due to the possible use of non-conventional weapons by clandestine terrorist groups requires the exploitation of any new technology for a rapid, efficient and safe in-situ decontamination of people and objects from highly toxic compounds, to save lives and to minimize damages.

The combination of these factors prompted the academic, civilian, military and private research institutions to address their interests towards emerging disciplines, such as nanosciences and nanotechnologies, to propose solutions and remedies to these sources of threat worldwide.

17.2 From Conventional Decontamination to Innovative Nanostructured Systems

Decontamination is required not only on the field, but also in laboratories, pilot plants and chemical agent storage and destruction sites. Conventional decontamination methods rely on the dissolution, adsorption, thermal decomposition or stoichiometric chemical degradation (oxidation, hydrolysis, combustion, etc.) of the toxic agents [7]. All these methods need huge amounts of reactants and/or energy and this poses several problems in terms of safety, environmental and economic sustainability, costs and disposal of the detoxified by-products [8]. In fact, conventional decontamination and abatement techniques rely on the use of super-stoichiometric amounts of strong oxidant agents (mainly containing active chlorine, such as hypochlorite chloramine-B or sodium dichloroisocyanurate, or peroxides, such as H_2O_2 or peroxoacids [9]) together with an aggressive alkaline action (by alkali-metal hydroxides or amines). CWA are indeed unstable under highly oxidizing conditions (in particular, in the presence of sources of active oxygen and/or free radicals) and at very high pH values (typically higher than 12) [10]. For these reasons, the attention of both open and patent literature moved, in recent years, from stoichiometric to catalytic chemical decontamination, based on reactive sorbent materials, which are able to convert highly toxic species into non-toxic (or much less toxic) secondary products under mild conditions (or even ambient conditions, in the best examples).

Furthermore, thanks to the recent advances in nanosciences and nanosized materials, nanostructured inorganic compounds (such as nanosized inorganic oxides or metal nanoparticles) are particularly suitable for these applications, as they show very high specific surface area, a large number of highly reactive and easily accessible active sites and high reactivity, because of their reduced size in the nanometer scale (1–100 nm) [11]. Nanosystems display electronic, photochemical, electrochemical, optical, magnetic, mechanical or catalytic properties that differ significantly not only from those of molecular units, but also from those of macroscopic systems. Thanks to these peculiar features, it is also possible to use nanostructured materials as catalysts for the chemical decontamination with enhanced capabilities. In addition, the use of reduced amounts of detoxifying catalytic materials is particularly advisable in on-field decontamination of large objects and terrain surface, where a total-loss use of the powder cannot be avoided [7, 12].

However, in many cases, the use of decontamination techniques based on nanostructured systems is not necessarily the best solution, fully free from drawbacks. In some research papers, the relevant cost and/or the technical difficulty to obtain the nanosized catalytically-active species has been underestimated and this is a main weakness when a large-scale use and production is foreseen. In other cases, noble metal-based catalytic centers (*e.g.*, Au or Pd nanoparticles; *v. infra*) are needed to carry out the catalytic transformation and degradation of the toxic agent. In other papers, the lack of a multidisciplinary research approach led to partial results that cannot find ready practical applications [13]. In addition, the recent ever-increasing concerns about the consequences on human health of nanosized inorganic systems [14] must induce the research group dealing with extensive use of nanosystems to evaluate carefully their toxicological and pathogenic impact on human body and the environment [15, 16]. The potential exploitation of nanosystems as methods for depollution and decontamination from hazardous and toxic agents must thus be carefully evaluated and the possible drawbacks balanced [17].

From the analysis of these advantages and critical issues in the existing open and patent literature, it appears evident that a reliable active catalytic system for the decontamination of CWA should be:

1. cheap, for a economically-sustainable production scale-up, even at large scale;
2. active, as the catalytic detoxification action has to be as immediate as possible;
3. selective, in order to avoid the side-production of degradation products which may be even more hazardous than the initial aggressive agent;
4. versatile, to be applicable against a very wide range of potential aggressives;
5. safe and non-toxic, towards personnel and the environment;
6. able to work under mild conditions, as close as possible to ambient conditions; and
7. robust, in terms of stability under use conditions in all the practical on-field scenarios.

17.3 New Strategies to Decontamination: Some Relevant Examples

Numerous examples dealing with the exploitation of metal nanoparticles, inorganic metal oxides, porous materials and layered solids and combinations of these components have been reported so far as catalysts for the efficient decontamination and/or abatement of CWA. In the following paragraphs, a selection of some relevant examples is reported, paying particular attention to the catalytic activity, selectivity and sustainability of the proposed system.

17.3.1 Nanosized Metal Particles for Decontamination of CWA

Metal nanoparticles are typically present in supported metal catalysts, where metal particles having size usually in the range 1–20 nm are deposited on the external surface and/or in the porous texture of high-surface-area solids. In particular, those of industrial importance are typically noble metal (*e.g.*, Pt for mineral oil reforming, Ru for ammonia synthesis, Pd for hydrogenation of organic compounds, Pt and Pd for the abatement of volatile organic compounds, VOC, etc.) supported on active carbons and on high surface area oxides (mainly alumina, silica, silica-alumina, zeolites and molecular sieves with high specific surface areas) [18].

Supported metal nanoparticles potentially represent a new class of nanostructured materials for decontamination or abatement of CWA. They can be easily incorporated

Scheme 17.1 Chemical Warfare Agents (CWA)

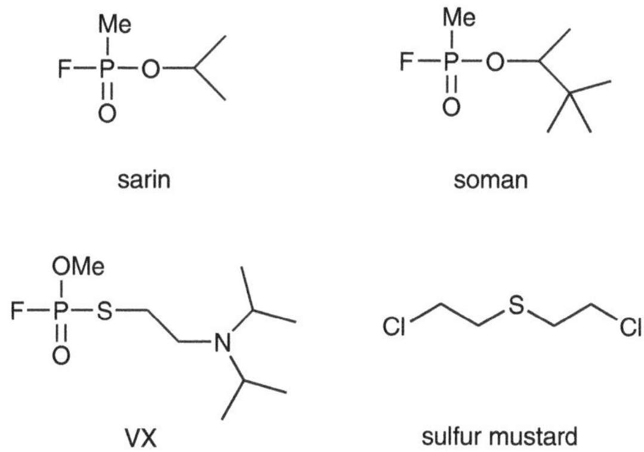

into textiles, air filtration systems or sorbents for contamination. These systems, thanks to different physical and electronic properties compared to their bulk metal counterpart, offer the possibility to exhibit new surface reactivity. The major obstacle to their use for a complete decontamination and/or abatement of CWA is that the metal surface must be able to cleave efficiently P-C, P-O and S-C bonds (Scheme 17.1), avoiding the formation of by-products which cannot easily removed and which can eventually poison the catalytically active metal surface [19]. In fact, compounds containing phosphorus and sulfur atoms are typical side-products that coordinate and bind irreversibly to the catalytically active sites of the nanoparticles, causing a gradual deactivation of the metal catalyst. Chen et al. reported that small Ni nanoparticles (with main diameters of 5.0 nm) supported on TiO_2 show higher efficiency than larger particles (8.8 nm) or annealed Ni films in the decomposition of dimethylmethylphosphonate, DMMP (a widely used simulant for organophosphorus nerve agents) into CO and H_2 at room temperature [20]. However, the reaction left high amounts of carbon and phosphorus adsorbed on the metal catalyst surface. This led to catalyst passivation and hindered any further reaction, making such system not useful for a real application.

Although a large number of studies involving catalytically active metal nanoparticles have been reported for the degradation of CWA, the most promising approaches require photochemistry. Recently, Au nanoparticles supported on TiO_2 have been reported to be able to perform degradation of a wide range of organic compounds, including CWA, under visible light at room temperature. Since its discovery, the photocatalytic properties of titania (mainly as anatase phase) under UV light have been widely studied as a new promising technology for the degradation of organic compounds, including highly toxic chemicals. However, the major limitation of titania-base materials is their low efficiency under common visible light irradiation, due to the wide absorption of the active anatase phase in the UV region, centered at 370 nm (with a band gap of 3.2 eV). On the other hand, when Au metal particles fall in nanometer size range (3–20 nm) a broad absorption, generally centered at 520–650 nm, denoted as surface plasmon band deriving from the collective excitation of electrons confined in the metal nanoparticle, is observed. Recent reports have shown that titania materials containing small gold nanoparticles (with mean diameters <10 nm) possess a visible light photocatalytic activity arising from photon absorption by the gold surface plasmon band and subsequent electron injection into the valence bond of titania [21]. This cooperation between titania support and gold nanoparticles, which act as light harvester and catalytically active sites, respectively, renders such material able to decompose a wide range of organic compounds including CWA [22–25]. Garcia et al. showed that Au/TiO_2 systems containing gold nanoparticles ranging 2–5 nm in size, having different metal loadings (from 0.4 to 1.5 wt.%) are able to decompose *soman*, VX and sulfur mustard compounds under visible light at room temperature [26]. In particular, 20 mg of titania (P25, Degussa) containing 0.7 wt.% of Au is able to decompose 100 μL of a dichloromethane solution containing 0.77 wt.% of CWA in 2 h under visible light irradiation. In all the cases, at the end of the reaction time, only harmless products are formed and only minor amounts of strongly absorbed and unrecoverable organic materials were still present on the solid. Unfortunately, no details about the reusability of the material were reported, even though this kind of solid is likely intended for a total-loss use. However, no change in the Au nanoparticle size distribution after the photocatalytic experiments was observed by high-resolution transmission electron microscopy, HR-TEM. The authors confirmed the crucial role of visible light performing comparative experiments under UV light, obtaining lower activities. Moreover, preliminary experiments performed using Au/TiO_2 under dark conditions or using TiO_2 alone under visible light, did not cause any decomposition of the organic compounds.

The same research group reported in a different paper the synthesis of mesoporous titania materials containing Au nanoparticles (with loadings in the range 0.4–0.7 wt.%) and their use as efficient photocatalysts for the visible light-driven decontamination of the CWA *soman* [27].

It is worth highlighting that over these catalysts the catalytic tests were performed on genuine CWA samples, rather than on simulant agents. In fact, simulant compounds often show a comparable, but not fully identical, reactivity with respect to real agents and these differences cause, in several cases, problems in the effective on-field use of the decontamination device.

17.3.2 Nanostructured Inorganic Oxides for the Decontamination of CWA

Inorganic oxides (binary compounds of metals in high oxidation state with oxygen) are typically shaped, at nanometric scale, in two forms: nanopowders and nanoporous materials. Nanopowders are generally defined as powders having an average particle size of less than 100 nm. The specific advantages of nanopowders include superior phase homogeneity and lower densification temperature. Beyond achieving nanoscaled particles, the control of other powder characteristics such as particle distribution, phase purity and morphology are equally important in obtaining promising performance. Conversely, nanoporous materials belong to a subgroup of porous materials displaying voids ranging from 1 to 100 nm. The materials can be crystalline (as zeolites), non-crystalline (as most of the siliceous mesoporous materials) or completely non-ordered (with pores without any a translational repetition along the three space dimensions).

In the field of CWA decontamination, abatement or destruction, both nanopowders and nanoporous materials are promising candidates, as they show very large specific surface areas (from 500 up to 1,500 $m^2 g^{-1}$ and over) and a huge number of defective sites with enhanced adsorption and catalytic properties [28]. Oxides from metals such as Zn, Ti, Fe, Mn, Mg, Al, Zr, Cu, thus with a oxidizing and/or an (Lewis or Brønsted) acid character, are typically the most interesting solids.

Prasad showed that it is possible to decontaminate (2-chloroethyl)phenylsulfide (CEPS, simulant of sulfur mustard) with total (100 %) abatement and good selectivity to non-noxious by-products, using mixed metal oxide nanocrystals of Al_2O_3-Fe2O3, Al_2O_3-V_2O_5 or Al_2O_3-CuO obtained by aerogel synthesis process [29]. Analogously, zinc titanate ceramic nanofibers, in particular, with a ZnO-TiO_2 40–60 % composition, were proposed as active filters in face masks or protective clothing and have displayed a good detoxification activity in the presence of paraoxon (as nerve CWA simulant) and of (2-chloroethyl)ethylsulfide (CEES; another sulfur mustard simulant) [30] (Scheme 17.2).

A relevant number of research teams focused their attention on titania-based reactive sorbent powders or fibers, in order to exploit a photocatalytically-driven removal of the CWA [31, 32]. In these examples, no metal nanoparticles were necessary to carry out the reaction (as shown in the previous section), but the decontamination properties were enhanced by adding additional catalytic sites. For instance, some strong acid centers can be introduced on TiO_2 by sulfation treatment with H_2SO_4 and sulfated titania photocatalyst proved to be 50 % more effective in adsorption capacity and 20 % more active in CEES conversion than the unmodified pure TiO_2 [33].

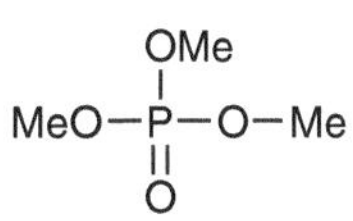

dimethylmethylphosphonate, DMMP

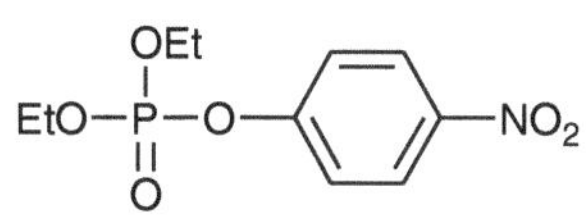

paraoxon

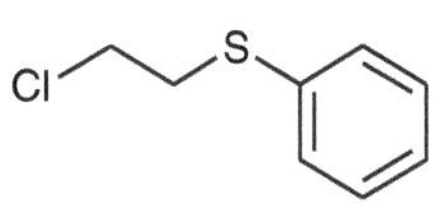

2-chloro ethyl phenyl sulfide, CEPS

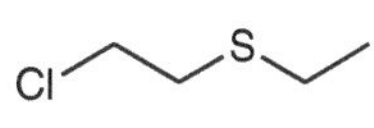

2-chloroethylethylsulfide, CEES

Scheme 17.2 Some common simulants of CWA used in catalytic tests

Interestingly, the use of solid catalysts under heterogeneous conditions permits the peculiar combination of active species that typically cannot be compatible with one another. Solid materials can, for instance, accommodate and stabilize the simultaneous presence of acid and basic sites together on the same support. Nanosized MgO can, in fact, show a high local concentration of Lewis acid (Mg^{2+} centers in the lattice of MgO) and Lewis basic (O^{2-} centers) in close proximity, without any mutual neutralization. This feature was exploited in the highly efficient adsorption and degradation of paraoxon over a polymer-supported MgO nanocrystals [34].

In addition, nanoporous materials with a high specific surface area is an optimal locus to insert or deposit catalytically active sites with redox properties (in particular, with oxidizing capabilities). A vanadium-doped mesoporous silica material, V-MCM-41, was particularly active in the catalytic oxidation with O_2 of the mustard gas simulant CEES, even if in the presence of a sacrificial aldehyde [35]. Such catalyst was, under controlled conditions and reaction times, rather selective to the desired non-noxious by-product sulfoxide, instead of the undesired sulfone.

This last example shows that aerobic oxidations, under mild conditions, based on easily accessible oxidants, such as molecular oxygen or air, constitute a promising, yet challenging, approach to the safe, efficient and cheap decontamination of CWA.

17.4 Conclusions

Thanks to the recent advances and efforts in the design, development and large-scale preparation of innovative, efficient and selective catalytic materials for the adsorption, decontamination, abatement and detoxification of CWA, a broad series of new tools can be soon available to national decision-makers who are active, at governmental level, in the field of abatement of toxic chemicals either stockpiled in obsolete arsenals or potentially employed in the clandestine production of CWA.

The new decontamination devices can play a relevant role in improving the global security in terms of:

1. reduced risks of illegal uses (better prevention),
2. reduced vulnerability for the civilian population (better protection), and,
3. improved risk management (minimization of negative consequences).

Acknowledgements M.G. and M.C.R. gratefully thank Prof. J. Banoub, Prof. E. M. Essassi and Prof. G. Giorgi for the kind invitation to the NATO-ASI held in Pontignano, Italy. M.G. and C.E. also thank NATO through the Science for Peace and Security Project "Nanostructured Materials for the Catalytic Abatement of Chemical Warfare Agents – NanoContraChem" (SFP.984481) and the 7th FP EU Project "NANoREG, Regulatory testing of nanomaterials" (NMP.2012.1.3-3) for financial support.

References

1. Nwanna GI (2004) Weapons of mass destruction. Library of Congress, Washington, DC
2. NATO, North Atlantic Treaty Organization (2003) AJP-3.8 – Doctrine for the NBC Defence of NATO Forces
3. Koenig KL, Boatright CJ, Hancock JA, Denny FJ, Teeter DS, Kahn CA, Schultz CH (2008) Health care facility-based decontamination of victims exposed to chemical, biological, and radiological materials. Am J Emerg Med 26:71
4. Technical Secretariat of the Organisation for the Prohibition of Chemical Weapons, convention on the prohibition of the development, production, stockpiling and use of chemical weapons and on their destruction, C.N.246.1994.TREATIES-5, issued on 31 August 1994, and following additions
5. http://www.opcw.org. Accessed 20 Aug 2014
6. Organisation for the prohibition of chemical weapons, OPCW Press release 12/2013, The Hague, 16 Sept 2013
7. Kim K, Tsay OG, Atwood DA, Churchill DG (2011) Destruction and detection of chemical warfare agents. Chem Rev 111:5345
8. Guidotti M, Rossodivita A, Ranghieri MC (2012) Nano-structured solids and heterogeneous catalysts: powerful tools for the reduction of CBRN threats. In: Vaseashta A, Braman E, Susmann P (eds) Technological innovations in detection and sensing of CBRN agents and ecological terrorism, NATO SPS series – A. Springer, Amsterdam, pp 89–97
9. Netherlands Organization for Applied Scientific Research, present state of CBRN decontamination methodologies, TNO Report TNO-DV, 2007, A028, p 10
10. Swedish National Defense Research Establishment (1992) A FOA briefing book on chemical weapons, Stockholm
11. Psaro R, Guidotti M, Sgobba M (2008) Nanosystems. In: Bertini I (ed) Inorganic and bio-inorganic chemistry vol. II, part of Encyclopedia of Life Support Systems (EOLSS). EOLSS Publishers Co. Ltd., Oxford, pp 256–307
12. Reynolds JG, Hart B (2004) Nanomaterials and their application to defense and homeland security. J Mater 56:36
13. Aas P (2003) The threat of mid-spectrum chemical warfare agents. Prehosp Disaster Med 18:306
14. Gil PR, Oberdoester G, Elder A, Puntes V, Parak WJ (2010) Correlating physico-chemical with toxicological properties of nanoparticles: the present and the future. ACS Nano 4:5527
15. Gatti AM, Montanari S (2009) Nanocontamination of the soldiers in a battle space. In: Linkov I, Steevens J (eds) Nanomaterials: risks and benefits. Springer, Dordrecht, p 83
16. Guidotti M, Ranghieri M, Rossodivita A (2010) Nanosystems and CBRN threats: a resource worth exploiting, a potential worth controlling. In: Trufanov A, Rossodivita A, Guidotti M (eds) Pandemics and bioterrorism, vol 62, NATO SPS series – E. IOS-Press, Amsterdam, p 117
17. Galluzzi L, Chiarantini L, Pantucci E, Curci R, Merikhi J, Hummel H, Bachmann PK, Manuali E, Pezzotti G, Magnani M (2012) Development of a multilevel approach for the evaluation of nanomaterials toxicity. Nanomedicine London 7(3):393
18. Feldheim DL, Foss CA (2001) Metal nanoparticles: synthesis, characterization, and applications. Marcel Dekker, New York
19. Guo X, Yoshinobu J, Yates JT (1990) Decomposition of an organophosphonate compound (Dimethyl methyl phosphonate) on the Ni(111) and Pd(111) surfaces. J Phys Chem 94:6839
20. Zhou J, Ma S, Kang YC, Chen DA (2004) Dimethyl methylphosphonate decomposition on titania-supported Ni clusters and films: a comparison of chemical activity on different Ni surfaces. J Phys Chem 108:11633
21. Tian Y, Tatsuma T (2005) Mechanisms and applications of plasmon-induced charge separation at TiO_2 films loaded with gold nanoparticles. J Am Chem Soc 127:7632
22. Primo A, Corma A, Garcia H (2011) Titania supported gold nanoparticles as photocatalyst. Phys Chem Chem Phys 13(3):886

23. Kowalska E, Mahaney OOP, Abe R, Ohtani B (2010) Visible-light-induced photocatalysis through surface plasmon excitation of gold on titania surfaces. Phys Chem Chem Phys 12(10):2344
24. Arabatzis IM, Stergiopoulos T, Andreeva D, Kitova S, Neophytides SG, Falaras P (2003) Characterization and photocatalytic activity of Au/TiO_2 thin films for azo-dye degradation. J Catal 220:127
25. Uddin MJ, Cesano F, Scarano D, Bonino F, Agostini G, Spoto G, Bordiga S, Zecchina A (2008) Cotton textile fibres coated by Au/TiO_2 films: synthesis, characterization and self cleaning properties. J Photochem Photobiol A 199:64
26. Neatu S, Cojocaru B, Parvulescu VI, Somoghi V, Alvaro M, Garcia H (2010) Visible-light C–heteroatom bond cleavage and detoxification of chemical warfare agents using titania-supported gold nanoparticles as photocatalyst. J Mater Chem 20:4050
27. Alvaro MA, Cojocaru B, Ismail AA, Petrea N, Ferrer B, Harras FA, Parvulescu VI, Garcia H (2010) Visible-light photocatalytic activity of gold nanoparticles supported on template-synthesized mesoporous titania for the decontamination of the chemical warfare agent Soman. Appl Catal B Environ 99:191
28. Ozin GA, Arsenault A (2005) Nanochemistry a chemical approach to nanomaterials. RSC Publishing, Cambridge
29. Prasad GK (2010) Decontamination of 2-chloro ethyl phenyl sulphide using mixed metal oxide nanocrystals. J Sci Ind Res 69:835
30. Ramaseshan R, Ramakrishna S (2007) Zinc titanate nanofibers for the detoxification of chemical warfare simulants. J Am Ceram Soc 90(6):1836
31. Panayotov DA, Morris JR (2009) Uptake of a chemical warfare agent simulant (DMMP) on TiO_2: reactive adsorption and active site poisoning. Langmuir 25:3652
32. Grandcolas M, Louvet A, Keller N, Keller V (2009) Layer-by-layer deposited titanate-based nanotubes for solar photocatalytic removal of chemical warfare agents from textiles. Angew Chem Int Ed 48:161
33. Han ST, Zhang GY, Xi HL, Xu DN, Fu XZ, Wang XX (2008) Sulfated TiO_2 Decontaminate 2-CEES and DMMP in vapor phase. Catal Lett 122:106
34. Sundarrajan S, Ramakrishna S (2007) Fabrication of nanocomposite membranes from nanofibers and nanoparticles for protection against chemical warfare stimulants. J Mater Sci 42:8400
35. Livingston SR, Landry CC (2008) Oxidation of a mustard gas analogue using an aldehyde/O_2 system catalyzed by V-doped mesoporous silica. J Am Chem Soc 130:13214

Chapter 18
Strategic Missile Forces in Ukraine: Brief Survey of Past and Present Environmental Problems

Igor Winkler

Abstract Various environmental problems related to the former strategic missile forces located in Ukraine are outlined and analyzed. As the combat units and equipment have been disarmed and mostly utilized according to international agreements, various auxiliary units, equipment, parts, stockpiles of the rocket fuel and oxidizer remained in Ukraine outside the agreements. Unavailability of the utilization technologies caused various environmental problems linked with the outdated materials leakage, emission and other types the environmental contamination.

General survey of the current problems related to this issue is given and some mitigation steps, already taken and prospective are analyzed.

Keywords Environmental hazard • Rocket fuel • Rocket oxidizer • Strategic missile forces

18.1 Introduction

A powerful group of the strategic missile force has been inherited by Ukraine after the collapse of the USSR. It consisted of one full strategic 43rd missile army located mainly in the Western-Southern Ukraine with some divisions located in Belarus and Russia.

It is well known that activities related to strategic and conventional missile forces development and their service and auxiliary units are quite unsafe for the environment [1–3].

The following environmental effects should be taken into consideration in the analysis of this influence:

I. Winkler (✉)
Department of Analytical Chemistry, Yu. Fedkovych National University of Chernivtsi, Chernivtsi, Ukraine
e-mail: igorw@ukrpost.ua

J. Banoub (ed.), *Detection of Chemical, Biological, Radiological and Nuclear Agents for the Prevention of Terrorism*, NATO Science for Peace and Security Series A: Chemistry and Biology, DOI 10.1007/978-94-017-9238-7_18

- allotment of land resources required for the missile and auxiliary units allocation;
- possible radioactive contamination in case of nuclear missile units development;
- unavoidable contamination of soil, groundwater and open water bodies with rocket fuels and oxidizers emergency spills and regular technological leakages.

Strategic missile units require rather significant land resources for their exclusive use. Such land areas are usually located not far from the populated area, as a rule this land should be of good quality and potentially available for the regular agricultural or other civil use. Before the missile unit is located within some area, it undergoes intense engineering and earthworks. Numerous ground and especially underground facilities should be constructed, various communication tunnels and storage installations should be built, and all these activities require thorough drainage network to keep them in the working conditions and to pump out the groundwater. As a result, regular profile of the groundwater undergoes serious deformation. Usually the groundwater level inside the missile unit area goes deeper while in the neighbourhood it can approach the surface and cause some swamping. Naturally, these changes disturb normal life conditions of the local plants and animals. Reclamation of the lands used previously for the strategic missile units is quite complex and costly. In reality, such areas often become abandoned and unsuitable for any civil or agricultural use.

Nuclear warheads of the missiles are well protected and usually cannot cause any serious leakage of radioactivity unless they get very seriously damaged but the technological processes of the warheads disassembling and utilization produce significant amounts of highly toxic and radioactive wastes requiring complex, long, and expensive storage involving potential environmental danger of serious contamination. Since no nuclear warheads remain in Ukraine and there is no nuclear utilization/disarmament facilities working in the country, this effect will be out of further consideration.

Many military missiles use the liquid fuel jet engines. Many kinds of rocket fuels are composed using highly toxic heptile (Fig. 18.1) as a fuel and concentrated mixtures of nitric and some other acids with nitrogen oxides and sometimes hydrogen peroxide as an oxidizer.

Heptile is a volatile, a highly toxic compound with a strong teratogenic and carcinogenic effect [4, 5]. It is decomposes slowly in the environment and it gets cumulated along the trophic chains [5]. Products of biodegradation of heptile are also highly toxic and capable of intensifying its toxic effect. Since heptile is easily dissoluble in water, any leakage causes contamination of the groundwater and other

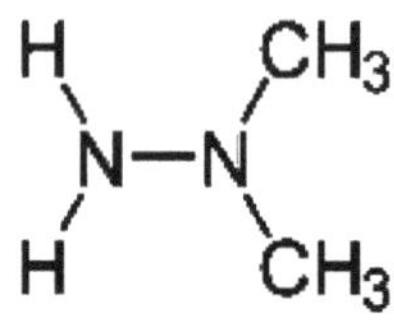

Fig. 18.1 Heptile – basic component of many types of rocket fuel

water sources very soon. Therefore, this compound can be classified as an agent of very high environmental hazard.

Several 'heptile' accidents happened during regular active operations of the 43rd strategic missile army. For instance, a critically defective missile had been disarmed and then destroyed by explosion in its launching silo in 1978. Then it was sealed with concrete blocks. However, in the 2000th metal hunters came to the spot that has been in the abandoned area at the time and removed the sealing constructions from the silo provoking serious leakage of heptile and massive contamination of the groundwater in the neighbouring area. Various forms of heptile intoxication affected dozens of inhabitants of the nearby settlements [6]. Besides, a serious outbreak of oncological diseases was reported in the entire region after dislocation of the heptile-fuel strategic missiles [6].

This paper deals with a general survey and analysis of some environmental effects caused by equipment and installation of the former 43rd strategic missile army located mostly in Ukraine.

18.2 Brief Story of the 43rd Strategic Missile Army Development During Soviet and Post-Soviet Times

The 43rd strategic missile army had been formed in 1960, during the peak of the 'cold war', in 1962 it took active part in the 'Cuban Missile Crisis".

Since its foundation, the army had been equipped with the middle-range liquid fuel heptile missiles SS-4 and SS-5. A process of gradual substitution of these missiles with new mobile solid fuel missiles SS-20 started in the 70th but it did not affect the stockpiles of the old rocket fuel and oxidizer much since the same compounds were used in a different missile complex SS-19. SS-19 came on duty in the late 70th – beginning of 80th and remained active until the collapse of the USSR. Later, all SS-20 were dismantled under the Strategic Arms Limitation Agreements enacted during the Soviet time.

A part of the missile army dislocated in Ukraine took the jurisdiction of Ukraine in 1991 when the inventory report disclosed an arsenal of 176 strategic missiles: 130 liquid fuel missiles SS-18, SS-19 and 46 solid fuel missiles SS-24. It should be noted that SS-18 is considered as one of the biggest missiles (see Table 18.1) and huge amounts of fuel, oxidizer and other compounds had to be produced, transported and stored in order to maintain normal functioning and service of this arsenal.

Table 18.1 Some technical parameters of the strategic missile SS-18 ("Satan" Mod. 1-6) and SS-19

Technical parameter	SS-18	SS-19
Weight, t	210–211	105
Length, m	33–34	27
Diameter (max), m	3	2.5
Fuel weight, t	~150	~72

Fig. 18.2 Abandoned installation of one of the 43rd army regiments near Vinnytsia, 2011

The decision to dismantle all strategic missile forces had been taken by Ukrainian authorities in 1991 and implemented in a number of legislations enacted in 1991–1994. As a result, the process of missiles dismantling started and 40 missiles SS-19 were disarmed and dismantled by 1994. All nuclear warheads were dismantled by 1996 and sent to Russia for utilization. The rest of missiles SS-19 and SS-18 were dismantled by 1997 and all 40 SS-24 were dismantled by 2000. The 43rd strategic missile army was officially dismissed on August 20, 2002 (Fig. 18.2).

It was declared that all equipment, headquarters, launching silos, storages and other installations were reclaimed except those transformed into museums or education centres.

However, many objects were abandoned without any signs of reclamation, while others were either forgotten or neglected (see Fig. 18.1). The latter also happened to many storage facilities where thousands tons of old rocket fuel and oxidizer were accumulated.

18.3 Current Environmental Problems Related to Old Stores and Equipment of the Missile Units

Ukraine encountered serious problems related to decontamination and utilization of old rocket fuel and oxidizer stockpiles that remained after official completion of the strategic missiles army dismantling.

Financial support allocated in the framework of the dismantling program was aimed mostly on warheads deactivation and destruction of silos and other direct battle equipment and ammunition. Technical equipment storages were of secondary importance. Nevertheless, about 1,171 t of rocket fuel and 3,090 t of oxidizer have been pumped out of the storages and sent to Russia for utilization in 1993–1995. At the same time, dozens of other storage areas scattered around many regions of Ukraine were left beyond this program.

It should be understood that it is quite uneasy to utilize old stockpiles of heptile and oxidizer in an environmentally safe way.

There is no chemical or other ‘civil’ technology requiring heptile as the source material. It can only be used as a fuel for conventional or converted military rockets launch. But old heptile is often polluted and unsafe, which impedes this way of utilization.

Chemical neutralisation of heptile is a complex and quite costly process that produces other toxic compounds. That is why this method is considered unacceptable.

Pyrolysis of heptile is much safer and provides a way to decompose this substance to less hazardous oxides of carbon, nitrogen and water. However, this process requires temperature over 1,000 °C to ensure complete decomposition of heptile. Also, it should be kept within the high temperature zone for 1.5–2.5 s to achieve required depth of decomposition, which is quite a complex technological solution. Currently, there is no plant in Ukraine that is capable of realizing this technology.

Rocket oxidizer mixtures (sometimes referred to as ‘Melange’) contain highly concentrated nitric acid, which is a compound of high demand for the chemical and many other branches of the industry. On the other hand, additional components of ‘Melange’ prevent direct utilization of this mixture instead of the regular HNO_3. There are some technologies available ensuring separation of nitric acid, but the final cost of this acid is significantly higher than the cost of the acid available on the market. As a result, thousands tons of ‘Melange’ remain in old cisterns for years without any prospect of utilization.

As of 2009, Ukraine had about 16,000 t of ‘Melange’ when a specialized OSCE-funded project aimed onto its utilization started. About 12,000 t were considered outdated and non-conditional [7]. This program continues quite successfully and the last 125,000 t of ‘Melange’ are expected to be sent for utilization by the end of February 2014.

The problem related to heptile is still open since the technology of its utilization is more hazardous and dangerous. Gradual transferring of the heptile stockpiles for use in the civil space launches still seems the easiest and most feasible option. However, this ‘utilization’ is also far from environmental safety as any space rocket launch emits tons of hazardous gases (nitrogen oxides and unburnt heptile (see dense red clouds under the heptile-rocket on launch in Fig. 18.3).

Fig. 18.3 Dense red clouds of highly toxic nitrogen oxides and heptile around the space rocket on launch

18.4 Conclusion

As a part of the Soviet Union, Ukraine was satiated with various kinds of dangerous military equipment and materials. While the main attention has been naturally given to the nuclear warheads dismantling after the collapse of the USSR, the problem of toxic rocket fuel and oxidizer utilization remained in the background. Ukrainian economy was in transition, and its own resources were insufficient to resolve this problem, even using the EU and USA funding it took more than 20 years to accomplish the complete utilization of comparatively less dangerous 'Melange'.

The problem of neutralization of heptile stockpiles still awaits its solution since there is no reliable and cost effective technology available yet. It is obvious that this problem will require even higher potential investment and wide involvement of various international institutions in order to utilize thousands tons of heptile stored currently in Ukraine and many other countries.

References

1. Kim HI, Lince JR, Eryilmaz OL, Erdemir A (2006) Environmental effects on the friction of hydrogenated DLC films. Tribol Lett 21:51–56
2. Westing AH (2013) The Gulf war of 1991: its environmental impact, Springer briefs on pioneers in science and practice, vol 1. Springer, Heidelberg, pp 51–75
3. U.S. Air Force strategic deterrence capabilities in the 21st century security environment: a workshop summary. National Academies Press, 47 p
4. Carlsen L, Kenessov BN, Ye S, Batyrbekova A (2008) QSAR/QSTR study on the environmental health impact by the rocket fuel 1,1-dimethyl hydrazine and its transformation products. Environ Health Insight 1:11–20
5. Rozkov A et al (1999) Laboratory study of bioremediation of rocket-fuel polluted groundwater. Water Res 33:1303–1313
6. 46th strategic missiles regiment: the “strange diseases”. http://pioneer-club.at.ua/publ/1-1-0-6
7. http://rus.newsru.ua/ukraine/12feb2009/melange1.html

MIX
Papier aus verantwortungsvollen Quellen
Paper from responsible sources
FSC® C105338

If you have any concerns about our products,
you can contact us on
ProductSafety@springernature.com

In case Publisher is established outside the EU,
the EU authorized representative is:
Springer Nature Customer Service Center GmbH
Europaplatz 3, 69115 Heidelberg, Germany

Printed by Libri Plureos GmbH
in Hamburg, Germany